The Biological Origin of
Human Values

The
BIOLOGICAL
ORIGIN
of
HUMAN VALUES

George Edgin Pugh

Basic Books, Inc., Publishers

NEW YORK

Library of Congress Cataloging in Publication Data

Pugh, George Edgin, 1926–
 The biological origin of human values.

 Includes bibliographical references and index.
 1. Values. 2. Psychobiology. I. Title.
BF778.P83   153.8'3   76-43470
ISBN: 0-465-00687-6

To her whose dedication
and values inspired

Contents

PART III

VALUES FOR PERSONAL DECISIONS AND SOCIAL POLICY

Acknowledgments

THE RESEARCH program that led to this book would not have been feasible without the generous cooperation of many busy people who lent their time to provide guidance through a maze of unfamiliar disciplines. In the field of psychology, I particularly want to express appreciation to Professor S. S. Tomkins and Professor Julian Jaynes of Princeton University. In the field of primatology, I am particularly indebted to Dr. W. Mason of the University of California. Appreciation is also due to Professor Walle Nanta of MIT for his assistance concerning the neurophysiology of the brain.

I wish to recognize a critical contribution by Mr. Randy Simpson of the Office of Naval Research who recognized the relationship between Milton Marney's philosophical work on "normative systems" and my work on the design principles for automatic decision systems, thereby initiating the chain of ideas that lead to this book. I am indebted to a number of readers who read and commented constructively on early versions of the manuscript: Dr. Edward Dunn, Dr. Hugh Everett III, Dr. John Fielding, Mr. Robert Gessert, Dr. Robert Kerchner, Mr. Neil Killalea, Dr. Gary L. Lucas, Dr. David Nobel, Ms. Barbara Slaughter, and Dr. Tom Wyatt. I particularly want to express my appreciation to Dr. Joseph Firestone, for his very careful review of the manuscript from a philosophical and sociological perspective, and to Mr. Herb Reich, whose excellent editorial judgment was responsible for major changes in the final version.

I am deeply grateful to Dr. Ben Alexander, chairman of the board of General Research Corporation, who, during some of the worst years in the company history, had the foresight and confidence to provide support where the whole bureaucracy failed. Finally I want to express my heartfelt thanks to Professor Roger Sperry, Professor Ralph W. Burhoe, and Professor Edward O. Wilson, whose generous support and practical assistance made it possible to carry the project to completion.

Preface

I N MOST SOCIETIES, the transmission of information on values has been, historically, one of the most important functions of education. In the United States, the responsibility for such education has been left primarily to the church and the family; but with the decline in established religion, much of the population fails to receive any instruction at all concerning their own human values. If there is any field of study that should pass the test of relevance, it should be the relationship between human values and human decisions. It is my hope that with the introduction of a more scientific approach to the theory of values, it will be possible to introduce courses on the subject into the college curriculum.

Although the theory of values as developed here may seem like a logical outgrowth of recent developments in sociobiology, it actually had a completely independent origin in cybernetics and decision science. The present theory makes use of a very general construct, which I call a value-driven decision system, and shows how this concept provides a unifying framework that links both behavioral science and human values to the evolutionary design of the brain as a biological control system. The remarkable correspondence of the ideas developed in this way with the basic findings of sociobiology is one of those coincidences that seem to occur in science when the time is right for a new idea. The present theory helps to explain how the genetically inherited behavioral tendencies identified by sociobiologists can operate through the rational mind to motivate conscious human behavior. The findings of sociobiology in turn help to confirm the genetic origin of such behavioral tendencies, which was one of the major premises of the present theory. The theory as developed fills some important gaps not only in behavioral science but also in our theoretical understanding of human values and the goals of social policy.

To make the ideas more widely available to the relevant disciplines, the book is written in a style that avoids all technical terminology, including unnecessary use of mathematics. Although this involves some loss in the precision with which the ideas can be developed, the advantage of reaching a wider audience seems to far outweigh this disadvantage. Because the book is written in a form suitable for a nontechnical audience, it necessarily contains a certain amount of elementary material that will already be familiar to technical readers. Most of this material, however, also serves the less obvious function of restating familiar concepts so that their compatibility with the new conceptual framework becomes more apparent. Because of the orderly way that the development in each chapter builds on the preceding chapters, the book has a flavor much like an elementary college text that could have been titled "An Introduction to the Theory of Human Values."

A natural question that arises in connection with a book of this type is the degree to which it presents new ideas or summarizes old concepts. A large part of the book deals with a poorly defined middle ground where the distinction between common knowledge, common sense, and new ideas is surprisingly obscure. Where ideas are clearly attributable to preexisting sources, I have tried to identify them. There are, however, a few key concepts which to the best of my knowledge are original: in particular, in Part I, the recognition of the importance and generality of the value-driven decision system as a cybernetic design concept; in Chapter 6, the interpretation of the physiology of the brain in terms of the design requirements for a value-driven decision system; in Chapter 7, the resolution of the free will paradox and the clarification of the mystery of consciousness; in Chapter 8 the recognition of a causal link between normal and abnormal behavior in the infant primate; in Chapters 10 and 11, the identification of an evolutionary design concept for the motivation of cooperative human social behavior, which manifests itself in human emotions and facial expressions; and in Chapter 14, the system-design interpretation of human social ethics. Most of the material in Part III is far from new from a philosophical perspective. Although the present theoretical framework is quite consistent with some existing philosophical views, it is also incompatible with others. The material in Part III is included not because it introduces new philosophical concepts, but to show how the scientific value-theory perspective can be related to existing ideas in sociology, ethics, religion, and philosophy.

The development of these ideas has been an exciting process, rather analogous to assembling a large jigsaw puzzle. So far, only the main outlines of the picture are available. I hope that my readers will also find it interesting and be able to fill in some of the missing pieces.

The Biological Origin of
Human Values

Introduction

> . . . human value priorities stand out as the
> most strategically powerful causal agent now
> shaping events on the surface of the globe. More
> than any other causal system with which science
> now concerns itself, the human value factor is
> going to determine the future.
>
> ROGER SPERRY (3) *

PROLOGUE

THE FAILURE to consider human value factors has been responsible for
some of the worst mistakes of modern social history. The pollution of the envi-
ronment and the loss of green space are two widely recognized examples. But
there have been other mistakes of comparable importance: Modern mass
production evolved without regard for its effect on the dignity and morale of
factory workers; and our large metropolitan areas developed without regard
for the human implications of vast depersonalized societies. It is now widely
recognized that we need better ways of anticipating the human value implica-
tions of new technology and social policy decisions.

Our inability to anticipate valuative implications can be traced to some
very basic conceptual problems. There is no generally accepted definition of
"human values," so we do not really know what we mean by the term. More-
over, we lack a scientific understanding of both the origin and the structure of
those enduring human values that we wish to protect.

It seems probable that our ignorance of human values may also be re-
sponsible for serious mistakes in the decisions that determine our personal

---

* Parenthetical numbers throughout the text allude to the list of references at the end of
each chapter.

life-styles. We are finding it increasingly difficult to establish and maintain a satisfactory relationship with the opposite sex. There is a growing sense of personal frustration and inability to find meaning in life. As will be shown in the chapters that follow, these difficulties may be symptomatic of our imperfect understanding of basic human values.

Common sense tells us that values are important in all human decisions. Each decision involves some kind of value judgment: Which alternative is likely to provide the most "desirable" outcome? Which policy will produce the "best" results? What is the "right" course of action? Ultimately our decisions are determined by what is "desirable," "best," or "right" in terms of our own system of values. When faced with similar alternatives, people with different value commitments will make different choices. To make better decisions we need a better understanding of the fundamental value criteria that determine what is "desirable" and what is "right."

What is the ultimate source of our basic human values? For generations wise men and philosophers, as well as many religious and moral leaders, have believed that certain fundamental and enduring human values are intrinsic to human nature. According to the philosophers these fundamental human values provide the ultimate foundation for *all* value judgments. A decision is desirable, right, or good if it is consistent with these fundamental values. A literal interpretation of the philosophers' view would imply that these intrinsic human values are genetically inherited, as an "innate" component of human nature.

But the prevailing theories of psychology and behavioral science have failed to support this idea of "innate" human values. It has been widely believed that human behavior can be attributed almost entirely to environmental "conditioning experiences" and that human nature contains very little that can properly be described as "innate."

Nevertheless, behavioral comparisons among different species have shown conclusively that many of the behavioral characteristics of a species are genetically inherited. Somehow, each individual of a species is endowed with certain specific behavioral "tendencies" that cause it to behave in a way that is characteristic of its own species.

In 1975, Edward O. Wilson published an important book, *Sociobiology: The New Synthesis* (4), which summarizes much of the recent work on comparative behavior. He develops a theoretical perspective regarding the genetic origin of social behavior that is uniform over all species ranging from the primitive sponge to the higher primates. His work leaves little doubt about the genetic origin of basic behavioral tendencies that distinguish the social behavior of different species. He also observes that behavior in the lower animals such as sponges and insects is very stereotyped and predictable, so that individuals of the same species react to a given stimulus in a very predictable and reproducible way. In contrast, the behavior of the higher vertebrates seems to be less directly linked to the genetic design. Although genetic differences between species remain intuitively obvious, the wide diversity in the response of individuals to specific stimuli makes an objective analysis of the genetic determinants of behavior much more difficult.

Wilson's work reinforces a growing conviction that the genetic determinants of behavior must be assigned greater importance even when dealing with human behavior, and that an understanding of the genetic foundation of behavior may be essential to the development of scientifically sound social policy. Although the existing work on comparative behavior has clarified our understanding, it leaves a number of very important questions unanswered: How can such "tendencies" be built into the human mind? How can we reconcile such genetically inherited "tendencies" with our subjective belief in the rationality of our own behavior? What are the practical implications of this new point of view for our understanding of ethics and social policy?

This book introduces a decision-science model of conscious behavior that shows how genetically inherited behavioral tendencies can be linked to actual behavior in the higher vertebrates. It explains the specialized "behavioral tendencies" of each species, as well as our enduring "human values," as manifestations of a built-in *value system*, which is an essential part of evolution's basic "design concept" for a biological "decision system." The theoretical perspective is based on an analogy between the design principles for artificial computerized "decision systems" and the principles that evolution seems to have used in biological "decision systems" such as the human brain. The theory suggests that our fundamental human values are not an accident or a mystery, but rather a natural and almost inevitable consequence of evolution's basic "design concept" for the brain.

Although the theory has grown out of practical experience in the design of artificial decision systems, it consists primarily of some very simple common-sense ideas. Because the subject of human values should be of interest to a wide and varied audience, the ideas are presented in a nontechnical form that does not require familiarity with either computers or formal mathematics.

THE HUMAN DECISION SYSTEM

The present theory was inspired by some striking similarities between the principles used in computerized decision systems and the principles that nature seems to have used in the design of biological decision systems—such as the human brain. The purpose of a computerized decision system is to help human planners with difficult decision and planning problems. In some cases the systems serve only to assist planners by suggesting potentially desirable courses of action. In other cases they operate independently, making a sequence of routine decisions that can be implemented automatically without human intervention. Artificial decision systems have been developed to deal with a wide range of practical problems, such as scheduling production in a factory, developing flight plans and schedules for an airline, routing delivery trucks, developing war plans, and even developing plans for school desegregation.

When the decision problems to be addressed are simple, the designer of such an automated system can select from a number of different system-design concepts. But when the decision problems are complex, the designer almost always finds that the most practical approach involves a rather standard design concept known as a "value-driven" decision system.

A value-driven decision system operates in a way that seems very similar to the human mind. It compares alternative courses of action and selects one that seems "best" in terms of a built-in system of values. Much like the human mind it makes valuative and judgmental decisions. It is capable of finding "creative" solutions to complex problems. The behavior of such value-driven decision systems is very *different* from most standard computer systems. Most computer systems are not designed to exhibit good judgment or creativity. They are designed to do *exactly* as they are told, so that they can serve as simple and predictable tools for men. Such traditional computer systems exhibit no creativity; they are machinelike robots, and they bear almost no resemblance to the creative human brain.

In contrast, the value-driven decision systems are deliberately designed to use "judgment" and to make decisions. These artificial decision systems can be almost as unpredictable as the human brain.* When the brain is analyzed as a value-driven decision system, then our fundamental "human values" seem

---

* These systems are more commonly known simply as "automatic" decision systems, because they can operate without outside intervention. But biological decision systems (such as the human brain) are also automatic in exactly the same sense. For this reason we will usually refer to the computerized decision systems as "artificial" to distinguish them from the more familiar "automatic decision systems" that are of "natural" or biological origin.

to follow naturally, as an almost inevitable consequence of the "design require-ments" for such a value-driven system. They are an essential part of nature's system design for the brain.

Of course, the biological design principles used in the brain are not really identical with those used in artificial decision systems, but there is strong evi-dence that the evolution of the brain has been guided by the same basic princi-ples of efficiency that dictated the value-driven design concept for the artificial decision systems. Thus, both the biological and the computerized decision systems seem to use the same basic "design concept"—the "value-driven" decision system.

Experience in the practical design of such decision systems has provided some very fundamental lessons about values. The primary values, which pro-vide the system with its *ultimate* criteria of decision, must be built into the sys-tem by the designer; and the resulting behavior of the system can be under-stood *only* in terms of this built-in value structure. These primary values are an essential part of the "system design." If the human brain is to be in-terpreted as a value-driven decision system, then biological evolution must have played the role of system designer. It follows that evolution's "behavioral plan" for each species must be defined in an underlying "system of values" which is an essential part of the design of the brain. Assuming that this is true, then there must be an underlying system of "innate" human values that mo-tivate people to behave like people.

This decision-science perspective differs substantially from previous or-thodox theories of psychology and behavioral science. For this reason, much of the book is devoted to a revised interpretation of the behavioral data which shows how this decision-science concept, of an "innate" or built-in value struc-ture, can be reconciled with the existing behavioral data. The analysis of the behavioral studies, in terms of this new point of view, provides surprisingly strong support for the value-theory perspective. Thus, the value-theory be-havioral model seems to provide a more realistic model of conscious behav-ior that may be generally useful in neuropsychology, sociology, and general behavioral science.

Decision systems normally use *two* types of values: "primary" values and "secondary" values. The primary values for any decision system are those that are built into the system by the designer. They define the system's ultimate criteria for evaluating decisions. Secondary values may be developed by the decision system itself as a practical aid to problem solving. The secondary val-ues normally reflect the primary values. Both types of values are used as cri-teria for decisions. The analysis of the human brain as a decision system in-dicates that our subjective valuative sensations (for example, the unpleasantness of pain and hunger, the pleasant taste of good food, or the pleasure of a sexual experience) are manifestations of a built-in *primary* human "value system." These primary value sensations are products of biological evolution. They are built into the hardware of the modern human brain, almost exactly as they evolved during primitive human evolution. To understand the human decision system we must try to understand these innate or "instinctive"* values.

Human beings also make use of secondary values. Whereas pain and hunger are manifestations of the "primary" human value system, the values we attach to money and to our moral, ethical, and social principles are all examples of "secondary" human values. According to the decision-theory perspective, the built-in *primary* human value system is the real source of *all* human values. It provides the ultimate criterion for all personal decisions.

Although our "innate" values provide the ultimate basis for our commonsense value judgments, our intuitive understanding of the "values" themselves nevertheless remains both vague and confused. In our everyday language, the term "human values" has many different meanings. If we were to ask what value criteria are most important in personal decisions, each person would give a different set of answers, but upon reflection most people would agree that at least the following elements are important: happiness, companionship, love, self-respect, pride, dignity, joy, freedom from pain, freedom from fear, and freedom from hunger. On the other hand, if we were to ask what characteristics we "value" in *other* people, we would get quite a different set of answers. The answers might include honesty, wisdom, courage, strength, sincerity, kindness, humor, and integrity. Finally, if we were to ask what "values" are important in the operation of our society the list would probably include liberty, justice, freedom from foreign oppression, equality of opportunity, and freedom from poverty. This diverse set of commonsense "human values" seems disorganized, and the relationship of the values to practical human decisions seems obscure (1, 2). One of the objectives of the book is to clarify the origin and structure of such commonsense values and to define the relationship between the different types of values in our conscious decision processes.

Traditionally, the study of valuative questions has been the province of ethics, religion, and philosophy; but we have advanced into a scientific age and there is a need for scientific understanding. Our analysis method, therefore, is very different from the traditional philosophic and religious approach.

* Psychologists have properly become very wary of using the word "instinctive" to describe any aspect of human behavior, so that the use of the word in this context is likely to be controversial. As will be shown later, however, the innate values exhibit characteristics such that the word "instinctive" seems to be more applicable to these values than to any other aspect of human behavior.

The analysis is built on the foundation of modern decision science and it serves to integrate human values into a "science of decision."

The traditional approach to the subject of human values has been obvious and direct (the author simply begins with a discourse on the subject of "values"), but the route to scientific understanding is not so direct. From a scientific point of view we can understand human values only within the context of the evolutionary role of the values. If we are to gain a scientific understanding of human values we must begin our study where the values originate, in the evolutionary origin of the human decision process.

THE IMPORTANCE OF HUMAN VALUES

Human values provide the guiding criteria for all personal decisions. They are therefore the fundamental driving force of human history. If we wish to achieve any real changes in either the momentum of history or the trend of our personal lives, we should probably begin with a reevaluation of basic value priorities. The whole structure of modern society is the result of an accumulation of human decisions. It is therefore a reflection of our historical value commitments. To improve our ability to make decisions in a changing environment we need a better understanding of those basic human values that determine what is "desirable" and what is "right."

Every great civilization has been guided by certain basic valuative concepts. These concepts have provided a common basis for interpreting world events. They define the goals and aspirations of the society and thus control the movement and direction of history. They also provide the cement that holds the society together and allows it to function as a cohesive cooperative unit. A society without unifying valuative concepts is like a ship without either engine or rudder. It may drift, buffeted by random internal and external forces, but it lacks energy and direction.

The need for a better understanding of human values has probably never been more acute than at present, in our rapidly changing society. As a result of scientific and technological progress, many traditional value commitments no longer seem adequate or relevant. Although scientific progress has tended to undermine the traditional ethical and religious perspective, it has failed to produce any generally acceptable replacement for the traditional ideas. The lack of credible value criteria is apparent in the increased crime, violence, and

general amorality of the cities. It is evident in the sense of meaninglessness
and despair that so many feel in their personal lives. It is reflected in the wide-
spread concern that public policy has become disconnected from human ob-
jectives and that our social institutions are pursuing abstract and meaningless
economic goals, while ignoring fundamental issues of real human value.

Roger Sperry, a professor of psychobiology at the California Institute of
Technology and one of the foremost scientists in the field of brain research,
has become a vigorous advocate of a scientific approach to the problem of
"human values." More than any other scientific writer, Sperry seems to have
recognized that human values originate in the basic design of the human
brain. When human values are recognized as an essential part of nature's sys-
tem design for the brain, they become a natural subject for scientific study.
Sperry states his case very clearly:

> I tend to rate the problem of human values Number One for science in the
> 1970's, above the more concrete crisis problems like poverty, population, energy,
> or pollution on the following grounds: First, all these crisis conditions are man-
> made and very largely products of human values. Further, they are not correcta-
> ble on any long-term basis without first changing the underlying human value
> priorities involved. And finally, the more strategic way to remedy these condi-
> tions is to go after the social value priorities directly in advance, rather than wait-
> ing for the value changes to be forced by changing conditions. Otherwise we are
> doomed from here on to live always on the margins of intolerability, for it is not
> until things get rather intolerable that the voting majority gets around to changing
> its established values. It is apparent, further, that other approaches to our crisis
> problems already receive plenty of attention. It is the human value factor that has
> been selectively neglected and even considered, in principle, to be "off limits" to
> science. (3, p. 32)

Of course, the "social value priorities" mentioned above are secondary
human values. The secondary values are products of the human mind and are,
therefore, subject to change on the basis of rational thought. They are the
"human value priorities" which need to be brought into harmony, both with
the modern technological environment and with the innate (or primary)
human values.

Because the innate (or primary) human values are built into the human
mind as part of our genetic inheritance, they are not subject to change either
by rational persuasion or by social pressure. The primary values originally
evolved in a primitive prehuman society very different from our modern
urban environment. There is reason to believe that much of the discontent
and alienation we find in modern society may be the result of our failure to en-
sure that the modern social environment remains compatible with our ancient
and "innate" human values.

In principle, of course, it might be possible to change even the primary human values: by brain surgery, with neurologically active drugs, or through genetic engineering. But the ethics and morality (as well as the actual feasibility and desirability) of doing so is subject to serious doubt. Our present approach is to treat these primary human values as constants of our analysis. Our objective is to seek a renewed harmonious relationship between the secondary value priorities and the primary or "innate" human values, within the context of a more humane version of technological society.

One might expect that a cold and objective scientific analysis of "human values" might lead to a personal ethics that is narrow, selfish, and devoid of altruism or human feeling. Exactly the opposite seems to be the case. The new perspectives seem generally to support the altruistic principles of Western religion and modern humanism. Indeed, many of our traditional social values such as "freedom," "justice," "equity," and the concept of "right and wrong" seem to follow as almost automatic consequences of theory.

HUMAN VALUES VERSUS SCIENTIFIC KNOWLEDGE

Human knowledge traditionally has been separated into two broad categories. On the one hand, there are disciplines concerned with what we "ought" to do, such as ethics, religion, philosophy, law, and policy studies. On the other hand, there are many disciplines (such as physics, biology, and mathematics) that are purely factual or scientific. The first are concerned with what is desirable, what is good, or what is right. The second are concerned simply with what is. The first are sometimes identified as the "prescriptive" disciplines because they try to prescribe or recommend courses of action. The latter are sometimes known as the "descriptive" sciences.

Traditionally, there has been little interaction between the prescriptive and descriptive areas of study, and indeed the subject of human values has been widely believed to be beyond the limits of scientific study. Nevertheless, whenever we consciously make a decision, we draw on both areas of knowledge. We draw on our factual or descriptive knowledge to predict the consequences of a proposed action. We draw on our prescriptive knowledge to *evaluate* the consequences so we can decide which course of action is "best."

Although the presciptive and the descriptive disciplines have often seemed to be independent, there is an obvious link between them which is the decision process itself. The field of "decision science" therefore lies at the interface between the prescriptive and the descriptive disciplines. Ultimately a mature science of decision should provide a link between the descriptive and the prescriptive disciplines so that scientific methods (based on fundamental knowledge of the human decision system) can assist in deciding (in terms of fundamental human values) what is desirable, or what is right. Of course, at present we can only begin to lay the foundation for such a science of human decision. But one of the main objectives of this book is to highlight the link between the prescriptive and the descriptive disciplines and to show how this link can be used to produce practical improvements in both personal and public planning.

To demonstrate this link between science on the one hand and the value priorities appropriate to personal behavior and public policy on the other, the book deliberately makes the transition across the "forbidden" territory. It begins with the development of a scientific theory; it ends with a discussion of implications for human goals and objectives. This violation of a long-standing scientific taboo is one of the most fundamental messages of the book. Because of the complexity of policy issues and the embryonic state of a "science" of human values, specific policy implications must obviously be treated with a great deal of caution. The real reason for dealing with policy issues is simply to exhibit the logical connection between the theory and policy.

ORGANIZATION OF THE BOOK

The book is divided into three parts:

Part I: The Human Decision System, develops the theory of decision systems and provides a decision-oriented theory of human behavior. It describes evolutionary forces, based on principles of efficiency in the design of automatic decision systems, that may have led to evolution's present "design concept" for the human brain.

Part II: Structure of Human Values, deals with the evolutionary factors that seem to have determined the present structure of instinctive human values. It provides a revised interpretation of human behavior within the context of the theory of values. It includes a preliminary analysis of the innate human "value structure" and of the evolutionary role of this value structure in motivating effective human behavior.

Part III: Values for Personal Decisions and Social Policy, deals with the potential implications of the theory for human policy. It shows how our innate human values seem to explain many of the traditional principles of personal ethics and social policy. It also develops some illustrative general ethical and social principles that seem to arise from the theory and discusses some implications of the theory with regard to personal and social issues of current interest.

Part I provides the scientific foundation for the theory. It is based on recent developments in cybernetics and decision science. Although it is written so that it does not require a technical or mathematical background, it is nevertheless the most technical of the three parts.

Part II draws on work concerning primate behavior and human psychology and develops a new interpretation of these results in terms of the decision science theory. Because this interpretation is quite different from some older behavioral theories, readers with prior training in the behavioral sciences should read the theoretical justification in Part I carefully before attempting to read Part II. The psychological concepts developed in Part II provide the theoretical basis for the discussion of ethics and social policy in Part III.

Part III is the least technical of the three parts. It is also the one with the most direct application to real human problems. In addition to a discussion of basic principles, Part III also discusses a number of illustrative issues in both personal ethics and social policy. Because of the embryonic status of work in "scientific value theory" this illustrative material must not be taken too seriously. It is intended to show, first, that the theory has potential practical applications and, second, that it appears to produce a shift toward value priorities which seem both more humane and more realistic.

REFERENCES

1. Perry, Ralph Barton. *General Theory of Value*. New York: Longmans, Green, 1926.
2. Pepper, Stephen C. *The Sources of Value*. Berkeley: University of California Press, 1958.
3. Sperry, Roger. "Messages from the Laboratory." *Engineering and Science*, January 1974. (Special issue on behavioral biology at Caltech.) Published by California Institute of Technology.
4. Wilson, Edward O. *Sociobiology: The New Synthesis*. Cambridge, Mass.: Belknap Press of Harvard University Press, 1975.

PART I

THE HUMAN DECISION SYSTEM

Chapter 1

Background and Preview

The purpose of life is to stay alive
and that is impossible,
but it is the ultimate purpose nevertheless,
our everyday assumptions say.
Thus great minds struggle valiantly
to conquer disease,
so that people may live longer,
and only madmen ask why.

ROBERT M. PIRSIG (8) *

THE SEARCH FOR A CONCEPT

FROM THE BEGINNING of history people have speculated on the mystery of human nature. The human mind is capable of rational thought, yet human beings often behave irrationally. We recognize, even in our own behavior, the influence of the subconscious, the irrational, and the emotional. We have an intuitive understanding of concepts such as pleasure or pain, good or evil, and right or wrong. Yet efforts to be scientific about these concepts usually produce only obscure philosophical discussions.

For several decades, these basic issues have received relatively little attention from the scientific community. At the same time, there has been great progress in related fields of automation, artificial intelligence, decision science, neural physiology, and behavioral science. Recent work in these areas provides the basis for a new conceptual synthesis which interprets the brain as a biological "decision system" and thus provides a new theoretical model of conscious behavior.

We should not expect too much of any single theory of behavior. No

---

* Paraphrased from Chapter 7.

theory can be more than a simplified model of the mechanisms for certain types of behavior. The existing concepts in behavioral science include a number of models or mechanisms that are helpful in understanding certain aspects of behavior, including the following:

1. *Feedback control:* An error signal is used to provide sensitive adjustment of a physical response. (For example, when we reach for a pencil, we see the distance between our hand and the pencil, and move the hand so as to reduce the gap.)

2. *Homeostatic behavior:* The organism responds to external stimuli so as to maintain certain critical parameters, such as body temperature or chemical balance in the blood, within acceptable limits for the welfare of the organism.

3. *Tropisms:* These are ritualized muscular response patterns that are genetically inherited and are triggered by specific stimuli.

4. *The conditioned response,* as developed by Pavlov: An inherited response to a specific stimulus can be conditioned so that it occurs automatically in response to quite a different stimulus that has become associated with the primary stimulus as a result of experience.

5. *Operant conditioning:* A random, perhaps subconscious, muscular response is reinforced and becomes greatly strengthened as a result of reinforcement (or reward) that consistently and closely follows the response.

Each of these concepts provides a useful way of thinking about certain aspects of human behavior. The boundaries between the concepts can be vague, and often more than one concept is applicable in explaining specific aspects of behavior. In view of the evolutionary origin of biological behavior, this is not surprising. The actual mechanisms underlying biological behavior are complex, and do not correspond exactly to any simplified model. In some cases the biological behavior may involve mechanisms for which we have no suitable model at all.

None of the generally accepted concepts seems to provide an adequate explanation of intelligent human behavior. As will be shown in later chapters, the concept of the value-driven decision system appears to provide a more suitable basis for understanding these aspects of behavior. Of course, the older concepts, particularly the concept of operant conditioning, have sometimes been stretched to provide apparent explanations of conscious behavior. But in terms of what is now known about the brain, the resulting model of intelligent behavior seems to be totally inadequate.

Our present approach is motivated by the need for a behavioral model that can be used within a science of decisions. The desirability of a decision-oriented behavioral model is apparent even within the field of psychology. Most people begin their study of psychology in the hope that it will help them to understand themselves and thus help in their day-to-day decisions or choices. From their point of view, the behaviorist psychology (of operant conditioning) suffers from a very critical defect. It does not include the concept of "choice." Because the behaviorist theory does not include the "decision" concept, it cannot be easily applied to provide guidance concerning personal decisions.

GOAL SEEKING AS A THEORY OF BEHAVIOR

The "decision-system" concept interprets human behavior as a purposive or goal-seeking activity. This basic concept of behavior as an activity involving goal-oriented decisions is probably the oldest and most intuitive of all theories of behavior, but during the early 1900s this intuitive theory of behavior began to fall into disfavor. Early successes in the physical sciences had demonstrated the importance of quantitative and objective experiments as a foundation for scientific knowledge. The behavioral scientists attempted to rid their discipline of nonscientific and nonquantifiable "mentalistic" concepts such as desires, goals, or objectives. To provide a foundation for a more "quantitative" science, they tried to explain behavior in terms of simple cause and effect (or stimulus and response) relationships such as one might use to explain the operation of simple nineteenth-century machines. This mechanistic approach is represented most forcefully in the work of J. B. Watson (13, 14, 15) of Johns Hopkins and more recently in the writing of B. F. Skinner (9, 10) of Harvard. The dominance of this narrow behaviorist view was not achieved easily. Numerous workers believed that the concept of purposeful behavior should be retained even in a "quantitative" theory of psychology. This view was perhaps most effectively expressed in 1932 in E. C. Tolman's *Purposive Behavior in Animals and Men* (12).

Recent developments, both in behavioral science and in computer technology, have now reversed the trend toward a narrow behaviorist view. Many of the old "mentalist" concepts have begun to appear in a very quantitative and scientific form in the structure of computer systems. Objectives or goals are explicitly written into the mathematics of computer programs, and the "behavior" of the programs is determined by these specified "objectives." This development has widened the scientific perspective so that once again many behavioral scientists are talking about purposeful or goal-oriented behavior.

Of course, the concept of purposeful behavior never really disappeared from the psychological theories. There is universal agreement that behavior is influenced both by the basic biological drives such as hunger, thirst, and sex puter examine thousands of alternatives unless the computer has some way of recognizing a "desirable" alternative when it finds one. If the computer is provided with a way of measuring the "desirability" of the alternatives, it can print out on a sheet of paper the best alternative (or a few of the best alternatives) for consideration by the human planner. Such results from the computer can be very helpful and will often allow the planner to make much better decisions. However, if the computer does not have a way of measuring the "desirability" of an alternative, it can only print out the thousands of unevaluated alternatives. Such a mass of unevaluated results would only make more work for the planner. It is obvious that the definition of suitable values or decision criteria is central to the design of a good decision system.

Experience with artificial decision systems has shown that the design of really satisfactory decision criteria can be a very difficult problem. It is often the most difficult aspect of the system design. As will be shown in the later chapters, nature seems to have encountered many of the same problems during the evolution of biological decision systems.

THEORETICAL FOUNDATIONS

Our analysis of the human decision system will rely heavily on certain well-established background concepts. These include the ideas of biological evolution, the concept of biological systems, and the ideas of information theory and data processing. A few words are necessary to explain the way these concepts will be used; otherwise there is a risk that the logic of the development will be misinterpreted.

BIOLOGICAL EVOLUTION

Because of the long time span of evolution, most of the products of the evolutionary process exhibit extreme efficiency and specialization of function, as if they had been deliberately and rationally designed. Therefore in speculating about human behavior and the human brain (or any other biological system) it can be fruitful to approach the subject *as if* we were dealing with a logical "system design." We can ask what functions are needed and how they can be most efficiently performed. Because biological organisms tend to be efficiently designed, such speculation can often provide concepts that are useful in understanding the biological organisms.

For simplicity of language, we will often speak loosely of the "design concept" or the "design objectives" of biological systems, as if they were the product of deliberate purposeful design. This figure of speech, however, should not be misinterpreted as attributing a grand design or pervading purpose of evolution. The essence of the theory of evolution is that the process takes place without purpose, as a result of random genetic variations and "natural selection" of the most survivable species. As Jacques Monad has so clearly pointed out in *Chance and Necessity*, the efficiency in the design of biological organisms, as well as the purposeful behavior of animals and men (as they apparently pursue the goals of survival and reproduction) are all natural consequences of the purposeless evolutionary process (6).

The brain, like any other biological organ, must be thought of as an instrument of biological survival. The evolutionary process has been in progress on earth for more than 2 billion years. During that time natural selection has been incessantly winnowing the living organisms. As a result, the organisms that survive today are efficient and highly specialized in the arts of survival. Much of their behavioral specialization results from the intricate design of the brain as a control and decision mechanism. We therefore should not be surprised to find that the human brain itself reflects a very sophisticated and logical "system design."

As our understanding of biological organisms has progressed, we have found more and more reasons to be surprised and impressed with the complexity and efficiency of biological designs. Even a casual review of the obvious physical endowments of the human species (the eyes, the ears, the sense of smell, the overall physical coordination) reveals an efficient "system design," as if very detailed consideration had been given to the design and function of all components.

An analysis of the hereditary mechanism itself can only produce amazement at the efficiency with which information is coded, the reliability with which it is transmitted, and the effectiveness of sexual reproduction as a way of achieving both stability and adaptability in the genetic inheritance of a species. For example, the design principles of the human eye are essentially the same as those used in fish, birds, and reptiles. We must, therefore, puzzle how the stability of this excellent optical design was maintained while evolution nevertheless proceeded with extensive redesign and restructuring of so many other system components. Recent research has shown that the genetic code itself includes a large degree of separation of function (5), so that those parts of the code which affect the design of the eye may be almost unaffected by mutations that determine the design of other organs. Such independence of function in the genetic code permits almost independent evolution of the many components and subcomponents that comprise the integrated biological system. Although we do not know in detail how this independence of function and resulting efficiency in the genetic design process) is achieved, the evidence for the efficiency is overwhelming.

THE SYSTEMS APPROACH

Complex computer systems are almost always developed through the use of "system design" methods. The basic system design is conceived initially in terms of the functions to be performed and procedures for performing them. The "system design" defines the general characteristics and requirements of the system, including the requirements for storing, processing, and transmitting data. The conceptual design process can be completed without asking whether the system will be implemented with an old-fashioned vacuum tube computer or a modern solid-state system.

In our description of the human decision system we will follow a similar approach. We will be concerned with understanding the basic conceptual or logical design of the system. Because the logical design does not depend on the specific physical components used in the implementation, it is possible to develop an understanding of the logical design without even knowing the underlying physical and biological principles. This is fortunate because some of the most important physical principles of the brain remain unknown.

The question arises, does such an understanding of the "system design" constitute a theory? In part, the question is one of definition, but there are also some important principles involved.

In the past, scientific advances have usually been the result of careful study of the most elementary processes. The "scientific" method has been to reduce complex systems to their most elementary components. Finally, having achieved an understanding of the elementary components, these are recombined to synthesize an understanding of the more complex systems of the everyday world. This basic scientific method (sometimes known as the reductionist approach) has been very successful.

However, there is a growing recognition that it is not the only way to achieve scientific knowledge. It is often expedient to analyze complex systems from a broader perspective, without waiting for completion of the reductionist approach. The history of science has shown that knowledge obtained in this way often continues to be valuable even after the reductionist research is complete.

For example, the entire field of classical physics has been subsequently "explained" in terms of the theories of relativity and quantum mechanics, but most engineering continues to be done in terms of the simpler classical laws. Both inorganic chemistry and organic chemistry have been "explained" by quantum mechanics. Although these new explanations have improved our understanding, most day-to-day work proceeds using simplified ideas about the behavior of molecules that are not stated in terms of quantum mechanics. It is clear that there is a need for different levels of theory in the explanation of natural phenomena.

Ideally we would like the theories at the more aggregated levels to be completely and rigorously "explained" at the more detailed level. It seems clear that we are approaching a time when almost all of the basic biological processes will be "explained" in terms of the laws of physics and chemistry. But many of the problems of interest to the biologist, the behavioral scientist, and the social scientist are so complex that there is a need for understanding at a broader system level. Even if detailed explanations were available in terms of the elementary laws of physics, the relationships would be so complex that they would be of little practical value.

For these reasons there has been a growing interest in the development of a general scientific methodology for dealing with complex systems. This movement (1, 2, 3, 4) has been classified under the heading "general systems theory." So far it has not proved possible (and it may never be possible) to say much that is worthwhile concerning all systems. Nevertheless, most biological organisms and many social systems fall into a special class for which unifying concepts and principles seem to apply. These systems can be broadly characterized by the fact that they seem to behave purposefully—as if they are seeking to maintain certain system norms or to achieve certain goals. In the face of changes in the environment these systems tend to behave adaptively. Although the behavior changes, it can still be explained in terms of essentially the same set of norms or goals. The theory developed here provides a foundation for analyzing human behavior in terms of the design concepts for purposive systems.

A GENERAL THEORY OF PURPOSIVE SYSTEMS

The possibility of a unifying scientific methodology for dealing with this class of systems seems to have been first clearly perceived by Smith and Marney (11) in their pioneering research on the foundations of decision science. Smith and Marney call such systems "normative systems" because of their tendency to maintain (or seek) certain norms. They point to the possibility of a general theory of normative systems and have made substantial progress in the development of concepts for such a theory. It is, therefore, appropriate to mention a few key insights from their work which provide a foundation for the present analysis.

Since the goals or norms that motivate purposive systems tend to be the *most constant* and thus most predictable characteristic of their behavior, it follows that a useful description of such systems must begin with a definition of the goals. The goals or norms of such systems thus constitute an *essential* part of the system description. To understand the system's behavior we must begin with an understanding of the "value structure" against which the system will make its decisions.

Although these ideas may seem obvious in terms of our commonsense view of the world, they represent an important break with scientific tradition. Traditionally, scientists have made a sharp distinction between factual cause and effect relations, which were presumed to be the province of science, and valuative judgments (concerning good or bad), which were presumed to be out of bounds for science. The proposed approach to the analysis of "normative systems" includes the valuative aspects as an essential part of the scientific description.

This break with tradition may be necessary if progress is to be made in certain areas that have previously seemed resistant to the scientific method. The scientific method has been extremely successful in fields such as physics, chemistry, and even neurophysiology, where cause and effect are directly related.* The scientific method has been less successful in dealing with other fields such as human behavior, decision science, and social science in which the valuative issues are intertwined with the factual. In order to integrate the rapidly expanding factual information into the valuative fields there is a need for a common language for describing the results and a common *precise* understanding of valuative concepts.

It seems probable that the "normative system" concepts will help to provide this much-needed communication bridge. Smith and Marney make the important point that all actual systems are finite. The finite data-processing limitations of such systems impose practical design constraints that are crucial to understanding the systems. Our analysis of the human decision system very strongly supports the importance of this view.

PREVIEW OF THE THEORY

LIMITATIONS OF FINITE SYSTEMS

All biological decision systems and all computerized decision systems are *finite* data-processing systems. They have limited capacity for data input, data storage, and data analysis. Any finite decision system that must deal with a complex decision environment will necessarily incorporate compromises in its

---

* At least in the sense of the statistical predictability of quantum mechanical laws.

design. It must make decisions based on limited input data and limited analysis of the data, even though the environment may involve almost unlimited variability. As a result, compromises must be made in the rigor of the decision process. The compromises may affect the way the problem is defined and the way the environment is analyzed and interpreted. To understand a practical decision system we must begin by understanding the main compromises inherent in the design.

The procedures appropriate to such a finite decision system are qualitatively different from traditional rigorous mathematical techniques. Allen Newell (7) describes the methods as "weak methods" because they use "weak" input information to produce "weak" results. I prefer to describe them as robust and efficient methods because they produce "useful" but "nonrigorous" results with a limited analysis of incomplete information. The finite system is denied the luxury of rigor, precision, and certainty. It must deal with uncertainty, with approximations, and with simplified representations of reality that may be of uncertain validity.

Such finite systems necessarily use simplified heuristic methods that are likely to be unsatisfying to a mathematician or logician. The design of a practical decision system is more like a problem in engineering design than a problem in mathematics or logic. For this reason, the experience in the design of artificial decision systems is particularly useful in understanding evolution's design for the biological decision systems.

THE WORLD MODEL

To predict the consequences of different action alternatives, both the artificial and the biological decision systems use a model of the environment. This introduction of an explicit and detailed model of the problem environment within the control system (as an essential component of the control or decision process) is one of the most important distinguishing characteristics of the value-driven decision system. The model is used to calculate or simulate the consequences of different decision alternatives. The system performance depends very strongly on the accuracy and reliability of this model. The mental model used by the human decision system can be as unsophisticated as the child's visualization of his room or his "model" of his mother that he uses to predict her response. Conversely, it can be as sophisticated as quantum mechanics or relativity. The ensemble of such concepts used by the human decision system constitute a total "world model" (a world view) which serves as a tool for the decision process.

In most computerized decision systems the designer of the system also designs the "world model" that will be used by the decision system. In such systems, the world model is an integral part of the original system design. Biological decision systems differ from artificial ones in that the systems themselves have the responsibility for developing most, if not all, of their own world model. Many of the decisions that must be made by the human decision system are really decisions about how to develop and refine the world model.

VALUES AS A CRITERION OF DECISION

After the world model has been used to predict the outcome of different action alternatives, a decision system must have some way of evaluating the outcomes, to decide which alternative is "best." In most artificial decision systems the need for criteria of decision is satisfied through the use of numerical "values" that are assigned to each alternative. The designer of the system specifies a procedure by which such "values" are to be calculated. The decision system then uses this procedure to calculate a numerical "value" for the outcome of *each* alternative that is considered. A "decision" is made simply by selecting the alternative with the highest computed "value." The specified value algorithm, therefore, determines the choices. In effect, it drives the decision system. Decision systems that use this basic method of decision will be referred to as "value-driven" decision systems.

This usage of the word *value* is consistent with present conventions in decision science, where a value is simply a quantitative measure (or numerical quantity) that is associated with alternatives for the purpose of making decisions. If a procedure is defined for calculating the value of all alternatives, then a decision can be made simply by selecting the alternative with the highest computed "value." Conversely, if one can observe the choices made by such a decision system, over many combinations of alternatives, one can often make deductions about the way the values are assigned. In principle, then, a behavioral model for a biological decision system can be developed by scientifically studying behavior to deduce the underlying structure of values that guides the decisions.

In the remainder of this book our use of the word *value* will follow the convention of decision science. It will be used to refer simply to a quantitative criterion of decision. This usage seems to differ from the traditional usage where value implies a judgment about desirability—good or bad, right or wrong, useful or useless. As will soon become apparent, however, our formal definition of the word *value* is a natural generalization of the traditional usage.

Not all decision systems are value driven. The selection of the value concept as the design principle is itself one of the fundamental compromises that must be faced in the design of a decision system. Experience with artificial decision systems suggests that as the decision environment becomes more complex, it becomes progressively more difficult to obtain satisfactory behavior with any of the simpler design concepts.

Our analysis suggests that evolution may have encountered very similar difficulties with its earlier design concepts, so that it settled on the value-driven method for its more sophisticated decision systems.

DESIGNING A VALUE STRUCTURE

The development of a value structure for an artificial decision system is usually an iterative process. The designer makes an initial judgment about the appropriate decision criteria. He then tests his initial value structure by observing the operation of the system. If he is happy with the resulting decisions, he has a satisfactory design and can stop. More frequently, however, he will find that the resulting decisions are not entirely satisfactory. He must then try to improve the decision criteria to make the resulting decisions more consistent with his own objectives. This type of iterative development of a value structure can be tedious for a human designer, but it is just the type of design refinement that is most efficiently accomplished within the evolutionary process.

During the design of an artificial decision system, the resulting system decisions are tested against the true objectives of the system designer. During the evolution of a biological decision system, the resulting system behavior is tested against the evolutionary goal of survival of the species. The evolutionary process of survival of the fittest thus plays the role of "designer" for the biological systems, and the biological value structure is tailored in this way to motivate behavior conducive to survival of the species.

THE INNATE HUMAN VALUES

From an evolutionary point of view, the innate human "values" are simply inherited decision criteria that were built into the human species during the process of evolution. There is no need to assume any inheritance of learned or acquired characteristics. Our presumption is that these built-in valuative sensations are a result of physical linkages in the neurons of the brain, which are inherited in exactly the same way as any other physical characteristic. As with any genetic inheritance, these neural linkages are the result of random mutations and the survival of the fittest.

Human behavior is influenced not only by these inherited primary decision criteria, but also by secondary decision criteria that are *learned* as a result of experience with the environment. The secondary decision criteria are what we most frequently think of as "values." They include learned goals or objectives as well as moral and ethical principles. Although these secondary decision criteria are vitally important to human behavior, they are not the ultimate source of our value criteria. The secondary values are developed individually and culturally as an outgrowth of more basic innate human values. To really understand human behavior (including the secondary values) we must begin by trying to understand the built-in primary value structure. Because the innate value structure is fixed by our genetic inheritance, it is one of the most invariant characteristics of human nature. It imposes practical limits on modern man's ability to adapt to the changing physical and social environment.

The innate human value structure is one of the most important parts of our subjective * mental experience. The innate built-in values are experienced as good or bad valuative sensations, such as tactile pleasure or pain, comfort or discomfort, joy or sorrow, and good or bad taste. These primary human values include both the "emotions" and what have been traditionally known as "biological drives." Thus the innate values are not constant over time, but they respond to physical and social stimulation in accordance with complex built-in rules. The natural preferences that result from this innate value structure are concerned with preferences for different types of activity, not with specific or ultimate goals. Thus the human value structure is concerned with the quality of the journey, not the ultimate destination.

Experience with artificial adaptive systems suggests that the ensemble of primary values required to explain the behavior of animals (and especially human behavior) is much richer and more elaborate than is usually described in the current literature. The orchestration of the primary values which takes place unconsciously and without rational thought is largely responsible for our complex emotional personalities. Thus, the theory of the primary values that will be developed here is considerably more complex than the traditional description of psychological needs and biological drives.

To understand instinctive human values, as we find them today, we must consider the evolutionary environment in which the values evolved. The values were selected to contribute to the survival of the species, not necessarily the survival of the individual. The innate human values include many elements which motivate behavior supporting the welfare of the tribe or band, as well as other elements supporting the welfare of the family and the individual. Thus, the instinctive values include a mix of both altruistic and selfish elements.

Despite the strong evidence that man is fundamentally a social animal, there has been a surprising tendency, both in psychology and formal philosophy, to overlook the importance of such innate social drives in human behavior. Social animals are designed by evolution so that as individuals they "enjoy" the social activities that are required by their intended role in their so-

---

* The word "subjective" is used to refer to our personal mental experience which is observable only within our own minds, as opposed to "objective" or external experience that can be shared with others.

ciety, even when these activities contribute nothing to the *physical* survival of the individual. Humans are no exception to this general rule.

DESIGN COMPROMISES IN THE HUMAN VALUE STRUCTURE

The evolutionary design objective (survival of the species) is a straightforward and unchanging objective that has been constant throughout biological evolution. It may seem paradoxical that the innate human values (based on this simple fixed objective) are so complex and variable. Why are there so many separate value components? Why are the values so variable?

Experience in the design of artificial decision systems has shown that it is possible to compensate for the data-processing and analysis limitations of a finite system by providing a more elaborate and detailed value structure. The complex human value structure can be understood as a design compromise, which served to match the decision criteria to the limited intellectual resources of the biological decision system. As the human brain increased in size, during the final stages of the evolutionary process, the available analysis resources have increased, so that the built-in value structure may be more detailed than is really necessary at present for the human decision system.

Obviously the innate human value structure is not an ideal or perfect approximation to the evolutionary objective. It is simply the best compromise the evolutionary process has been able to produce. From the point of view of the evolutionary designer, the innate human values are simply surrogates or substitutes for the real evolutionary goals. However, from a human point of view, these same valuative criteria are the ultimate human values. They are the ultimate measure of all human activity and the source of all that makes human life worthwhile.

RATIONAL THOUGHT AND IRRATIONAL VALUES

The innate human values respond to the physical and social environment in accordance with predetermined nonrational laws. Our emotions and biological drives are generally not subject to rational control. They respond instinctively * in accordance with procedures which are a part of our genetic inheritance. For example, the hunger sensation is an innate response to a deficiency in nourishment, pain is an innate response to dangerous heat or physical pressure, fear can be an innate response to specific stimuli such as a sudden noise or a strange person or animal. Because these values respond to internal and

---

* The innate human values can appropriately be described as "instinctive" values because they respond automatically to specific types of circumstances without rational reason or forethought. They probably fit the classical concept of "instinctive" behavior as accurately as any element in the human behavioral system.

external stimuli in accordance with genetically determined rules, it seems appropriate to refer to these values as "instinctive." The resulting behavior, of course, is far from instinctive, but it is motivated by instinctive values.

It is no accident that these ultimate valuative sensations are not based on rational thought. If the primary values were based on reason, they would be subject to change on the basis of rational thought. If the human decision system were able to tinker with its own primary "value system," it might randomly modify the ultimate criteria of decision, so that the resulting decisions would be inconsistent with the evolutionary objective. If a biological decision system is to serve the evolutionary objective, it must *not* be permitted to modify its ultimate criteria of decision.

Experience with artificial decision systems confirms that a decision system cannot be allowed to change its ultimate valuative criteria. Of course, in a few cases artificial decision systems are designed to learn from experience and thus improve their *secondary* decision criteria. Such changes in the *secondary* decision criteria can be evaluated against the primary or ultimate values, but there is no standard against which changes in the *primary* values could be evaluated. Improvements in the *primary* values can be achieved *only* by intervention of the system designer.

The paradox of man's apparent ability to think rationally and his inability to behave rationally seems to originate in the conflict between rational thought and the irrational and variable innate value system. Decisions must ultimately be evaluated in terms of their effect on the fluctuating instinctive value system. But the fluctuations in the instinctive value system cannot be accurately predicted. The human decision system is, therefore, faced with the added burden of trying to understand and predict its own valuative responses. This is far from easy because the valuative system is complex; and the value structure changes with age as the system matures.

Despite its imperfections, the built-in human value structure is fundamental to the design of the human decision system. The relationship between the value structure and human behavior is quite complex. To understand the relationship, it is necessary to analyze the innate human value structure from the broad perspective of an evolutionary system design. Part II of the book is devoted almost entirely to such analysis of the human value structure.

THE ROLE OF SECONDARY DECISION CRITERIA

One of the major objectives of this book is to trace the origin of our fundamental human values. Although this objective necessarily focuses attention on the innate human values, we should not underestimate the role of secondary values and other secondary decision criteria.

These secondary decision concepts can take the form of rules of thumb, wise proverbs, social conventions, moral or ethical principles, and even habit. The secondary decision criteria serve as a practical aid in making daily decisions. They help the decision system to make decisions more efficiently or more reliably. The purpose of the secondary criteria is to help in selecting decision alternatives that can be evaluated as "good" when measured in terms of the *primary* value structure. In principle, therefore, the secondary criteria should reflect the objectives of the primary values.*

Some secondary values are developed as original discoveries or conclusions of an individual, but the more familiar moral or ethical values and social taboos are the collective work of a society. Because the secondary values are in one way or another the product of rational thought, people tend to be consciously aware of the secondary values. When values are discussed in ordinary conversation, the word *value* usually refers to secondary values such as ethical and moral concepts. When people think of making decisions based on "values" they are almost always thinking about secondary values. Conversely, because the primary values are innate, and subjectively seem absolute, they tend to be ignored in our ordinary discussion of values.

The day to day behavior of an individual is likely to be determined at least as much by the secondary as the primary decision criteria. When secondary decision principles have been adopted, they are used almost intuitively without thought or question, much as if they were primary values. But this does not detract from the even more fundamental role of the primary values—for the primary values determine what types of secondary criteria the individual will be motivated to adopt.

In Parts I and II our attention will be focused almost entirely on the primary values. Part III of the book is concerned with the relationship between the primary and secondary values.

SYSTEM VALUES VERSUS GOALS OF THE DESIGNER

The distinction between primary and secondary values must always be made in the context of a *specific* decision system. The same value component may be a primary value for one decision system and a secondary value for another.

This *relativity* of values is particularly pertinent when we consider the

---

* The distinction between primary and secondary values in the present theory should not be confused with the traditional distinction between intrinsic and instrumental values. The traditional distinction is a subjective concept, derived from the introspective experience of the human decision system. The present distinction between primary and secondary values is an objective system design distinction that can be applied to any value-driven decision system. As will be shown in Chapter 5, when the present theory is interpreted from a subjective perspective, it leads to a very concrete version of the traditional distinction between intrinsic and instrumental values.

relationships between the decision system and its designer. The system de-
signer will usually specify objectives (or primary values) for the decision sys-
tem which do not correspond exactly to the true objectives of the designer.
The values that he specifies for the decision system are tailored to the intellec-
tual capacity of the system. They may be simplified representations of his real
objectives, or they may be simply surrogate values designed to make the
decision system behave as he wishes. Thus the primary values specified for a
decision system may be secondary values from the point of view of the de-
signer.

In the case of biological decision systems, evolution was the designer.
The ultimate evolutionary goal was survival of the species, but the biological
decision systems are designed to use primary values that are surrogates for the
ultimate evolutionary objective.

COMMON MISCONCEPTIONS ABOUT VALUES

As was mentioned earlier, the concept of behavior as a goal-oriented or
purposeful activity is far from new, but the present theoretical approach in-
volves a number of departures from common or traditional views. In the past,
the development of a consistent theory has been impeded by some of the fol-
lowing misconceptions about primary values or goals:

1. The assumption that primary values should be simple, that is, that they should
   include only a small number of independent components.
2. The assumption that primary values should be constant, and should not vary
   from day to day with biological needs.
3. The assumption that primary values should be rationally consistent, or that
   they could be deduced or improved by reason.
4. The tendency to relate values directly to "biological needs" and to overlook the
   social or altruistic component of instinctive drives or values.
5. The assumption that primary values should refer to ultimate goals or destina-
   tions, rather than the activity involved in the journey.

Most of these misconceptions seem to be related to a central misconcep-
tion—that human values are in some sense absolute. As mentioned earlier,
the primary human values are really only surrogates for the ultimate evolu-
tionary objective. They are surrogate values that have been tailored by evolu-
tion to match man's limited intellectual capability.

ELABORATION OF CONCEPTS

In this brief review of the theory, we have confined ourselves to an ex-
tremely simplified and idealized version of the ideas. It will not be surprising
if the reader at this time fails to see much similarity between the idealized con-
cepts and the practical operation of the human brain.

Moreover, because the vocabulary used to describe the concepts in this chapter parallels the traditional usage, it is easy to misinterpret the discussion as a vague qualitative concept similar to early "mentalist" theories. To remove these problems of oversimplification, it will be necessary to develop two fairly extensive sets of ideas:

1. The oversimplified skeleton of the theory must be filled in to include the realistic limitations of a biological decision system.
2. Basic design principles for automatic decision systems must be developed to show how evolutionary pressures could have operated to produce the present design of the brain. We need to know why nature would have chosen to design the brain as a value-driven decision system. We need to know why the innate value system is so complex, and why it involves such elaborate time dependence.

The next six chapters are devoted primarily to the development of these ideas. The design principles and evolutionary forces are developed in Chapters 2, 3, and 4. The idealized theory is more carefully and accurately stated in Chapter 5. Finally, in Chapters 6 and 7, the idealized skeleton is fleshed out to provide a more relistic conception of the human brain.

COMPUTERS AND THE MIND

In retrospect, World War II constituted a technological turning point. It was during this period that man first moved from the "machine age" to the "age of automation."

It was soon recognized that the principles of control and information processing used in automation might also be relevant to understanding the mechanisms of control in the human body and other biological organisms. The present study rests on the assumption that biological systems operate within the same laws of information theory and system design that apply to artificial systems. This hypothesis is supported by a considerable body of evidence.

THE PRINCIPLES OF CYBERNETICS

In 1948, Norbert Wiener published his pioneering book on a subject which he called "cybernetics." In introducing the term *cybernetics* he said,

> We have decided to call the entire field of control and communication theory, whether in the machine or in the animal, by the name *Cybernetics*, which we form from the Greek χυβεργη'τηδ or *steersman*. In choosing this term, we wish

to recognize that the first significant paper on feedback mechanisms is an article on governors, which was published by Clerk Maxwell in 1868, and that *governor* is derived from a Latin corruption of χυβεργη'τηδ. We also wish to refer to the fact that the steering engines of a ship are indeed one of the earliest and best-developed forms of feedback mechanisms. (16, pp. 11–12)

Norbert Wiener's basic hypothesis, that the principles of control in animals and machines would prove to be similar, has now been tested by theory and experiment for more than twenty-five years.

Scientists have written computer programs that exhibit many of the properties of learning, reasoning, and problem solving that we usually attribute to higher intelligence. Biologists have established that the individual cells of the nervous system (the neurons) possess electrical properties that allow them to operate both as information conductors (transmitting signals between different parts of the body), and as miniature circuit elements that can accomplish data-processing and logical operations much like the circuit elements of a computer. Scientists studying the operation of the brain have identified numerous data-processing activities that make sense only as methods for increasing the efficiency of data storage and communication in the brain. The cybernetic analogy between the processes of communication and control in automation and the control processes in living organisms has proved to be extraordinarily fruitful. Consequently it now seems probable that all the major functions of the brain will ultimately be understood in terms of theories of communication and control.

Control systems that deal with complex problems almost always exhibit a hierarchical structure. Tactical decisions are delegated to the lower echelons of the hierarchy while strategic decisions are centralized in the top echelons of the hierarchy. This hierarchical division of responsibility is very similar to what exists in the familiar command structure of an army or corporation. Most previous work in biological cybernetics has been concerned with the lower echelons of the control hierarchy, where the nervous system seems to function "automatically" independent of our conscious control. These processes include the transmission of nerve impulses from the brain to control muscle response, the generation of nerve impulses which convey sensory information (vision, hearing, sense of touch, smell, etc.) to the brain, the preliminary processing of such sensory information so that it can be more efficiently transmitted or stored, the operation of reflex actions which proceed without conscious control, and the role of "learning" in simple neural responses such as the conditioned reflex. Although the present development is also concerned with cybernetics, it is concerned with the higher control processes which seem to be under conscious control. Whereas most previous work has tended to work upward from the bottom levels of the control hierarchy, this effort approaches the problem from the top down.

THE ANALOGY WITH COMPUTERIZED SYSTEMS

To develop an understanding of some of the problems involved in the design of a decision system, we will depend heavily on the analogy with the artificial, or computerized, decision systems. These computerized systems exhibit many of the properties usually associated with intelligent adaptive behavior. They make choices, in terms of an internal value scale. When confronted with changes in the environment, they exhibit very systematic and purposeful behavior.

Such computerized decision systems are usually designed to operate on a standard general-purpose computer. It may seem strange that such a standard computer can be made to operate like a sophisticated decision system. In most applications, the computer performs only routine tasks that can be planned in every detail by a human designer. However, the standard general-purpose computers are designed so that they can be programmed to follow almost any process of logical analysis that the human mind can conceive. The way the computer will actually behave depends on a very detailed set of instructions called a "computer program" that is prepared by a human designer. When these instructions are stored in the computer they literally become a part of the computer system and they control the way the system operates.

In order to make the computer operate as a value-driven decision system, the programmer must prepare a very extensive set of instructions that incorporates all the logic of the system. When such a computer program is stored in the computer, the computer will obediently follow the instructions for a value-driven decision system. Thus when we speak of an artificial decision system, we are usually referring to a complete operating system that includes both the physical computer and the special instructions that are stored in the computer.

The amount of flexibility and "intelligence" that can be built into such a computer system of course depends on the capacity of the computer as well as the patience and creativity of the people who develop the computer program. With the miniaturization of computer circuit elements and the resulting availability of computers with millions of circuit elements, it has been possible to build more flexibility into computers and computer programs. As this has happened, the sharp dichotomy between the "machinelike" computer systems and the "flexible" human brain has become much less clear.

Of course, for most applications computers are still concerned with rather routine data processing where there is little need for creativity in the operation of the computer. Indeed for such applications it is usually desirable for the computer to operate in a simple predictable machinelike way, and most computer programs are designed for such "predictable" operation.

The artificial decision systems, in contrast, are usually designed to provide a high level of judgment and flexibility. The designers and the programmers for such systems will often seek to provide the maximum "intelligence" that is practical within the capacity of the computer. The resulting "intelligent" behavior is therefore primarily a consequence of the instructions in the computer program.

Although it is traditional to distinguish between the computer and the computer program, this distinction is not at all fundamental. Modern computer systems always obtain their flexibility from extensive sets of instructions. Some of the instructions are built directly into the "hardware," other instructions called the "firmware" are permanently stored, and still others called the "software" are temporarily stored during the operation of a specific program. But the design principles for a computer system are quite independent of whether the required instructions are contained in the "hardware," the "firmware," or the "software." Thus, the distinction between the computer and the computer program is becoming blurred.

For the purpose of the present discussion, we will think of the computer *plus* the instructions as a single integrated system. In making comparisons between artificial decision systems (those developed within a computer) and biological decision systems (such as the brain) we will usually consider the systems as a whole. What parts of the system are in permanent hardware and what parts are easily modified software is, of course, an interesting question that has to be asked separately for the biological and the computer systems. In existing computerized decision systems almost all of the logic is contained in the software. Presumably in the biological decision systems a much larger fraction is built into the biological hardware.

METHOD OF PRESENTATION

Although the basic concepts involved in the design of the human decision system are simple, they differ in some subtle but important ways from more traditional concepts. Because of these differences, it is necessary to develop the ideas very carefully. This can be done most easily by discussing artificial decision systems first.

If we were to begin with the human system, we would suffer from two very serious handicaps. First, the human brain is so complex that it is very hard to say anything simple that is really correct. Statements intended only to illustrate principles would have to be so qualified that the train of logic would be lost. Second, because of the extensive body of common prejudices and misconceptions, the meaning of words could be easily misinterpreted.

For these reasons we will develop the basic principles in terms of examples selected from experience with artificial decision systems. During this development, in Chapters 2, 3, and 4, some effort will be made to show the applicability of concepts to biological systems. The reader is asked to be patient during this development. Although the relevance of the discussion may not always be obvious, the resulting principles are essential to the later development. In Chapters 5, 6, and 7 the principles are finally applied to provide a better understanding of the human decision system. Thereafter, the relevance of the material to the subject of human decisions and human values should be obvious to the reader.

REFERENCES

1. Bertalanffy, Ludwig Von. *General System Theory.* New York: Braziller, 1968.

2. Buckley, Walter. *Sociology and Modern Systems Theory.* Englewood Cliffs, N.J.: Prentice-Hall, 1967.

3. Buckley, Walter. *Modern Systems Research for the Behavioral Scientist.* Chicago: Aldine, 1968.

4. Laszlo, Ervin. *The Systems View of the World.* New York: Braziller, 1972.

5. Maniatis, Tom, and Mark Ptashne. "A DNA Operator-Repressor System." *Scientific American* 234, no. 1 (January 1976): 64–76.

6. Monad, Jacques. *Chance and Necessity.* New York: Knopf, 1971.

7. Newell, Allen. "Artificial Intelligence and the Concept of Mind." In *Computer Models of Thought and Language.* Edited by: Roger C. Schank and Kenneth Mark Colby. San Francisco: W. H. Freeman, 1973.

8. Pirsig, Robert M. *Zen and the Art of Motorcycle Maintenance: An Inquiry into Values.* New York: Morrow, 1974.

9. Skinner, B. F. *Science and Human Behavior.* New York: Free Press, 1953.

10. Skinner, B. F. *Contingencies of Reinforcement: A Theoretical Analysis.* Englewood Cliffs, N.J.: Appleton-Century-Crofts, 1969.

11. Smith, Nicholas M., and Milton C. Marney. *Foundations of the Prescriptive Sciences.* McLean, Va.: Research Analysis Corporation, March 1972.

12. Tolman, Edward Chase. *Purposive Behavior in Animals and Men.* New York: Century, 1932.

13. Watson, John Broadus. *Behavior: An Introduction to Comparative Psychology.* New York: H. Holt, 1914.

14. Watson, John Broadus. *Psychology from the Standpoint of a Behaviorist.* Philadelphia: Lippincott, 1919.

15. Watson, John Broadus. *Behaviorism.* New York: Norton, 1925.

16. Wiener, Norbert. *Cybernetics: or Control and Communication in the Animal and the Machine,* 2nd ed. New York: MIT Press and Wiley, 1961.

Chapter 2

The Artificial Decision System

We normally consider that our senses
are "windows" to the world . . .
we see with our eyes, hear with our ears.
But such a view, . . . is not entirely true
for a primary function of sensory systems taken
as a whole is to discard "irrelevant" informa-
tion.

ORNSTEIN (6) *

THE DESIGN PRINCIPLES used in artificial decision systems may, at first, seem too formal or sophisticated to be applicable to a biological system such as the human brain. But we should not be too quick to assume that good design principles would be ignored by evolution, just because they seem formal or sophisticated.

Biological evolution has shown remarkable success in rejecting inefficient solutions and in finding efficient solutions to difficult technological problems. Very frequently the evolutionary solution and the human technical solution prove to be the same. Historically, our understanding of biological systems has paralleled the development of technical solutions to similar problems. We understood the skeleton and muscles when we learned the physics of levers. We understood the respiratory and circulatory systems when we learned the principles of chemical energy and the technology of pumps. We gained an understanding of the eye when we learned the technology of the camera. We gained

* Chapter 2.

understanding of the nervous system when we understood information feedback and techniques of coding information.

This experience suggests that we should gain a better understanding of intelligent human behavior as we learn the information-processing requirements for efficient decision making. The success of evolution in finding efficient biological designs suggests that we should seek our understanding of human behavior in those principles of decision making that are both simple and efficient. Research in physiology has already demonstrated numerous ways in which principles of efficiency in data processing have been incorporated into the design of the human nervous system. Therefore, we should not be surprised if nature also exploits principles of efficiency in the design concepts for biological decision systems.

OUR CHANGING CONCEPT OF THE BRAIN

The brain, like any biological organ, must operate within the laws of physics and chemistry. Its biological design must be such that it could have evolved through random genetic variations and natural selection. Because of this evolutionary consideration, early attempts to understand the brain centered on unspecialized concepts.* The brain was viewed as an almost unstructured mass of randomly interconnected neurons. The remarkable properties of the brain were assumed to be an inevitable (but mysterious) consequence of this interconnected neural mass.

This simple and attractive hypothesis is no longer consistent with experimental facts. With increasing information, our concept of the brain has incorporated more and more functional specialization. In the emerging modern concept, the brain looks more like a detailed engineering design.

This revised concept is entirely consistent with experience in the development of artificial systems. Although it is theoretically possible to design artificial systems without much built-in structure that can learn everything from experience, this is an inefficient approach when other alternatives are available. The more a "system design" can be specialized to its specific task, the more efficiently it can perform. The more relevant facts and information about the problem that can be built into the system in advance, the less it will have to learn from experience—and the more efficiently it will operate. The issue of

* For example, Ashby (1), Yovits and Cameron (15).

efficiency in data processing is not just a matter of 10 or 20 percent effects in system size, performance, or cost. The effects of efficient design on size and cost will typically involve factors of 10 to 100 or even 1,000,000. The use of efficient rather than inefficient procedures can make the difference between feasibility in a system of moderate cost and infeasibility even in systems where cost is no object. On the basis of this experience we should expect efficiency to be a *very* important consideration in the evolution of adaptive biological systems.

For readers not familiar with the evidence concerning nature's use of data-processing principles, the next section develops a few illustrative examples. The examples are included as background, so that our application of similar data-processing principles to develop theoretical concepts concerning the design of the brain will seem more natural to nontechnical readers.

INFORMATION-PROCESSING PRINCIPLES IN THE BRAIN

The higher biological organisms are equipped with millions of sensory elements. Relevant information from these sensors must be available for processing at the higher decision levels. Moreover, the brain must be able to retrieve sensory experience stored in its memory so that comparisons can be made between new and old experiences. If *raw* unprocessed data from all sensors had to be processed or stored in the brain, the data-processing and storage requirements would be astronomical. Therefore, principles of efficiency dictate that the brain must provide preprocessing of sensory data to put it in an efficiently coded form. They also dictate a need for selection mechanisms to avoid unnecessary storage of irrevelant data. As expected, both types of mechanisms seem to be included in the nervous systems of the higher animals.

CODING EFFICIENCY

As the sensory information approaches the brain (and as it approaches the higher decision and storage centers within the brain) there is a progressive reduction in the number of neurons involved in transmitting the information. This reflects an improvement in coding efficiency as raw sensory information is processed into more meaningful and more compact forms.

Probably the most spectacular example of this principle has been found in the processing of visual information. Raw visual data contains a great deal of redundant pictorial information. Television engineers have known for years that the essential pictorial information is contained primarily in the edges of the objects in a picture. This fact is used by artists and cartoonists, who can convey the essential information from a photograph with just a few pencil lines. Using a similar principle, it is possible to transmit only the boundary data (where there are sharp changes in color or darkness) and simply infer the colors of the interior of objects from the color of the boundary. When this method is used it permits much more compact storage of pictorial data and allows television pictures to be transmitted through communication channels of lower capacity. Even further improvements in efficiency are possible if information is transmitted only at those times when there are changes in the picture.

These techniques are considered quite sophisticated and at present their use is very limited. Nevertheless, evolution has provided for just such processing in the optical pathways that lead to the brain. Although the sensory neurons in the retina of the eye respond directly to the received light, the neurons further along the optical nerve are so connected that they respond primarily to the differences in color and brightness between neighboring pictorial areas. Indeed neurons are found that respond only to those pictorial boundaries with a specific orientation (horizontal, vertical, and intermediate angles; see Figure 2.1). Others are found to respond only to moving objects.

As signals move along the optical pathway toward the brain, neurons with more and more sophisticated responses are encountered. The pictorial representation changes from an inefficient literal duplication of the picture toward a more efficient abstract or symbolic representation. Specific neurons have been identified in the optic nerve of the frog that respond only to small dark moving objects about the size of a bug flying in range of the frog's tongue!

In the human brain there is evidence that the principle of boundary information is carried much further than just the identification of edges. In cases of eye damage or damage to the optical nerve, when parts of the picture are missing, the victims are usually unaware of the missing parts of the picture. The picture that they "see" seems to be completed across the visual gap. The picture is completed, not just with uniform colors, but with a continuation of checked patterns, lines, or whatever is required to fill the gap. The only symptom of the disability occurs when small objects fall *entirely* within the area of the visual gap. Such objects simply are not perceived.

The techniques used in the optic nerve to provide efficient coding of information are not perfect. They can be fooled by certain contrived situations. This is the basis of many famous optical illusions. Two such illusions (Figures 2.2 and 2.3) are included here to show that our visual perception is a symbolic rather than a literal representation of reality.

Figure 2.2 shows that the optic nerve includes neurons that respond differently to different gradations in the light intensity. Those that respond to

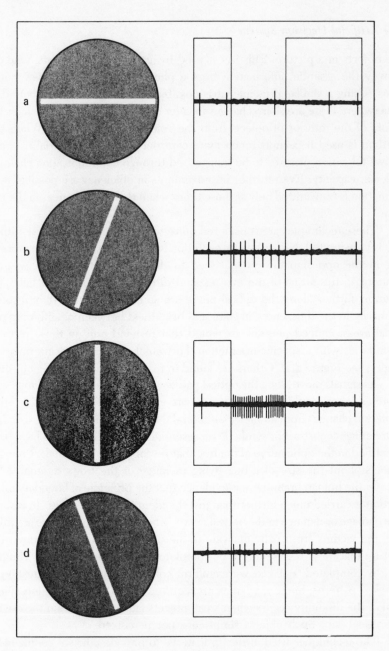

FIGURE 2.1
Illustrative Response of Neuron to Slit Orientation

A single complex cell in the cortex shows varying responses to a slit projected onto the cell's visual field in the retina in an experiment by Hubel and Wiesel. The cell only responds to the image of the slit when it is in a particular orientation (vertical) with respect to the retina. Various positions of the slit are shown at left; at right the responses of the cell.

From *The Conscious Brain* by Steven P. Rose. Weidenfeld & Nicolson, Ltd., 1973.

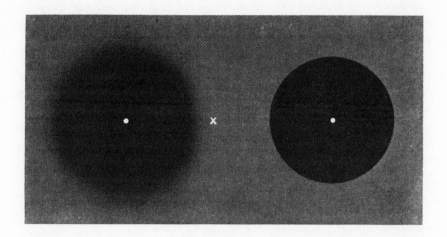

FIGURE 2.2
The Disappearing Circle

Look steadily at the left dot; the hazy circle disappears. You can make it reappear by looking at the X. The right-hand circle will not disappear if you stare at it, although it is as light in the center as the other circle. The visual system, like other sensory systems, is maximally sensitive to sharp changes. Lacking sharp changes either in position or time, the image on the left simply fades away.

From *The Psychology of Consciousness* by Robert E. Ornstein. W. H. Freeman and Company. Copyright © 1972.

FIGURE 2.3
Your Visual Blind Spot

To find the blind spot, close your right eye and stare at the circle. Hold the book 9 to 10 inches from the eye, and move it toward you and away until the square disappears. Notice that this "hole" is always present in your visual world. If you stare at the square and close your left eye, the dot will disappear.

From *The Psychology of Consciousness* by Robert E. Ornstein. W. H. Freeman and Company. Copyright © 1972.

gradual shadings cease functioning when exposed to a constant unchanging stimulation. Those that respond to sharp edges continue to operate even though there are no apparent changes in the stimulation.*

* Actually, minute rapid eye movements produce changes in the position of the image on the retina that serve to keep these neurons active. If the picture is projected so that it moves in synchronism with these eye movements, then these neurons also will cease to function.

Figure 2.3 illustrates the phenomenon of field continuation. Even the normal eye includes a small blind spot away from the center of vision. The figure illustrates the disappearance of images that fall in the blind spot.

Many of the techniques of advertising display (bright colors, sharp lines, moving displays, bright lights, etc.) are designed to defeat our built-in mechanisms for rejecting irrelevant information. As we step into a brightly lit supermarket we may experience a sensation almost like pain. We receive an overload of visual information that does not automatically filter out. We may feel impelled to close our eyes, or to deliberately limit our field of visual attention. Ordinarily the coding and preprocessing of visual data protects the higher levels of the brain by limiting the quantity of visual data to manageable levels.

It is particularly noteworthy that this sophisticated coding or compression of visual data is actually built into the optical hardware of the brain. It is not a learned or adaptive behavior, but is part of the original genetic design.

REJECTION OF IRRELEVANT DATA

The brain also exhibits remarkable selectivity in the choice of information to forward to the higher levels. We know that as we fix our attention on specific aspects of our environment (or as we concentrate on problems) we become less aware of irrelevant sights or sounds. Scientific experiments have shown that such extraneous sensory signals not only seem to be reduced, but that they are in fact reduced. This was first shown in 1956 in a series of experiments by Hernandez-Peon, Scherrer, and M. Jouvet (3). A sensing electrode was placed in the brainstem of a cat, in the region where signals from the ears are processed before passing on to the upper brain. The cat was exposed to a clicking noise, and corresponding electrical pulses in the brainstem were measured by the electrode. But if the cat was shown mice in a glass container (or other objects of great interest) the measured electrical pulses were greatly reduced. Thus, the brain includes distinct physical mechanisms for selectively controlling information that comes to the attention of the higher centers of the brain.

SELECTIVE STORAGE OF DATA

To limit data storage requirements we might expect that the sensory memories would be stored only in their most compact and selectively limited form. This assumption also seems to be borne out by physiological evidence.

In preparing patients for brain surgery it is sometimes necessary to use miniature electronic probes to map the region of damaged tissue or tumors. The probes do only minimal damage to brain tissue, and because the brain has no sense of touch or pain, the tests can be made without discomfort to the patient. The surgeon uses the patient's response to such electrical stimulation as a guide in diagnosing the area of damage.

Sometimes the electrical stimulation triggers the release of stored sensory information. The patient experiences a vivid recall of past experience, which unfolds almost as if he were reliving the experience. Usually this replay of the experience will continue as long as the stimulation is continued. But the patient is also conscious of his current environment, so he can describe the experience. If there is music in the experience, he can sing or hum along, just as if he were watching a movie. If the stimulation is stopped the experience ends. If the stimulation is resumed, it is as if a movie film had been rewound. The experience restarts from the beginning. Wilder Penfield of the Montreal Neurological Institute has reported on about forty cases of this type (7), and reports that the memory flashback seems always to be *limited* to those aspects of sensory experience that fell within the patient's consciousness at the time. Thus, it appears that the memory traces may indeed be maintained (as they are received by the higher brain) in the most compact and selective form.

APPLICATION TO THE DECISION PROCESS

The foregoing examples are illustrative of a rapidly growing body of data which shows that data-processing and data-handling procedures in the brain are highly developed and very specialized. It is obvious that evolution has given meticulous attention to the procedures for processing and storing sensory information. But the stored information can contribute to survival only if it permits better and more efficient decisions. Therefore, it seems most unlikely that evolution would have failed to provide similarly efficient decision procedures.

The psychological model of conscious human behavior that is now most widely accepted is the model of operant conditioning. According to this model, specific random behavioral responses to a stimulus can be reinforced by systematic rewards, so that after many rewards the response will occur (in a very pronounced form) with almost 100 percent reliability. Although there is no doubt that such conditioning processes occur in the nervous system, it seems most unlikely that this is the mechanism of conscious decisions. If we were to take operant conditioning seriously as the main human decision process we would have to believe that nature carried its sophisticated design only to the threshold of the higher decision centers and then settled for a very crude and inefficient central control. On a priori grounds this seems most unlikely.

The available evidence shows that evolution is capable of very sophisticated cybernetic designs. Clearly the more efficient decision mechanisms should not be neglected in any study of the brain simply because they involve specialized cybernetic designs. Since evidence shows that the principles of information theory are indeed respected and effectively exploited by biological systems, we will now ask what we can learn from our automation experience.

BASIC PROBLEMS IN ADAPTIVE BEHAVIOR

Most recent work in artificial intelligence has been concerned with the duplication of rather sophisticated intellectual processes, such as communication in English or other natural language, solving puzzles, proving theorems, or playing games such as chess or checkers. Impressive progress has been made in all of these areas,* and some of the concepts from this work will be of interest to us later. But it seems unlikely that such sophisticated processes can provide the key to evolution's basic design concept for efficient adaptive behavior. Such processes appeared late in the evolution of biological systems, and many of them are unique inventions of modern man. It seems more likely that the human animal superimposed such intellectual accomplishments on a more basic mode of adaptive behavior that is common to all higher animals.

Our initial interest therefore will be focused on the everyday aspects of conscious adaptive behavior: the maintenance of adequate nourishment, the avoidance of dangers, the care and feeding of young, and so on. We will be concerned with the design of systems capable of making good "commonsense" decisions with regard to such issues.

Of course, the experience with practical automation has not been concerned with such everyday problems. Instead it has been concerned with systems to assist human executives with decision and planning problems. Although this work in automation has not dealt with the exact problems faced by biological organisms, the general character of the problems (involving alternative responses to a variable environment) is quite similar to traditional problems of adaptive behavior. We will begin by considering some lessons learned in the development of such systems.

---

* For example, Minsky (4), Chomsky (2), Newell (5), Simon (12), Winograd (14), and Schank and Colby (11).

THE NEED FOR VALUE-DRIVEN DECISION SYSTEMS

The first obvious question is, "Why should evolution have developed a complex behavioral mechanism, such as the value-driven decision system?" Why not use a "simple" stimulus and response system like the tropism? *

Some computerized planning systems are indeed constructed by designing a set of preplanned procedures, together with a list of rules of thumb (a decision table) which specifies the circumstances under which each procedure is to be invoked. Obviously this approach is similar to the preprogrammed muscular responses, called tropisms, that characterize the behavior of lower animals. But where the decision problems are complex and variable this design concept of the tropism is almost unworkable. When the simple decision table approach is used, the designer himself must preplan and specifically program every required response. Moreover, he must specifically define the kinds of stimuli that are to generate each such response. This places much too great a burden on the advance planning ability of the designer.

Often it is possible to reduce this burden on the designer by using control theory, or feedback principles. Using the feedback principle, the designer can simply specify some control variable, such as body temperature, that is to be kept at some specified control level. At the same time he will specify compensating actions, such as increasing or decreasing the rate of metabolism, that are to occur when the control variable moves outside the specified control range. Then when any disturbance moves the control variable outside the desired range, the system will automatically correct for the disturbance. Thus the feedback principle minimizes the need to plan future responses in great detail.

Such feedback systems can be linked together to provide rather sophisticated control systems. For example, the desired value of the control variable for one feedback loop can be specified dynamically as the output of another control loop. Through such techniques it is possible to provide rather complex system responses to a number of variables. This type of design concept is undoubtedly used in many of the biological systems that are related to the control of metabolic processes. The feedback principle is also widely used in both mechanical and biological systems to simplify the control of output devices such as muscles and automated tools. The possibilities for providing complex adaptive behavior for biological systems through the use of a hierarchy of such control systems has been discussed in some depth in a recent book, *Behavior: The Control of Perception* by William Powers (8).

* See Chapter 1, page 18 for a definition of the tropism.

Although this type of concept generally works quite well, and is undoubtedly used, at the lower levels of a control hierarchy, recent automation experience has shown that it does not really solve the decision problems that must be faced at the highest control level of a complex system. At the highest control level, procedures are needed to dynamically adjust the control levels for the relevant control variables. When we think realistically about control variables, such as the pressure between the fingers that is to be used in grasping a tool, or the dynamic positioning of a tool as it is used to work on some artifact, it becomes apparent that the dynamic adjustment of control variables can be a very complex problem.

Some technique must be provided that will generate an appropriate sequence of control variable settings to accomplish such tasks. This problem leads us back once again to the consideration of familiar techniques, such as decision tables, to dynamically adjust the required control variables. Indeed, such approaches (which combine feedback techniques with decision tables) can be quite successful in automated systems that deal with rather simple, essentially repetitive control problems. However, such decision-table control techniques are not suitable for the creative solutions of complex problems, such as the development of tools, or the solution of other problems that may not have been anticipated by the designer.

Many planning processes require a complex and interacting set of decisions that cannot be preplanned in a decision table. For example, consider the steps an architect must face in developing a plan for a home or the interactions involved in scheduling a fleet of commercial aircraft. The development of a decision table for such activities is extremely difficult, because there is a long sequence of decisions to be made. The correct decisions for the later stages depend in detail on decisions that have already been made. The designer of such a planning system would be faced not with just a single decision table but with a branching tree of them where each new decision leads only to the next decision table. For each branch in this ever-expanding set of alternatives, specific responses would have to be anticipated and programmed.

But the problem is really even more difficult. Even the best architect does not produce a satisfactory design on the first attempt. He tries many different design concepts. As he develops the details, some of the concepts prove to be unsatisfactory, others show promise but need to be refined. Thus the planning process involves the exploration of many alternatives. If a system designer is to develop satisfactory decision tables for such a process he must think through all possible alternatives in advance. This is simply too much to ask of the system designer, whether he is a human designer or the evolutionary process.

I vividly recall my first effort to deal with a really complex planning problem using the tropism, or decision-table design concept. In the late 1950s and the early 1960s the U.S. Defense Department was interested in automatic planning of bomber flight plans. The problem was to develop flight plans that would allow each bomber to strike several important targets and at the same time minimize the exposure of bombers to enemy defense systems. Some experienced military officers had already assembled a list of about thirty to forty rules to be followed by the decision system in the development of the plans. I will mention just a few to give a flavor of the problem:

> Aircraft will normally fly in a straight line between targets except where this would involve flying too close to enemy surface-to-air missile defenses.

> The angles of turn from one target on the route to the next should not exceed about 45°, to avoid excessive wandering of penetration routes.

> All targets of specified types are to be attacked with at least two weapons.

> No single bomber will drop more than one weapon on a specific target.

> All bombs carried by the aircraft must be used against some important target.

> Provision must be made to recover all aircraft and crews in friendly territory.

> To minimize exposure to defense radar systems aircraft will fly at low altitude at least to the first target and as far thereafter as fuel reserves permit.

To design a computerized system capable of generating appropriate flight plans, a *complete* and *internally consistent* set of rules was needed. Careful examination of the specified rules showed numerous situations in which they were contradictory and inconsistent. An effort to remove the inconsistencies resulted only in the addition of numerous exceptions and numerous new rules. As the list of rules grew, the sequence of branches that would have resulted in the computer code became more and more unworkable—and the possibility of actually producing a successful computer program to implement the rules became less and less.

Such rules had worked reasonably well as a guide for human planners, because when the rules were in conflict (or inapplicable) the human planners could use common sense. But the computer had no "common sense" to fall back on. When inconsistencies in the rules were encountered the computer was sure to do one of a number of unfortunate things: (1) make an arbitrary—probably foolish—choice, (2) make no choice and fail to produce a plan, or (3) go into a computational loop and continue forever seeking a nonexistent solution. It was clear that unless the computer could be made to apply common sense to unexpected situations the whole approach was doomed to failure.

Our effort then began to focus on the commonsense aspect of the human decision process. Typically, when faced with a dilemma, the human planner

ternatives. The design of a good value structure is far from trivial, but it is much easier than trying to anticipate all future alternatives. The value-driven design concept is uniquely simple and robust. It can operate at a very low level of cybernetic resources or at a high level. As the level of cybernetic resources increases it provides the possibility of extremely intelligent and adaptive behavior.

The foregoing discussion demonstrates the basic advantages of the value-driven decision system. The field of decision science is relatively new, so the power and generality of this basic concept as a formal design principle is only now beginning to be recognized. Nevertheless, as an intuitive concept it has been widely used in the design of practical computerized decision systems. In the following chapters it should become apparent that the same design principles have also been used by evolution in the design of biological decision systems. But to understand the biological systems we need a better general understanding of the structure and operation of value-driven decision systems.

BASIC STRUCTURE OF A DECISION SYSTEM

To use the value-driven decision method certain basic functional components are needed. In one form or another the following elements are included in almost all value-driven decision systems:

1. A data collection procedure to supply information needed to define the environment as it affects action alternatives.
2. A model of relationships in the environment which defines action alternatives and their consequences.
3. A procedure for exploring available action alternatives and estimating their consequences.
4. A method for assigning values to the estimated consequences.
5. A decision mechanism for selecting the alternatives that show the best value.

A decision system may also include other elements such as procedures for creation, improvement, and refinement of a model. However, at least the five elements above are required. The existence of the five separate elements is sometimes obscured in a simple decision system. For example, simple decision systems can be built around formal mathematical optimization methods such as linear programming. In such systems many of the separate functions may seem to merge into a single algorithm. For readers familiar with linear programming, the following paragraph describes how all five functions are combined within the linear programming approach. Other readers may choose to skip the paragraph.

The linear programming method itself is a procedure for systematically exploring alternatives (item 3). The procedure is called an optimization method because it ensures that the best of all feasible alternatives will be found. The payoff (or cost) which is defined as a linear function of the activities provides a method of assigning a value to every possible alternative solution (item 4). The linear programming procedure is designed so that new alternative solutions are considered only if they result in an improvement in value, and the process stops when no further improvement is possible, thus providing the decision mechanism for selecting the best value alternative (item 5). The coefficients of the activities, which define the rate of consumption or production of resources by each specified activity, provide the required model of the environment (item 2). The data collection process (item 1) of course corresponds to the input data which defines the current values of the constraints and other coefficients needed to define the current environmental situation.

The linear programming decision system, however, represents almost a degenerate limiting case. Because of its simplicity, it is widely used, but it is not at all typical of systems dealing with complex problems. The model of the environment that can be incorporated within a linear program is so stylized that it is hardly recognizable as a specific model of the problem. The format is too rigid to allow incorporation of richer, more realistic models such as those needed by advanced biological decision systems.

As the decision problems become more complex the following trends seem to occur in the structure of a practical decision system:

The model of the problem becomes much richer and the correspondence between the model and the real world problem becomes much more obvious and direct.

The rules for manipulating the model to consider alternative feasible courses of action become much more detailed and correspond much more closely to the real world alternatives.

The procedures for calculating the consequences of the decisions become more complex and begin to resemble a direct simulation of the environmental system.

It is no longer feasible to provide a systematic search through the possible solutions to guarantee achievement of the best solution. As the problem complexity increases, the systematic search methods of linear programming give way to less efficient branch and bound techniques. As the complexity increases further, the formal requirement of an optimum solution has to be dropped and the search procedure begins to resemble a more random exploration of alternatives. Of course, the system efficiency still depends very strongly on the extent to which good principles can be found to guide the search.

The exploration of alternatives is terminated when the cost of exploring additional alternatives becomes comparable to the benefits that can be expected from further improvements in the solution.

The value criterion needed to determine which alternatives are best tends to become more complex.

The types of decision problems facing the higher biological organisms are typically very unlike the familiar problems of linear programming. In our further consideration of artificial systems we will be speaking primarily of more complex kinds of decision problems that more closely resemble the problems of the biological organism. Although the formalization of concepts is still in an embryonic stage, there has been enough practical experience with such systems to define important design principles relevant to our understanding of biological decision systems.

EFFECTS OF ACCURACY OF THE MODEL

One of the most important distinguishing characteristics of the value-driven decision concept is the inclusion within the decision system of a detailed symbolic model of the problem environment as an essential component of the control or decision process. The model is used to project or estimate the probable outcomes for different courses of action. The quality of the decisions that can be produced by such a system are completely dependent on the ability of the model to correctly project the probable outcomes. If the model is inaccurate or incomplete so that it does not correctly project the outcomes, the performance of the decision system will be correspondingly degraded. Experience in the development of artificial decision systems has shown that it is usually far better to obtain an approximate decision with an accurate model than it is to obtain a mathematically optimum decision with an inaccurate model.

Although the models that are used in such computerized systems are always (like the mental models in the brain) symbolic rather than physical, they must nevertheless incorporate enough realism to be accurate in predicting outcomes. For those who are used to working with simplified mathematical equations, it may be difficult to visualize the accuracy with which symbolic models can provide a correspondence with reality. In one recent application involving an air combat decision system, the symbolic model in the computer was actually used to generate a video display, much like a moving picture, that provided a visual representation of an air battle involving hundreds of aircraft.

This video representation was developed so that users could confirm the validity of the symbolic model used in the computer by personal visual observation.

It appears that evolution has also discovered the importance of an accurate and reliable model in order to obtain good performance in its decision systems. The elaborate processing of information in the optic and audio nerves, as well as the large amount of the cerebral cortex devoted to the analysis of visual images, testifies vividly to the importance evolution has placed on an accurate and up to date model of the environment. Indeed, it appears that one of the most demanding and important evolutionary functions of the human brain is concerned with the development and updating of this mental model of the world environment.

The development of the cybernetic mechanisms within the brain that enable it to spontaneously generate and maintain this mental model is undoubtedly one of the most spectacular achievements of biological evolution. It is an achievement that we are not even close to duplicating in computerized systems. Because of our present inability to duplicate this feat in computerized systems, the models that are used in artificial decision systems must still be laboriously specified by human designers. A recent review of the present practical limits of decision automation concluded that this impediment in the specification of computerized models constitutes one of the most important barriers to the automation of complex decision processes (9).

One of the important consequences of using detailed and realistic models is that the analysis of alternatives by formal mathematical techniques such as linear programming or control theory usually becomes completely impractical. So the system designer is usually driven to heuristic and semi-random search techniques similar to those we use to consider alternatives in our commonsense decision-making processes.

EFFECTS OF ACCURACY OF OPTIMIZATION

During the early history of decision science, research workers were much concerned with the mathematical tools of the trade. Because of their training in mathematics they placed great emphasis on the logic and precision of the decision process. They were influenced by the traditional mathematical dichotomy, of "right" versus "wrong" answers, and tended to concentrate on formulations that made it possible to obtain truly "optimum" decisions. As a consequence there was a tendency to oversimplify both the model of the problem and the desired goals so as to fit the problems into a mathematical framework that would permit rigorous optimization. Such oversimplification in problem formulation was the cause of numerous fiascos, and many workers began to lose faith in the principle of optimization as a practical decision tool.

There is now a growing recognition that practical decision problems do not typically require "optimum" solutions. It is far more important to provide a good description of the problem and objectives (even if less rigorous optimization methods must be used). Most of the absurd results obtained in early analyses were the result of obtaining a rigorously "optimum" solution to the wrong problem.

As the size and complexity of decision problems has increased, it has been necessary to reexamine the economic desirability of rigorous "optimization." Even for problems where a true optimum is theoretically achievable, one must ask whether the benefits justify the cost of calculation. In many cases an approximate solution can be obtained quite cheaply, whereas the precise optimum is very costly. The solution techniques for complex nonlinear problems often use a method of successive approximations that may never achieve an exact solution, but they may very easily provide a solution that is within about 1 percent of the optimum.* For most practical decision problems this is more than adequate accuracy.

Practical decision issues are quite unlike a mathematics problem which has rigorous "right" and "wrong" answers. They are more like an engineering art in which there is a range of alternatives that can be graded in quality: good, better, best. For large complex decision problems, one can often find hundreds (or thousands) of alternative courses of action that are almost optimum. Since improving the accuracy of the solution costs time and money, the cost of accuracy in the optimization becomes an important factor to be considered in the decision process.

Work on approximate computational methods has led to the development of a wide range of techniques. These include special heuristic techniques, network optimization methods, generalized Lagrange multipliers, branch and bound techniques, and nonlinear programming techniques. The range of available techniques has developed so that an experienced analyst (given time and computer resources) can usually find a combination of techniques that will provide useful approximate solutions for almost any well-formulated decision-science problem.

But since it is not the purpose of this book to discuss such mathematical techniques, we will simply assume that the computerized decision systems

---

* Sometimes formal bounding techniques can be used to define an upper bound for the optimal solution.

include suitable procedures to allow them to explore alternatives and select one which is quite good, perhaps approximately optimum in terms of the specified decision criterion.

IMPORTANCE OF VALUES USED IN A DECISION SYSTEM

It is obvious that the quality of the behavior that can be expected from a decision system will depend on several factors:

The accuracy of the representation of the problem environment that is used.

The accuracy of optimization.

The specific value system used to provide the decision criterion.

It is perhaps less obvious that, of these three factors, system performance is typically *most* sensitive to the specified *value system*. The *extreme* importance of the value criterion for system behavior is probably the most significant single result to come out of the recent experience with artificial decision systems.

In the past, workers in the decision-science field have tended to concentrate their attention on the precision of the model and the accuracy of optimization. The crucial importance of the value issue has only recently been recognized. Our experience with the behavior of artificial adaptive systems makes it clear not only that the value system is an essential part of the system description, but also that it is the most *decisive* element in determining behavior. The development of a really satisfactory value system is far from trivial. This point is of great importance in understanding the design and behavior of biological decision systems. It helps to explain the extreme complexity of the built-in human value system. It also helps us to explain some of the nonfunctional behavior which will always occur as a by-product of any value system that is less than perfect.

We identified earlier (p. 54) five key functional elements that are common to almost all artificial decision systems. When we ask how these functions correspond to our subjective mental experience, we find that four out of the five functions are experienced subjectively as a part of conscious mental activity. The assignment of *primary* values, however, occurs *outside* the veil of consciousness. The primary values are orchestrated by automatic neural and chemical processes that are not a part of conscious mental activity and are not normally under conscious control. This primary human value system is the source of the fundamental or "instinctive" human values. The other four components identified in artificial decision systems appear in human beings as a part of the "rational" thought process.

The primary human value system is experienced by the individual in the form of specific valuative sensations of physical and psychological origin which include both our fluctuating emotions and our biological drives. The paradox of rational thought versus irrational emotions originates as a result of this automatic and nonrational assignment of primary values. It is no accident that the control of this primary value assignment lies outside the veil of consciousness. If the value assignment were part of conscious thought, it would be subject to conscious modification. This could destroy the entire intent of the system design.

It must be emphasized that the primary value system or decision criterion is always *built into* the system by the designer. It is not subject to review or modification by the decision system. The decision system may be allowed to modify or adjust secondary values to make them more consistent with the primary values, but the designer must not allow the system to modify its primary value system. If he were to do so, the values could be unpredictably changed, and the system would no longer behave in accordance with the goals for which it was designed. Such changes, if permitted, could completely defeat the objectives of the designer. Thus, from within any decision system, the primary values will appear to be absolute, or beyond question.

One might think that a designer could provide a higher valuative standard against which such changes in primary values might be evaluated, but if he were to do so, he would just be substituting a *new* primary value system. What had been the primary value system would become simply a set of secondary values to be tested and evaluated against the *new* ultimate standard. Thus, almost by definition, a decision system cannot be designed to improve on its own primary value system. The choice of this ultimate or primary value system is solely the responsibility of the designer.

THE CONCEPT OF DERIVED OR SURROGATE VALUES

Up to now we have discussed the derivation of values in only a very general and intuitive way. In Chapter 1 we noted that human secondary values can take many forms: habit, rules of thumb, wise proverbs, social taboos, and moral and ethical principles. In most cases, these secondary human values are developed through rather vague intuitive processes so that the logical relationship of the secondary to the primary human values tends to be obscured.

The process of value derivation is much more obvious when we are dealing with artificial decision systems. Derived or surrogate values are used in such systems to simplify the decision process. They make it possible to obtain reasonably satisfactory decisions from a much less complex decision process. The sole purpose of the surrogate values in such a system is to generate decisions that would be judged as reasonably "good" when evaluated in terms of the designer's real objectives for the system. The derivation of the surrogate values can be accomplished either by the designer or by the decision system itself.

A system that is designed to use an ad hoc value criterion that was developed intuitively by the designer is usually referred to as a "heuristic" system (or algorithm), and the value criteria specified for such systems are described as "surrogate values," "judgmental values," or "heuristic decision criteria." Although these decision criteria, from the point of view of the designer, may be only rough approximations to his real objectives, they nevertheless define the ultimate or primary decision criterion that will be used by the decision system. From the point of view of the decision system they become the primary values.

But a decision system can also be designed to automatically generate surrogate values that will serve to simplify its own decision processes. This automatic derivation of surrogate values is most commonly encountered in systems that use formal methods of mathematical optimization. In such systems the surrogate values arise automatically from the mathematics, and they are variously described as "shadow values," "marginal values," or "Lagrange multipliers." When a decision system automatically generates such surrogate values (either formally or informally from its own primary values), the resulting values are described as secondary values. Such secondary values can serve as a very important aid to the decision process. The human "secondary values" are analogous to surrogate values of this type. They are derived informally from the genetically determined primary human values.

This concept of derived values is absolutely fundamental to the theory of values, so it is important to make the principle as clear as possible. To illustrate the basic concept we will consider two different examples. The first example illustrates a situation in which the values can be formally derived by mathematical methods; the second example illustrates a situation in which the derivation of values requires more informal methods.

1. *Surrogate values that are formally derived.* Consider the case of an idealized manufacturing company which has no objective except to maximize

profit. The manager for such an organization (as a rational decision maker) should use profit as his one and only "real" value, but he may use many "surrogate" values. For example, he might notice that his production capacity is limited by a shortage of workers with certain specific skills. For each additional skilled worker, he might be able to increase his production by some amount, P, and he might calculate that for each increment, P, of additional production, he could produce D dollars in additional yearly profit. He might therefore conclude that each skilled worker is worth D dollars per year in profit. In economic terms this would be the "shadow value" of the skilled worker. The shadow value (or surrogate value) is an aid to his decisions. It helps the manager estimate how much he should be willing to pay for an educational program to train more skilled workers.

The shadow value is not a "real" value, it is a "surrogate" value. From the point of view of our idealized company manager, the worker has value *because* he can contribute to corporate profits. Although in this hypothetical example, profit is the only "real" value, the manager is nevertheless well advised to act as if the "shadow value" is real. In this simple case, it is possible to rigorously, or mathematically, calculate the shadow value of a skilled worker, so the idea that the surrogate value is "derived" from other values seems obvious. However the relationship is not always so simple.

2. *Surrogate values that are informally derived.* Sometimes surrogate values can be learned from experience but cannot be rigorously calculated. Consider the role of "values" in the strategy of chess.

In an idealized game of chess, the only real objective is to win, or at least achieve a stalemate. Other things have value only insofar as they affect the probability of a win or a draw. We know that a queen is a valuable piece and a bishop is a less valuable piece. Most experienced players use certain rule-of-thumb numerical value ratios to define the value of a bishop or a queen relative to the less valuable pawns. An experienced player will also assign intuitive values to certain strategic configurations on the board. The numerical values are not exact; they cannot be rigorously derived; and they may sometimes be misleading. Nevertheless, they serve as a useful guide to decisions. The experienced player uses these "surrogate" values almost as if they were real.

It is almost impossible to play chess without using "surrogate" values as a guide. If one were to try to play the game without using such values, the only way to evaluate a move would be to follow the consequences of the move, through *all possible* succeeding moves until a checkmate or stalemate occurred. For existing finite decision systems, this is clearly impractical. Although such an exhaustive examination of alternatives is sometimes used to solve very simple games (like tick tack toe) it is totally impractical in chess and most traditional games.* In practice, whether the game is played by a human being or even a very large computer, it is essential to utilize surrogate values. Although the consequences of a move may be projected several moves ahead, the outcome of the move (except in the end game) must ultimately be evaluated using the surrogate not the ultimate values.

An artificial decision system can be designed to learn such surrogate or secondary values from experience, or to use experience to improve initial surrogate values provided by the designer. For example, if the designer had estimated that a queen is only 5 percent more valuable than a bishop, a decision system could gradually learn that this estimate is low. In contrast, the system cannot improve its primary values on the basis of experience. If the designer were to tell the system that a win is only 5 percent better than a draw, the system could never improve on this estimate. Although primary values cannot be rationally deduced within a decision system, secondary surrogate values can be estimated from primary values.

The two examples above illustrate the role of derived values in practical decision making. Human judgment is based in large measure on the intuitive development of extensive networks of derived values. In the development of such a judgmental value network, the derivation process may be successively applied to generate new value criteria that are further and further removed from primary values. Just as secondary values can be derived directly from primary values, they can also be derived from other secondary value criteria. As the development of a judgmental network proceeds, the specific relationship of new value criteria to the primary values tends to be less and less clear. For example, a new value criterion may be evaluated in terms of *both* primary and secondary values.

The concept of value derivation is fundamental to the theory of values. It not only explains the origin of human secondary values, but it also allows us to understand how the genetically determined innate or primary values could have developed as a result of biological evolution. Evidently the primary or innate human values are related to the goal of species survival in much the same way that the values of the chessmen are related to the object of winning

---

* The values normally used in chess have absolutely no basis if we consider the game from a rigorous mathematical perspective. The values are relevant and useful only because the players are finite imperfect systems. Players of different skills might find that quite different values would be appropriate. Ideal or perfect players would play an optimum or perfect strategy, so the outcome of the game (win, lose, or draw) would be determined as soon as the status of the board was specified. Every possible change in the status of the board would either leave the outcome unchanged or change it between the three basic states: win, lose, or draw. Obviously, with such perfect players the traditional "values" have absolutely no meaning.

the game. From this evolutionary perspective, the primary human values are simply surrogate values derived from the ultimate objective of species survival.

The next chapter addresses two key questions:

1. Why has evolution chosen to use an innate value system involving so many separate components?
2. Why has it chosen to use values that fluctuate with time?

It will be shown that these characteristics of the value system reflect design compromises that were necessary to achieve satisfactory system behavior with limited cybernetic resources. In a decision system of finite capacity, it is possible to make great savings in the data-processing requirements by adding some complexity in the value system.

REFERENCES

1. Ashby, William Ross. *Design for a Brain*. New York: 1952; rev. ed., 1960.
2. Chomsky, Noam. *Syntactic Structures*. Atlantic Highlands, N.J.: Mouton, 1957.
3. Hernández-Peón, Raúl, Harold Scherrer, and Michel Jouvet. "Modification of Electrical Activity in Cochlear Nucleus during 'Attention' in Unanesthetized Cats." *Science* 123 (1956): 331–332.
4. Minsky, Marvin L., ed. *Semantic Information Processing*. Cambridge, Mass.: M.I.T. Press, 1968.
5. Newell, Allen. *Computer Simulation of Human Thinking*. The Rand Corp., P-2276, April 1961.
6. Ornstein, Robert Evans. *The Psychology of Consciousness*. New York: Viking, 1972. Paperback: San Francisco: W. H. Freeman, 1972.
7. Penfield, Wilder. "Speech, Perception, and the Uncommitted Cortex." Conference on Brain and Conscious Experience, Sept. 28 to Oct. 4, 1964. Reported in *Brain and Conscious Experience*. New York: Springer-Verlag, 1966.
8. Powers, William T. *Behavior: The Control of Perception*. Chicago: Aldine Pub. Co., 1973.
9. Pugh, George E. "The Limits of Decision Automation" in *Mathematical Decision Aids for the Task Force Commander and His Staff*. McLean, Va.: General Research Corp., Jan. 1976.
10. Rose, Steven. *The Conscious Brain*. London: Weidenfeld & Nicolson; New York: Knopf, 1973.
11. Schank, Roger C., and Kenneth Mark Colby. *Computer Models of Thought and Language*. San Francisco: W. H. Freeman, 1973.
12. Simon, Herbert A. *Models of Man*. New York: Wiley, 1957.
13. Von Neumann, John, and Oskar Morgenstern. *Theory of Games and Economic Behavior*. Princeton, N.J.: Princeton University Press, 1944.
14. Winograd, Terry. *Understanding Natural Language*. New York: Academic Press, 1972.
15. Yovits, Marshall C., and Scott Cameron, eds. *Self-Organizing Systems*. Proceedings of interdisciplinary conference, 5–6 May 1959. New York: Pergamon, 1960.

Chapter 3

Behavioral Experiments with Artificial Systems

Optimum is relative to the time and resources constraining the decision process. Even in hindsight, no universally "best" decision can ever be identified.

SMITH AND MARNEY (7) *

THIS CHAPTER describes some experiences with artificial systems that illustrate the need both for multiple value components and for values with a complex time dependence. In discussing such experiments we will be concerned only with the valuative issues and will not trouble the reader with the mathematical aspects of the problems.

Computerized value-driven decision systems exhibit very complex adaptive behavior. Changes in the problem environment give rise to apparently unpredictable but purposeful changes in behavior. Of course, when the value structure of a system is completely understood, it is usually possible to trace the reasons why specific decisions were made. Nevertheless, even with such complete understanding, the behavior of a system often surprises and indeed may seem to outsmart the designer.

The specialist working with artificial decision systems is in a unique position to study the effects of the value structure on behavior. He can know exactly what value structure he is using and can make controlled experiments in which only the value structure is changed. In such experiments, the model of the environment used by the system can be maintained absolutely without

* Chapter 14.

change, and the decision logic can remain unchanged except for the way the values are calculated. The resulting differences in behavior are informative and often spectacular. Such experiments provide important insight about the relation between values and behavior. Probably the best way to communicate such insight is to share some experiences with the reader.

THE PRINCIPLE OF MULTIPLE VALUES

The innate value structure that evolution has built into the human brain includes many components. There are many different biological drives, many distinct emotions, and many different valuative sensations (such as good and bad taste, smell, and tactile sensations). Why has nature chosen such a complex structure of primary values?

Experience with artificial decision systems has shown that multiple value components are almost always needed to provide satisfactory adaptive behavior in a complex environment. Systems that are designed to operate solely in terms of a single criterion will almost always behave in ways that seem silly and unacceptable to common sense. Almost all human decisions involve compromises between multiple goals and objectives. An automatic system will fail to produce decisions consistent with common sense unless it is capable of finding such compromise alternatives. Although the foregoing statements sound like a truism, they are much more important than was originally recognized. The following example illustrates the importance of multiple value components, even in an apparently simple and clearly stated problem.

During the late 1960s, the busing of students to achieve racial balance in public schools became an emotionally charged issue. Officials in the Department of Health, Education, and Welfare felt they could deal more effectively with the problem if they had a better understanding of the extent to which racial isolation in the schools could be avoided without excessive busing of students. To explore the problem, an automatic "student assignment system" was developed which operated as a value-driven decision system. The system was used to explore school desegregation and busing alternatives in forty major urban areas. To provide a realistic basis for the analysis, census data was obtained showing the racial composition and geographic location of school-age children. The enrollment capacity and location of schools in each area were identified and data was collected on the road network and travel speeds within each city.

Obviously, the objective was to achieve desired levels of desegregation with minimum inconvenience and minimum unnecessary cost. The system was designed to provide a great deal of flexibility in the primary value assignment. Naturally the value criterion included a component which depended on the desegregation achieved in the school system,* but it also included penalty points that could be subtracted for each student who had to ride a bus to school. The number of points subtracted could be a simple constant for each student bused, or it could depend on how much time the student spent on the bus.

For any specified form of the value criterion the system could provide a specific school assignment for the students from each census block. In each case the resulting assignment was approximately optimum for the specified criterion. Initially, the government officials believed that they were not interested in complex value structures with multiple components. They just wanted answers to some very simple questions. For this reason, the initial calculations used oversimplified value criteria.

The first question that was asked was "How much desegregation can be achieved without having any child ride the bus more than thirty minutes?" The value system for this calculation was very simple. It included only the points for desegregation, together with a very large fixed penalty for any child who had to ride more than thirty minutes. The resulting assignments answered the question. They showed that complete desegregation, in which all schools had the same percentage of minority students, was possible, even in the largest school districts, without violating the thirty-minute travel limit. However, as we had predicted, the resulting student assignments were absurd. The system perversely assigned almost *all* students to distant schools where busing was required! This occurred because the oversimplified value structure did not include any incentive to limit the number of students bused. The value structure failed to inform the system that it should avoid unnecessary busing of students. Although the original question had been answered, it was doubtful whether the answer was of any practical value.

To correct this difficulty the calculations were repeated, using a somewhat more realistic value system. The new value system was the same as before except that a *very* small point penalty was included for each additional student who had to ride a bus. Although the actual change in the value structure was small, the change in system behavior was immense. The number of students bused dropped from an average in excess of 95 percent to an average between 10 and 30 percent! The level of desegregation was just as high, and still no students were bused more than thirty minutes! The addition of this

---

* These points were given in proportion to the percentage of white classmates seen by each black in his assigned school.

new component in the value system completely eliminated unnecessary bus-
ing of students without degrading the desegregation goal. The new results
gave a much better understanding of the real policy problem.

A more careful inspection of the resulting school assignments, however,
showed that the assignments were still totally at variance with common sense.
An absurd percentage of the students who were bused were assigned to
schools where the travel time was close to the thirty-minute maximum. Ob-
viously, the still simplified value structure had not informed the system that it
should try to minimize travel time for the students bused! When a new value
component was added to include a small penalty for each minute a child spent
on the bus, the average travel time dropped dramatically. An entirely dif-
ferent set of school assignments was produced in which racial balance was still
maintained but the average travel time ranged from six to eight minutes, as
opposed to the fifteen to twenty minutes in the previous assignments!

To a superficial examination, the school assignments that resulted at this
point looked reasonable. However, a more careful examination showed that
they were still inconsistent with common sense. The assignments showed no
sense of equity or fair play with regard to travel time. Occasional assignments
were made in which one child would ride twenty-three minutes and another
would ride three minutes even though the same desegregation objective could
be achieved if both rode thirteen minutes! A change in the penalty for travel
time was made to make the travel penalty "nonlinear," so that additional min-
utes of travel would be penalized more heavily for long trips than for short
trips. This change removed unnecessary inequalities in travel time and cor-
rected the problem.

The school assignments produced by the system using this refined value
structure seemed generally consistent with common sense. However, the as-
signments involved more busing than most people would have been willing to
accept. Some students were bused even though they contributed only margin-
ally to the improvement of the racial balance. Thus these assignments (which
produced almost uniform racial composition in the schools) would have been
acceptable only to desegregation enthusiasts. To make the assignments more
generally acceptable, it was desirable to reduce both the number of students
bused and the average busing distance. To accomplish this, it was decided to
increase all the travel penalties while leaving the value for desegregation the
same. The resulting new assignments showed some loss in racial uniformity,
but they showed a substantial reduction in busing. The assignments produced
in this way were generally accepted as a good compromise solution (2).

Since the system behavior at this point seemed to be consistent with com-
mon sense, it was decided to use the results to develop detailed plans showing
the resulting school attendance boundaries for one city (3). As the maps were

being prepared, it became apparent that the system behavior was still not fully satisfactory. Some of the resulting assignments seemed to be unnecessarily complicated. We had failed to specify a preference for *simple* plans! Students from different locations within a neighborhood might be assigned to different schools, in a way that was both inconvenient and educationally undesirable, even though equivalent desegregation results could have been obtained by assigning the whole neighborhood to the same school. Clearly, we should have included a penalty to discourage different school assignments for the same neighborhood area. Unfortunately, because of technical difficulties, it did not prove practical to incorporate this additional value component in the automated system, so manual adjustment of the final computer plans was required to correct this remaining problem (4).

The foregoing experience illustrates some of the most important characteristics of the value-driven decision system. The value-driven technique is a very efficient and flexible method of making a system operate in a purposeful way. For any specified choice of the value function, the system could automatically produce comparable school assignments in any of the forty areas. Thus the system adapted easily, and it continued to behave sensibly in new environments. However, satisfactory system behavior required the designer to pay meticulous attention to the value structure.

We can now draw some important conclusions about the structure of a value system if it is to motivate sensible behavior. Presumably, these same criteria would apply in the evolutionary development of a biological value structure.

1. *All* relevant objectives must be included in the value structure. If *any* significant objective is omitted from the values, the behavior will almost certainly be absurd with regard to that objective.
2. Reasonable behavior that does not violate common sense can usually be obtained over a wide range of variation in the relative values, so long as the values are neither *zero* nor *infinite*. (Inexperienced designers frequently try to escape their responsibility for picking values by simply setting the values for some objectives to zero, or by specifying certain objectives as rigid constraints—which is equivalent to assigning an infinite value. Such oversimplification has often been responsible for the absurd behavior of artificial decision systems.)
3. Really satisfactory behavior requires careful attention to the relative values associated with each objective. To provide satisfactory behavior even with regard to minor objectives it is absolutely essential that these objectives be included in the value structure, but it may also be essential that the values be small.

The importance of multicomponent value-systems has recently received considerable attention in decision science literature (1). Writers have been concerned with how to estimate the proper weight for different objectives, and how to produce good compromise solutions. In other cases they are concerned with techniques to allow a scientist to estimate a decision maker's value structure from his history of decisions. Such techniques, as they become available, should be helpful in analyzing the value structure of biological organisms.

Before leaving the subject of multiple values, it may be appropriate to mention one of the earliest cases in which the multiple value problem was encountered in a rather dramatic form. In the late 1950s, there was a great deal of interest in strategic nuclear forces. In particular it was necessary to decide how many weapons and what types of weapons to procure in order to provide adequate deterrence.

To help in answering such questions, a variety of computer programs were developed which claimed to provide "optimum" assignments of weapons to targets. It was expected that these computer systems would tell the military planner what combination of weapons was most effective. The computer programs used for the work were early versions of the value-driven decision system, but the decision scientists of the time were not aware of the importance of the multiple-value concept. The computer programs were designed to maximize a single goal, namely the expected value of all the targets destroyed. Fortunately, the systems worked fairly well for most applications, but by the early 1960s the systems had developed a bad reputation and were falling into disuse because the "optimum" war plans they produced did not make sense under careful examination.

Any experienced military planner "knew" that important enemy targets should be attacked by several different *types* of weapons. In that way, if most bombers failed to penetrate enemy defenses, the missiles would still destroy the target. Conversely, if the missiles proved less reliable (or less accurate) than expected, the bombers could still destroy the target. This technique was known as "cross-targeting."

To the embarrassment of the decision scientists, the early computer programs stubbornly refused to put more than one *type* of weapon on a target. For each target, the program would find a "most efficient" weapon type and then just pile on as many weapons of that type as seemed to be justified. The problem was ultimately corrected by providing the computer programs with a more accurate "value function." The original value function reflected only the target value destroyed, assuming that all weapon types worked with *exactly* the assumed reliability. To correct the problem it was necessary to incorporate into the value function the probable value of targets destroyed over a wide range of possible weapon reliabilities. When this was done, the decision systems seemed much more intelligent and obliged with a very effective cross-targeting of their own invention.

We can summarize the present discussion by noting that multiple values are usually essential to satisfactory behavior. The development of a satisfactory value system to be used in a decision system is usually a trial and error process. During the course of biological evolution there has been sufficient time for a great deal of trial and error experimentation with values. The experience with artificial systems helps in understanding how evolutionary pressures must have worked to produce multicomponent value structures.

The need for multicomponent values highlights one additional advantage of the value-driven decision system. Whenever a problem involves compromises between *many* objectives, it is extremely difficult to provide consistent rules for selecting alternatives by any other decision method. With the value-driven method, choices can be made between alternatives simply by totaling the value contributions associated with each of the objectives.

THE PRINCIPLE OF TIME-DEPENDENT VALUES

Traditionally, goals, objectives, or values have been thought of as fixed or at least subject only to long-term variation. This traditional view is appropriate to broad social goals, objectives, and values. By establishing stable goals or objectives we also stabilize policy and thus make it more efficient. However, it does not follow that evolution's primary values for motivating individual human behavior must also be stable over time. The primary human values, such as the biological drives and emotions, fluctuate with time and circumstance.

The ultimate evolutionary goal, survival of the species, appears to be a fixed objective, so it may seem strange that evolution designed its decision systems using such a variable value structure. Recent experience in the design of artificial decision systems provides a simple explanation. Time-dependent values can provide a more practical design approach, even when the ultimate objectives are independent of time. To illustrate this point we will consider the case of a production-scheduling system that was designed to make decisions in the face of a changing and unpredictable product demand.

Around 1970 a large pharmaceutical manufacturer was studying a consolidation and modernization of its chemical manufacturing facilities. The idea was to replace existing outdated production lines with a single multipurpose manufacturing plant. The new plant would contain no permanent production lines. Instead it would contain standardized multipurpose equipment, or "production modules." When a particular product was needed, the necessary modules could be linked together to provide a temporary production line. These standardized production modules were to be very large, so that a year's production could be accomplished within a few days or weeks. By sharing such large-scale equipment among many products, the company hoped to achieve the economy of large-scale production, even for low-volume pharmaceutical products. The plant was designed so that many temporary production lines could be operated simultaneously. When the required production of a product had been completed, the modules would be disconnected, cleaned out, and made available for other products.

The proposal involved major changes in the manufacturing procedures of the company. While the concept looked good on paper, there was no experience with the operation of such a plant. The most serious problem concerned the scheduling of production. Paper and pencil efforts showed that it was very difficult to plan efficient schedules. Too many scheduling alternatives had to be considered. Since the economic advantages of the plant depended on efficient scheduling, an automatic scheduling system was developed to study the economics of the plant.

Let us consider the design problems that were faced in the development of the scheduler. In principle, there was no difficulty in specifying the "real" objectives for the scheduler. The scheduler should operate to minimize manufacturing costs and to ensure that the production demand would be met. The obvious "value structure" for such a system therefore would have included simply a penalty to reflect manufacturing costs and another penalty to reflect losses in customer good will and sales if production ever failed to meet demand. The theoretical objective of such a scheduler would then be simply to minimize these penalties.* In principle, therefore, the goals for the system could be specified in terms of a traditional value structure that did not change with time. A more careful analysis, however, showed that this simple value structure could not be used to guide system decisions without imposing an overwhelming computational burden.

The overall production cost for a schedule could be determined only after the completion of the schedule, so it could not be used as a practical criterion to guide the generation of a schedule. The situation was analogous to the procedure for playing chess. The ultimate purpose in chess is to win, but to help the player in estimating the desirability of individual moves it is useful to assign artificial values to the pieces. Similarly, to help the scheduler in evaluating individual scheduling decisions it was necessary to assign artificial time-dependent values which could be used to estimate the importance or

---

* More accurately, we should minimize the discounted value of these penalty costs, thus allowing for an appropriate rate of return on invested capital.

urgency of producing specific individual products. These artificial values were time dependent because they depended on the current inventory of each product.

An analogy with the value structure of biological organisms may be helpful to illustrate the problem. From a theoretical point of view, evolution could have designed individual biological organisms to make decisions only in terms of a single long-term goal—survival of the species. This theoretical goal would then have been constant over time. But such a fundamental goal would be a useful guide to behavior only if the organism were endowed with enormous information and intelligence. Any ordinary organism, faced with such a long-term specification of objectives, could only be confused. It would not know what activities to begin or when to begin them. If each individual decision had to be evaluated against evolution's ultimate objective (survival of the species) the organism would have to expend all its physical resources just thinking. To avoid this problem, evolution uses a much more practical design concept. Day-to-day decisions are motivated by a multiplicity of drives, or values, that are orchestrated over time to motivate behavior compatible with evolution's objectives. The ultimate evolutionary goal does not even need to be included within this value structure.

In the design of the scheduling system the same basic problem was encountered. While it seemed practical to design a mechanism to solve the scheduling problem one day at a time, it was not feasible or desirable to try to schedule many months (or years) in advance—as would have been necessary if we were to use the ultimate objective as a daily decision criterion. Even the choice of a production schedule for a single day involved a nontrivial computation. If certain products required all the high-pressure stills or all the vacuum driers, then no other products requiring these modules could be produced. Thus, only certain combinations of products could be simultaneously produced, so it was quite difficult even to find a good combination of products to make at any one time.

Ultimately, the following design concept for the scheduling system emerged. The scheduler would be driven on a day-to-day basis by "values" which reflected the "economic importance" of making each product at that particular time. For each day, the scheduler would consider all possible combinations of products. It would select the best feasible combination (i.e., the combination for which the total "value" of products in production was greatest). To motivate efficient system behavior over a period of days or months, the driving values would be carefully orchestrated. To design the value structure, a detailed analysis was made of the effect of the value structure on the economics of system behavior. Although we will not go into that analysis here, the general value structure that resulted from the analysis is of interest because of similarities to the biological drives that motivate the behavior of animals.

The "value" that was assigned for producing a product on any specific day depended on both the remaining inventory of the product and the rate of consumption. Products whose inventory was low (compared to the consumption rate) were assigned relatively high "values." To avoid frequent production changes (and the associated costs of cleaning equipment and setting up each new production line) the "value" for continuing the production of a product already in progress was increased by a substantial premium above the "value" used to put a new product into production. This premium for products already in production made it unlikely that a production run would be prematurely interrupted to make equipment available for other products. The value premiums were carefully designed so that resulting production runs would usually be close to the optimum economic run length.*

To clarify the principles involved, Figure 3.1 shows a sketch of a typical "value function" used in the scheduler. The sketch shows the "value" per hour of production for one product as a function of the product inventory. Note that two separate value curves are shown. The upper curve applies when the product has already been placed in production; it specifies the value of continuing production. The lower curve applies when the product is not in production; it specifies the value of initiating production. During normal operation we expect the inventory to oscillate between a "normal high inventory" (at the end of a production run) and a "normal low inventory" (just before the start of a new production run). These two nominal inventory levels are shown by the vertical lines in the sketch.

To illustrate the operation of this value structure, let us consider its operation during one production cycle. Suppose that the product is initially in production, then the value of continuing the production is defined by the upper curve. As production continues, the inventory increases, moving toward the right along the upper value curve. As the inventory grows, the "value of continuing production" decreases. When the normal high inventory level is reached, the value of continuing is just equal to the probable value of the equipment in other applications. Thus, if there is an average demand for equipment to make other products, the production of this product will stop. The equipment will be relinquished to other products that show a greater need (or current value) for the equipment. Of course, the production of this product could end earlier, if the value of the equipment in other applications

---

* If production runs are too long, large expensive inventories of products are produced. If runs are too short, the set-up costs per unit of production become excessive. Optimum run length occurs when there is a good economic balance between these extremes.

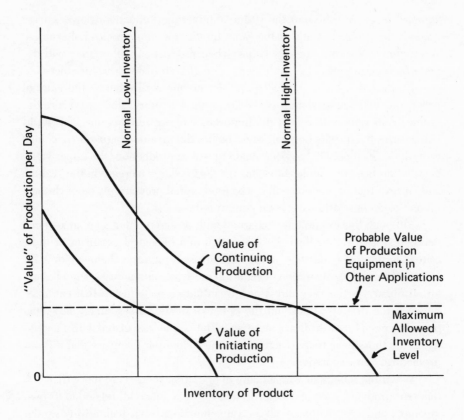

FIGURE 3.1

Illustrative Structure of "Values" Used to Drive Scheduler

The sketch illustrates the value structure for one product as a function of the inventory of that product.

is higher than usual. Conversely, the production of the product could continue, if the other demands for the equipment are less than average. But even if there is no other need for the equipment, the production will stop when the maximum inventory level is reached, for at this point the value of continuing production falls below zero. (Thereafter, the zero value which can be obtained by not producing is preferable to the negative value that would result from continued production, so the system acts as if it is satiated and production stops.)

After production has stopped, continuing sales of the product will result in a gradual reduction of the inventory, moving to the left in the figure. Since the product is now out of production the lower curve applies. Normally the production will resume when the inventory gets close to the "normal low inventory" level. At this point the "value of initiating production" (lower curve) exceeds the probable value of the same production equipment in alternative applications. Naturally the exact time when production will resume will depend on how urgently the equipment is needed to make other products.

If other demands for the equipment are unusually heavy the start of production will be postponed. However, as the inventory falls toward zero the "value" rises rapidly to reflect the importance of not running out. This rapid rise ensures that production will begin before the inventory situation becomes critical. Once a decision has been made to restart production, the upper value curve again becomes applicable. This results in a large increase in the "value" and ensures that production will not be interrupted prematurely (after the set-up and clean-out costs have been committed).

Although this example is concerned with design problems for an artificial decision system rather than the motivation of a biological organism, it illustrates a number of issues relevant in both areas. It indicates the need for elaborate and careful orchestration of values to motivate satisfactory system behavior. It illustrates the very direct and quantitative relationship that can exist between the driving values and the behavior of such a system. It also illustrates the need for an arbitrary increase in the "value" associated with an activity, at the time the activity is actually initiated, in order to ensure that it is not prematurely interrupted.

When the scheduler was actually in operation there were more than 100 different products, each of which had its own value function, related to its own inventory and sales. Although all the value functions were qualitatively similar to the sketch in Figure 3.1, the numerical values were related to the actual cost factors (set-up cost, storage cost, raw material cost, etc.) for each product. The system automatically selected the production schedule for each product and determined the allocation of about 300 production modules (of about 50 different types).

As with any such decision system, there was the usual trial and error period before all important factors were included in the driving value function, and there was a period of experimentation to determine how system behavior depended on certain key parameters of the value function. When such adjustments had been made the scheduler performed remarkably well. It was used during the plant design phase to simulate the operation of alternative plant configurations and to evaluate the economics of the proposed new facility.

The basic design used in this production scheduler illustrates many principles relevant to the cybernetic design of the higher animals. The specific problem dealt with by the scheduler resembles the animal's problem of obtaining nourishment. Just as the scheduler had to plan the production of products needed to meet the marketing needs of the corporation, the brain must plan the acquisition of food and nutrients needed to sustain the health of the organism. The scheduler is designed so that it "likes" (to produce) products that are in short supply in its inventory, just as the animal is designed so that it likes to eat when its level of nourishment begins to drop. Given a choice between alternatives, the scheduler will always select that mix of products which is most to its liking (those for which the total value is highest). The preference function (value function), which determines how much the system "likes" specific products at any time, is built into the system by the designer (by evolution). This function is not subject to rational control or modification by the system. The value function is innate to the system, and it provides the system with the motivation to produce needed products. However, decisions concerning the specific time for the production of a product, the specific products to make at any one time, and the specific methods of production are delegated to the "rational" decision process of the system. The rational thought process of the system accomplishes a systematic examination of available alternatives followed by the selection of a preferred alternative (one that is most to the liking of the system at the time). If the system could be asked why it preferred a particular alternative it would have no rational response. It could only say because on balance it likes it better than other alternatives (it has higher value). The reason for the value (or the liking) is unknown to the system; it is known only to the system designer.

It is worth emphasizing that (in this concept of a decision system) the animal does not eat because it needs nourishment, it eats because it likes to eat, and it eats what it likes at the moment. The close correspondence between what it "likes" and what it *needs* is an accomplishment of evolution (of the system designer); it is not a rational achievement of the decision system. Clearly, the selective process in evolution will show a preference for those value systems that provide a high correlation between "likes" and actual biological needs. In the design of the value function for the production scheduler, we were careful that the values correlated well with probable future system needs.

The word "like" was used in the foregoing passage to relate the results to our popular understanding of animal behavior. In the computer system, the "like" or "dislike" really refers only to the numerical magnitude of a value function that determines what the system will do when given a choice. This may be a good *objective* definition of the word "like"—even when we think of human behavior. However, the word is traditionally used more in a *subjective* sense to describe the internal sensation of preference which leads to an objective choice.

There is one other striking parallel that is evident in this analogy. The animal, like the production scheduler, normally encounters a set-up cost in getting ready to eat. He must find and perhaps prepare the food. When he has expended this effort it is desirable to eat enough so that he will not have to repeat the process too soon. As the animal actually smells the food, his interest or appetite seems to be heightened. Thus it is harder to distract the animal after he has begun to eat than it was before he started. This suggests that nature may have employed the same technique, that of a value premium, that was used in the scheduler to sustain production once it had begun.

TIME SEQUENCING OF MOTIVATING VALUES

The motivating values provided by evolution are often designed to occur in a sequence that will lead the animal through a complex series of activities that otherwise could not be accomplished. For example, in the mating process, the achievement of the desired evolutionary goal requires a specific sequence of steps. At the time the animal is engaged in the early phases of the activity he may be totally unaware either of the evolutionary goal or of later steps required. Thus the animal will engage in the preliminary activity because he enjoys or "likes" the preliminary activity, as a goal in and of itself, not because of any conscious expectation of the later steps.

Very similar problems in the sequencing of behavior were encountered in the design of the scheduler, and they were solved in a very similar way. The final design of the scheduler included some rather elaborate time sequencing of the values. Certain chemicals, called intermediate products, were required as "raw material" for other products. These intermediate products had to be produced before production of the final product was possible. To avoid unnecessary storage costs, the intermediate products should be produced *just* before they are needed. To motivate production of these products at the right time, their "value function" was based on a *projected* future inventory of the final product (which was calculated from the rate of consumption after allowing the lead time necessary to produce the intermediate). At the same time, the "value function" for the final product was set to zero until a suitable inventory of the intermediates was available. This *deliberate* sequencing of *values* automatically produced the necessary sequencing of the production.

THE FREQUENCY OF DECISIONS
IN ADAPTIVE BEHAVIOR

When the design of a decision system such as the scheduler is discussed, people often wonder how much the system behavior will be affected by the frequency with which decisions are reexamined. Since the motivating values depend on a changing product inventory, the values themselves are constantly changing. A decision made at one point in time can be different from the same decision made a moment later. However, the behavior will be almost independent of the decision frequency so long as decisions are reexamined often enough that the change in values between decisions is small. They can be reexamined once a second, once a minute, or once an hour, and there will be no significant difference in behavior. But if decisions are reexamined less frequently than once a day there will be a noticeable degradation in performance, and if they are reexamined only once a month the effectiveness of system performance will be largely destroyed.

This point is made simply to call attention to the fact that such a system can seem to behave purposefully (without any arbitrary decision points) so long as there is fairly frequent reexamination of activities in the light of changing values. When this is the case, it is not possible to guess the actual decision rate from the external behavior.

In summary, the use of values that are time-dependent is a cybernetic compromise which (1) permits more adaptive behavior than would be possible if responses were "wired in" and (2) avoids the fantastic computational burden that would result if the organism's individual decisions had to be measured against the ultimate evolutionary goal.

The use of such time-dependent value functions in artificial decision systems is still in an embryonic stage. Although in some cases formal mathematical analysis will lead to such time-dependent value criteria,* the procedures in most cases are more an art than a science. In the development of the scheduler it was possible to use economic principles to define many of the characteristics of the functions. Nevertheless, a number of the issues had to be decided arbitrarily using purely intuitive logic. There is obviously a need for additional research on mathematical methods for developing such "value functions."

---

* For example, see Pugh and Mayberry (5) and Pugh (6).

REFERENCES

1. Cochrane, J. L., and M. Zeleny, eds. *Multiple Criteria Decision Making*. Columbia: University of South Carolina Press, 1973.

2. Pugh, George E., et al. *School Desegregation with Minimum Busing*, A report to Asst. Secretary for Planning and Evaluation: U.S. Dept. of Health, Education and Welfare, 1971.

3. Pugh, George E., et al. "School Desegregation Alternatives in Prince George's County." *Lambda Report 71*, April 1972.

4. Pugh, George E., et al. "Computer Assisted Desegregation Planning for Prince George's County." *Lambda Report 73*, December 1972.

5. Pugh, George E., and John P. Mayberry. "Theory of Measures of Effectiveness for General Purpose Military Forces, Part I: A Zero-Sum Payoff Appropriate for Evaluating Combat Strategies." *Operations Research*, vol. 21, no. 4 (July–August 1973).

6. Pugh, George E. "Theory of Measures of Effectiveness for General Purpose Military Forces, Part II: Lagrange Dynamic Programming in Time-Sequential Combat Games." *Operations Research*, vol. 21, no. 4 (July–August 1973).

7. Smith, Nicholas M., and Milton C. Marney. *Foundations of the Prescriptive Sciences*. McClean, Va.: Research Analysis Corporation, March 1972.

Chapter 4

The Lessons of Automation

All cognitive agents are finite decision systems. They perceive the environment through sensors of finite capacity, process data at finite rates, use finite information-processing programs, store information within a finite memory, communicate through channels of finite capacity, and endure over finite lifetimes.

SMITH AND MARNEY *

THE PRECEDING CHAPTERS introduced some of the basic design principles for a value-driven decision system. This chapter discusses some additional principles which seem to have been important in the evolution of the human brain. The material in this chapter is somewhat more detailed and technical than the rest of the book. Although the material is needed to understand the evolutionary forces that may have molded the design of the brain, it is not essential to an understanding of the design itself. Some readers, therefore, may prefer to skip to Chapter 5 and return later to this chapter.

THE CYBERNETIC COSTS OF RATIONAL ANTICIPATION

Rational anticipation of future events seems to be a recent evolutionary development. The more primitive organisms such as frogs, fish, and snakes seem to operate with little or no anticipation of the future. Where the survival of such

* Paraphrased from Chapter 9 of Smith and Marney (1).

small-brained animals requires behavior which "anticipates" the future, the necessary "anticipation" must be built into the motivating values. Actual anticipation of the future is so costly in cybernetic resources that it probably is practical only for animals with quite large brains.

The design concept described in the previous chapter for the production scheduler provided absolutely no "rational" anticipation of the future. Decisions at any time were made without rational foresight, solely on the basis of immediate or current preferences. The capacity of the system for "rational" thought (i.e., deliberate comparison of alternatives) was limited to today's decisions, without any thought for tomorrow. The model of the environment used by the system did not even include the concept of a future or past; it included only today. The apparent "foresight" of the system in "planning for the future," which seemed superficially evident in the behavior of the system, was actually the result of careful orchestration of time-dependent motivating values. In this sense the scheduling system represented a design extreme. This design extreme, however, may approximate the designs used by nature in many of the lower animals. Certainly in humans, and probably in other intelligent animals, there is considerable rational anticipation of future consequences.

The designer of an automatic decision system must make some difficult design choices when he decides how much to rely on such rational anticipation. The costs of such anticipation in terms of cybernetic processing capacity can be immense. Again, the experience in the design of the scheduler is instructive. The decisions required of the scheduler for a single day were already quite complex. Despite the use of branch and bound techniques to reduce the number of alternatives that actually had to be explored, the scheduler would nevertheless search 200 to 2,000 production combinations before settling on the "best" one for each time period.

To allow the system to really think ahead, it would have been necessary to select some of the "best" candidate alternatives and then project the future consequences of each on a trial basis before making the final selection. To project future consequences it would have been necessary to run the scheduler forward in time on a trial run for *each* such candidate! Let us consider the computational costs of this type of "rational anticipation."

Suppose we wished to prepare an advance schedule for one year. As actually designed, using the usual daily decision rate, without rational anticipation, this would involve a computational investment I_1 given by

$$I_1 = 365 \, c_0$$

where c_0 is the normal cost of a single decision period. But if we decided to do a trial run of the consequences of each of n alternatives for the day as far as d

days ahead before making the final decision, the computational investment I_2 required would be given by

$$I_2 = 365 \times n \times d \times c_0$$

If we were to choose $n = 6$ trial alternatives and project each for $d = 90$ days the computational investment for the schedule would have been 540 times as large. If the original scheduler could produce a schedule with ten minutes of computer time, the "improved" scheduler would require almost four solid days of calculation! However, even this estimate is based on the assumption that we will not consider any further alternatives as we think ahead with each *trial*. Specifically, it assumes that the projection of the future, as we explore each of the six trial alternatives, is accomplished simply by following (at each future decision point) the *single* alternative that seems most attractive on the basis of purely local value criteria. Theoretically a better look-ahead procedure should consider other alternatives in the projection of *each* trial alternative. But such anticipation would be even more costly.

Suppose we were to consider such alternatives every tenth day in the ninety-day trial look-ahead, and we decide on each occasion to consider only two alternatives. Then the six alternatives considered during the first ten days would each multiply into 2^8 alternatives for the last ten days. The total investment, I_3, in the schedule would be given by

$$I_3 = 365 \times 6 \times 10 \times (1 + 2 + 2^2 + 2^3 + \ldots + 2^7 + 2^8) \times c_0$$
$$I_3 = 365 \times 6 \times 10 \times 511 \times c_0$$
$$I_3 = 365 \times 30{,}660 \times c_0$$

and the computational cost would be a factor of about 30,000 larger than for the simple scheduler. Compared to the ten-minute calculation for the simple scheduler, such an "improved" scheduler would run about seven months.

But even such a calculation would not allow us to rationally consider future uncertainty in the market for products. The rational anticipation considered above would correspond to the assumption that the environment (or market) is completely predictable. Since this is not the case, we might wish to consider alternative market futures. Suppose we add to the projections just two market (or environmental) alternatives to accompany each of the decision alternatives. The calculation will now have four branches (two environmental alternatives and two schedule alternatives) at each of the previous branch points. The computational investment, I_4, for such a schedule will become

$$I_4 = 365 \times 6 \times 10 \times (1 + 4 + 4^2 + 4^3 + \ldots + 4^7 + 4^8)\, c_0$$

Such a scheduler would require about 100 years of computation to produce the one-year production schedule. Although it might produce superb schedules it would be of no practical value. Obviously the purpose of this discussion has been to illustrate the enormous computational burden that can result from a "rational analysis" of future alternatives.

To develop a one-year production schedule we discussed four possible approaches, which are summarized in the table below:

TYPE OF SCHEDULER	RUNNING TIME
1. Simple scheduler, no look ahead	10 minutes
2. Look-ahead scheduler, nonbranching futures	4 days
3. Look-ahead scheduler, with branching futures	7 months
4. Same but also with branching future environment	100 years

Since the computation costs escalate so rapidly, such rational consideration of future alternatives must be done very sparingly. The examples deliberately limited both the number of branches and the frequency of branching; nevertheless, the required calculation got rapidly out of hand. To make rational anticipation practical, one must make careful judgments about what alternatives are to be projected and how far they are to be projected into the future. But the cybernetic resources required to make such judgments can also be substantial. Thus it is not at all surprising that the simpler animals have almost no capacity to anticipate future consequences.

In the design of an automatic decision system there is always a trade-off between the resources allocated to the rational consideration of alternatives and the resources committed to a more sophisticated value system. Theoretically the time phasing of production for intermediate products could have been accomplished simply by allowing the scheduler to consider a wider range of future alternatives, but the computational cost for this approach would have been astronomical. The final design therefore accomplished the same objective by using a more sophisticated value system.

The foregoing discussion deals with a number of practical lessons concerning advance planning. Since these are important to our later development, we will pause to review them.

1. Advance planning is fundamentally costly in cybernetic resources, because of the number of alternatives that logically should be considered.
2. Comprehensive planning which considers all relevant alternatives is almost never possible, so rules of judgment are needed to limit the alternatives explicitly considered.
3. The practical evaluation of alternatives that are projected into the future requires the use of values. The value of an alternative can be conveniently divided into two parts:
   a. The value or pleasure associated with the alternative up to the time when the alternative is discontinued.

b. An artificial value which is assigned on some basis to estimate probable future value if that alternative were continued further into the future.

This discussion shows that the cybernetic cost of rational anticipation is not confined just to the simulation of alternatives. Important costs are also involved in the development of judgmental methods for selecting and evaluating alternatives. Traditionally the accumulated wisdom of society including ethics, morality, and social norms has been concerned at least in part with providing such judgmental methods.

VALUE STRUCTURES FOR RATIONAL ANTICIPATION

A decision system that is designed to think ahead requires a more complex value structure than is necessary for one with no foresight. If a decision system cannot think ahead, it does not need to know anything about the "desirability" of its present state. It does not need to know whether it is comfortable or uncomfortable, happy or unhappy. It only needs to know what it would like to do next. This suggests that nature's value structure for the simpler organisms may be radically different from the value structure used in the more intelligent animals. Although it is possible to design a value structure that will operate successfully either *with* or *without* anticipation of the future, the design requirements for such a value structure are more stringent.

The production scheduler was designed on the assumption that it would not think ahead. Its value structure could *not* have functioned satisfactorily if the system had been allowed to think ahead. As the value structure was designed, values were specified in each time period for the production of each product in that time period. By convention, the value of *not* producing each product was zero. Given these conventions, the value structure required to motivate satisfactory behavior was rather clearly defined. It took the general form that was shown in Figure 3.1.

As originally designed, the system usually initiated production, as intended, near the "normal low inventory" level. But if we were to use this same value structure in a revised system, designed to think ahead, the system would stubbornly refuse to initiate production as planned.

If we examine the value structure in Figure 3.1 we can understand why. If production is initiated promptly at the low inventory point, the total value (or fun) realized during the production process, as the inventory moves from the normal low inventory level to the normal high inventory level, will be just proportional to the area under the upper curve. But if the production is postponed until the inventory drops to zero, the same amount of production can be accomplished at a lower inventory level where the value (or enjoyment) of the production is much higher. Consequently, a system which could think ahead could achieve a much higher total value by allowing the inventory to drop to a dangerously low level.

To use a biological analogy, the organism might decide to postpone eating until it was hungrier, so it could enjoy it more! If decisions on when to eat were governed strictly by how much we enjoyed the eating, this might indeed be the choice an intelligent organism would make. However, as we become more hungry we also become less comfortable. This requires us to balance the discomfort of hunger against the increased pleasure of eating. In practice therefore we usually prefer to eat fairly soon after we experience discomfort from hunger.

Apparently if a system can think ahead, a new term must be added to the value function to provide a negative value (or discomfort) when the inventory falls too low. Figure 4.1 illustrates how the value function might be modified for advance planning. In this revised value structure three separate curves are shown: one for not producing, one for initiating production, and one for continuing production. The risk of running out of inventory is now reflected entirely in a penalty for not producing, which goes sharply negative when the inventory level drops too low. This value component for *not* producing (analogous to the discomfort of hunger) can provide an advance planning system with a strong motivation to produce when the inventory is low, so it is no longer necessary to reflect this need in large positive values for the other two curves (which specify the value of production). This revised value structure retains essentially the same motivation as the previous value structure (difference between value of producing and not producing) for a simple non-lookahead scheduler; but the added penalty for nonproduction at low inventories will motivate a look-ahead system to avoid low inventories.

It is remarkable that with this modification for a look-ahead scheduler we are forced to a value structure that almost exactly parallels the evolutionary design! It includes the discomfort of hunger, a reward for eating which increases as the need for nourishment becomes more acute, and a premium for continuing to eat after the process is started.

Of course, the specific value structure developed here is not unique. For example, if we had developed the design originally for a look-ahead system, we might have explicitly included a set-up penalty in the value function, instead of a rather ad hoc premium for continuing production. But such a set-up penalty would not have worked at all in a more primitive non-look-ahead sys-

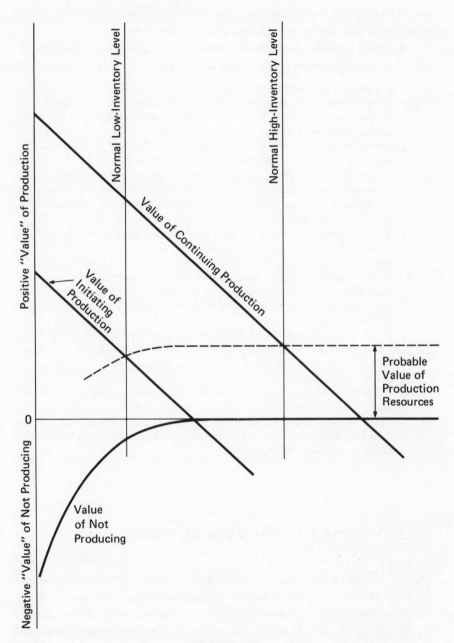

FIGURE 4.1
Illustrative Value Structure for Scheduler with Advance Planning Capability

tem that could not see beyond the penalty. Thus, the premium concept may be the more natural and robust evolutionary value structure.

It is now possible to summarize some of the fundamental principles involved in the selection of primary motivating values.

1. If a decision system had unlimited cybernetic capacity, the designer could simply assign values corresponding to his own *ultimate* objectives for the system. In this case there would be no difficult design problem in the choice of the primary motivating values.

2. For all real systems the cybernetic resources are limited and it is usually infeasible or impractical to use the ultimate goals or values directly. The ultimate values can be infeasible to use because they impose too large a burden on the process of rational anticipation. Instead the designer must select primary values for the system that (with limited anticipation) will motivate satisfactory behavior, in terms of his ultimate goals for the system.

3. When cybernetic resources are so limited that any rational anticipation is infeasible, the values must be designed so that decisions based on shortsighted preferences will lead to acceptable behavior. (This may be the predominant design principle for organisms of intermediate intelligence, such as fish, frogs, and snakes.) Organisms that operate on this principle have no need to evaluate the quality of their current situation. They only need to know what they would like to do next.

4. As cybernetic resources are increased so that some rational planning for the future is possible, value components such as comfort and discomfort or happiness and unhappiness are needed that will permit good choices between alternative future states.

5. From an evolutionary point of view, the ideal value system is one which will lead to adequate behavior in terms of strictly short-term preferences, but which will also lead to better behavior as the amount of long-term planning increases. (The value system for the more intelligent mammals seems to be based on this basic design concept.)

VALUES FROM A MATHEMATICAL PERSPECTIVE

The real purpose of the value concept in decision theory is to provide a simple decision rule. If a numerical value can be assigned to the "outcome" of any alternative, then choices can be made between alternatives simply by comparing the values. In mathematical terms, the most direct way to accomplish the assignment of values is to specify a "value function" which defines a scalar value over the space of possible *outcomes*. Such a function defines a specific value for every possible outcome.

When the concept of a "value function" is defined in this way it is necessary to ask what we mean by an *outcome*. In a mathematical sense the "outcome" of a decision includes all of the information about the consequences of the decision that is required to evaluate alternatives, using the value function to calculate values. The designer must decide how much information about the consequences of a decision must be available before a choice can be made.

One of the most important issues in the design of a decision system concerns how far the system should think ahead before making a decision. If a system is designed to think too far ahead, the cybernetic resources required for rational analysis will be excessive. If it is not designed to think sufficiently far ahead either the value function will become unduly complicated or the decisions will be unsatisfactory. Sometimes information about the state of the system only a moment after the decision is sufficient. Sometimes information is required about the entire trajectory of system states for a long time after the decision. In developing any specific decision system the designer has wide latitude in deciding how to define the "outcomes."

In the original production scheduler the "outcome" was defined simply as the state of the system immediately after implementing a particular production alternative for that time period. The value function for the scheduler was defined simply as a function of this state. A natural extension of this concept for a look-ahead system would sum or integrate such a value function over a trajectory of future states.

The evolutionary value functions appear to be based on a very analogous concept. For any specific state, S, of the system a value, $V(S)$, is defined. (Usually the state S is itself a function of the recent experience of the organism.) For an organism with an ideal look-ahead capability the preferred choice would be one that offers the highest integrated value, ϕ,

$$\phi = \int V(S) \, dt$$

For any real organism with a limited look-ahead capability, long-term values must not be allowed to dominate decisions, so some discounting of long-range values is appropriate. This can be accomplished by defining the value function $V(S)$ to include future values in a discounted form. For example

$$V(S) = V_0(S)\rho(t)$$

where the function $\rho(t)$ is some declining function of t such as e^{-at}.

Let us consider how this concept would operate for an artificial decision system with a very sophisticated look-ahead ability. Such a system will be constantly exploring alternative behavior trajectories through the foreseeable future. It will measure the desirability of the alternatives in terms of the integrated "value" or enjoyment of the whole trajectory. It will try to select a behavior trajectory for which this integrated value will be highest.

Any choice of a behavior sequence will result in some sequence of activities over time. Each activity will have associated with it a value per unit time (or a level of satisfaction). The preferred behavior sequence will be the one for which the accumulated value over the whole trajectory is the highest. This leads to the important observation that motivating values associated with various types of activities should be expressed as values *per unit time*. The ultimate objective of the decision system is to maximize the total accumulated value. A system with a very poor ability to look ahead will almost ignore future values and consider only the instantaneous value. A system with very good look-ahead capability will give almost as much weight to future as to present values. A system which optimizes the use of its cybernetic resources will vary the amount of look-ahead depending on the importance and urgency of a decision.

We can now ask what criteria a system designer can or should use to select good driving values. At present, the only practical approach seems to be for the designer to consider a wide range of alternative behavior sequences. He then designs the value function so that the difference in the integrated value calculated over alternative decision trajectories (using a reasonable range of look-ahead times) will correspond as accurately as possible to the difference in relative desirability of the decisions in terms of the designer's ultimate objectives for the system.

In the case of the production scheduler we tried to design the value structure so that the value for each alternative decision would be related as directly as possible to the probable "profitability" for that decision. In the case of biological decision systems we should expect that the value structures will have evolved so that those behavior strategies most conducive to survival of the species will normally show the highest integrated value. Conversely, this suggests that human behavior will usually correspond most closely to the evolutionary intent when it is most successful in realizing value from the innate value structure.

It is worth emphasizing that the primary driving values usually should not be identical with the designer's ultimate goals, both because of irreducible uncertainty about the future and because of the practical limitations of any finite system to anticipate future alternatives. For example, the value function in Figure 4.1 obviously does not correspond exactly with system profitability for any trajectory. The true maximum of system profitability would occur if the inventory always were allowed to fall to *exactly* zero before starting production. This would minimize inventory costs and no real losses would occur so long as the inventory did not actually run out. However, the "value" function in Figure 4.1 shows a strong penalty for allowing the inventory to get too close to zero. This penalty is a reflection of the uncertainty of future projections. If planning allows the inventory to get too low, there is a risk that it will actually run out.

As the look-ahead capability of a scheduler is improved and as the uncertainty in demand is reduced, this low inventory penalty could be reduced to allow the inventory to drop closer to zero. In the limit of no uncertainty in demand and infinite look-ahead capability the penalty could be reduced to zero. Thus, the proper value structure for a system will depend both on the amount of look-ahead to be used and the level of uncertainty within which the system must operate.

THE DISTINCTION BETWEEN HUMAN AND EVOLUTIONARY GOALS

The foregoing discussion emphasizes the distinction between the goals of the designer and the primary values that motivate system behavior. If the designer has done a good job, system behavior will correspond closely with the objectives of the designer. But from the point of view of the decision system, the correspondence is a coincidence. The goals of the system designer may not even be known to the system. The system makes its decisions and seeks its satisfaction solely in terms of its own primary values. If the value design is faulty, the system performance will be faulty—in terms of the objective of the designer.

This distinction, in which system goals are treated as entirely distinct from the goals of the system designer, has important implications for the theory of human values. To understand human objectives and human behavior we must deal with the built-in values that motivate the behavior of human beings as individual decision systems. Although these values are related to the evolutionary goal of survival of the species, the human goals are much richer and much more complex.

As with any genetic inheritance, the value structure varies in detail from one individual to another. An individual's personal innate value structure is his own private inheritance. This inheritance is the sole determinant of what makes life worthwhile. Human endeavor inevitably is directed to the achievement of objectives that are determined by the innate human values. The evolutionary function of human reason, ethics, and policy is to ensure that human activity is, in fact, wisely directed in support of these innate human values.

THE USE OF SECONDARY VALUES TO SIMPLIFY DECISIONS

In Chapter 1 we called attention to the distinction between a system's primary and secondary values. Human behavior is very strongly influenced by secondary values. For example, a person may work for money to buy food. The relief of hunger is the primary objective; earning money is a secondary goal. The secondary goal provides a more immediate objective, which focuses activity and helps in solving the problem. The value of money is a secondary rather than a primary value.

Artificial decision systems also make extensive use of secondary values. So far in the discussions of the student assignment system and the production scheduler we have made no mention of secondary values. Both were described simply as optimizing decision systems, driven by designer-specified primary value criteria. Although this description is accurate, it is not complete. In fact, both systems made extensive use of secondary values to increase the efficiency of their operation.

As we said earlier, most of the computational methods used in a computer would be irrelevant to a discussion of human behavior, but the use of secondary values seems to be an exception to this general rule. Secondary values are used in a formal decision system in much the same way as they are used in our "commonsense" reasoning.

In the scheduler the limited supply of production modules (stills, dryers, reactors, etc.) defined limits for the production that could be accomplished at any time. If more modules of certain types were available, more production could be realized, and a higher payoff in terms of the driving values could be achieved. Thus, additional production modules of each type would offer a quantitative value equal to the value of the increased production permitted by the additional module.

Such secondary values (known as shadow values) can be used to simplify the decisions that must be made. Without the shadow values it probably would be necessary to explore *all* possible combinations of products that might be put in production. The shadow values make it possible to make a guess about which products to produce simply by selecting those products for which the production "value" *exceeds* the "shadow cost." (The "shadow cost" is equal to the total *shadow value* of all production modules used to make the product.) Products for which the value of production is significantly *less* than the shadow cost almost certainly should *not* be produced. But there are usually a number of products where the value and shadow cost are almost equal. To find the best production decision, systematic (branch and bound) search techniques are used to explore the alternatives concerning production and nonproduction of these products.

In the scheduler, both the primary values and the shadow values change with time. The primary values change because they depend on the changing inventory of each product. The secondary or shadow values change because they are based on the primary values. For this reason they must be recalculated (or updated) for each day's decision.* The shadow values make it easier to make decisions a little bit at a time. It is much easier to ask, for each individual decision, whether the value justifies the "cost" than it would be to consider all the possible decisions at one time.

Shadow values play a very similar role in the student assignment system. In a school desegregation assignment the overall desegregation "value" achieved depends on the racial composition of schools. The shadow values in this case reflect the increase (or decrease) in the payoff that could be achieved by assigning one more black student or one more white student to each school. Schools that have too large a percentage of black students should be willing to pay a premium for white students; schools with too high a percentage of whites should be willing to pay a premium for black students; schools that are above allowable capacity should be willing to pay a premium to get rid of students; and those that are below capacity should be willing to pay a premium for either black or white students. By this technique it is possible to solve the problem simply by simulating a free market procedure for assigning students to schools. The students are assigned to schools a few at a time on the basis of the shadow values. For each alternative school assignment there is a travel penalty which depends on whether the school is within walking distance and, if not, on how long the student would have to ride the bus. To achieve the correct assignment each student is simply assigned to the school where the "value" in excess of travel penalty is greatest. As students are reassigned, of course, the shadow values change, but they gradually converge to correct values such that no further students should be reassigned. Thus as usual, the

---

* Although there are many different ways that the shadow values can be calculated, the methods are all equivalent to setting up a free market for the modules and allowing the law of supply and demand to set market prices. Modules that are not in short supply are automatically given a shadow value of zero. The market price for modules that are in short supply is determined by what the traffic will bear.

Probably the simplest method for doing this is an iterative one that simulates a free market. Using an initial estimate of the shadow values we make a trial decision which puts into production all products whose "value" exceeds the "cost." We then check to see which module types are used in excess of the supply and the shadow value for these is raised slightly, and the shadow values for the remaining modules are lowered slightly. Then with the adjusted shadow values we repeat the trial decision and once again change shadow values depending on which modules are in short supply. This process is repeated until fairly stable shadow values are obtained. The process converges on a valid (but not necessarily unique) set of shadow values for the time period.

shadow values reduce a very complex problem to a series of simple manageable decisions.

Obviously, biological organisms do not make decisions with anything like the numerical precision implied in this discussion of shadow values. Nevertheless, the intuitive concept of secondary values to assist in decisions is widely used by people, and it may even be used by some of the more intelligent animals.

THE PRINCIPLE OF INTELLECTUAL VALUES

A decision system needs some criterion for deciding when to stop worrying about a specific problem. This is particularly true for systems which produce only approximately optimal solutions. This issue can be resolved by the decision system itself if the cost of calculation (or of intellectual resources) is included as one of the components of value to be considered within the optimization process. When the rate of improvement in the solution drops so that the expected savings from a better solution are smaller than the projected increase in analysis cost, the analysis stops. In this way intellectual values are introduced as an intrinsic component of the decision process.

Most artificial decision systems do not go much further than this in the treatment of intellectual values, because their repertoire of intellectual behavior is very limited. In contrast, intelligent biological organisms have a very wide repertoire of intellectual behavior. They can rest, they can spend time learning, they can decide what to try to learn, or they can spend time developing decision principles to guide the intellectual effort itself. Chapters 12 and 13, which deal with the structure of innate human intellectual values, show how this requirement for guidance of intellectual behavior has been the source of many of our *cultural* ideas about intellectual values.

THE HIERARCHICAL DESIGN PRINCIPLE

When we decide to walk across the room, the action follows almost automatically. We do not think about which muscles to contract or how much to contract them. Such detailed decisions are made below the conscious level in a decision hierarchy. The top level makes only the basic policy decision. Lower echelons in the nervous system implement the decision. This very general principle of hierarchical design is also utilized in the design of artificial decision systems. Of course automatic systems that deal with very simple problems can sometimes deal with the entire problem within a single decision process, but as the problems become more complex it becomes essential to decompose them into a hierarchical structure.

Each of the artificial decision systems we have been considering required more than one level. The hierarchical structure simplifies the strategic decision level by delegating tactical decisions to a process that can be carried out after the strategic decisions have been made. A few examples will serve to illustrate the principle.

In the student assignment problem the strategic decision system was concerned only with the selection of school assignments for the students in a census "block group." In most cases, when all students are assigned to the same school, no further assignment decision is required. However, where the students from a single block group are divided between two or more schools, a decision must be made on how to divide the block group. This tactical decision requires detailed information on the internal structure of the block group which is irrelevant to the main assignment problem. Any processing of this detailed information during the main assignment would impose an unnecessary data-processing burden, so these decisions were reserved to a later processing step (which in this case involved manual processing).

The production scheduler was concerned only with what products should be made during each time interval and the number of modules of each type required. Tactical decisions concerning which specific modules of a given type to use and how they should be interconnected were postponed to later processors. In a fully automated plant, a third-level processor would have been required for detailed control decisions, adjusting valves and thermostats, to actually produce the products. Because these detailed decisions are irrelevant to the main scheduling process, the presence of such information would only encumber the system.

A similar division of responsibility was used in the strategic war planning system. The top-level system was concerned only with deciding which weapon types (from which geographical areas) to assign against each target. A second-level processor was used to decide which specific missiles and aircraft should be used against each target, and a third-level processor was used to generate detailed flight plans for each missile and bomber.

The hierarchical decision structure is almost universal in automated systems that deal with complex problems. Indeed, the pattern is quite parallel to the division of responsibility that one finds in human organizations. The hierarchical structure in human organizations (such as an army or a corporation) has usually been explained as a way of sharing planning loads which exceed the capacity of any single person. But the fact that the same type of structure is required in the automated systems makes it clear that the reasons for the hierarchical structure are much more fundamental.

In the systems discussed above the different levels of decision were accomplished at *different* times by the *same* computer. In these systems the hierarchical structure is not a way of sharing the decision load; it is a way of *reducing* the *total* load by using models at each level in the hierarchy which represent the problem at an appropriate level of detail. The use of different models with different levels of detail is *essential* to the efficiency of the decision process. The use of a simple model of low detail at the strategic level makes it possible to explore a much wider range of alternatives than would be feasible if each were developed in detail. Once the strategic decisions have been made, tactical planning which requires an increased level of detail can begin. Tactical planning requires the exploration of detailed alternatives within a basic plan, but it does not involve the exploration of any additional strategic alternatives.

The use of such a hierarchical structure is apparently essential for a system dealing with complex decision problems. The war planning system and the scheduler were both developed before the hierarchical concept was recognized as a basic design principle for such systems, and in both cases the client opposed the omission of detail at the strategic level. Although major efforts were made to avoid the hierarchical structure, there was simply no practical alternative.

The hierarchical principle also affects the processing of input data. Data which are to be used at the highest decision level must be stripped of all but the essential attributes needed at that level. The more detailed representation for lower levels of the hierarchy also requires a specialized structuring of the data, which must be done in advance to avoid encumbering the decision processes.

It seems evident that evolution also has been forced to the hierarchical design concept. The strategic decisions are accomplished at the conscious level. The processing of sensory data to prepare it for use at the strategic level is accomplished subconsciously elsewhere. Conscious decisions usually are concerned only with what to do, not the details of how to do it.

Evolution has also developed another important way of conserving cybernetic resources at the critical strategic decision level. Routine processes that are learned originally at the conscious level are gradually transferred to automatic or subconscious control. The ability to learn at the subconscious level allows the adult to carry out routine activities while giving full conscious attention to more complex problems.

HOMEOSTATIC BEHAVIOR IN A VALUE-DRIVEN SYSTEM

The preceding sections have been concerned with general design principles applicable to any value-driven decision system. How can these concepts be reconciled with earlier behavioral concepts such as feedback systems and homeostasis? The evidence seems convincing that all of the concepts are relevant in their own way as system design techniques. Sometimes the same cybernetic system can be described equally well in terms of several of the alternative concepts. A homeostatic system almost always includes feedback. A value-driven system can display homeostasis, and it will usually employ feedback principles. Although it is easy to incorporate homeostatic behavior within the value-driven decision system, the value-driven system can also display a wide variety of purposeful behavior which is not easily encompassed in the homeostatic model. Thus the value-driven decision concept has the advantage of greater versatility.

Homeostatic behavior can be motivated within a value-driven decision system simply by introducing an appropriate value system. The maintenance of a proper range of body temperature is probably the most commonly used example of homeostatic behavior. The solid curve in Figure 4.2 illustrates a value criterion that will induce a similar homeostatic behavior in a value-driven system. The example shows a range of acceptable skin temperatures from about 77 to 86 degrees. Although minor temperature preferences exist within this range, other elements of value could easily dominate the temperature preferences within this range. But as the skin temperature moves beyond this region, the value rapidly becomes very negative and strong preferences develop to return the temperature to the desired range. Obviously by increasing the strength of the temperature component of value relative to other values, these temperature preferences can be made arbitrarily strong.

Because of this ability of the value-driven system to exhibit homeostatic behavior, it seems likely that much of the reported "homeostatic" behavior in the higher animals is actually the work of a value-driven decision process. The value mechanism as a motivation for such behavior has the advantage that it provides a quantitative specification of the way temperature preferences are to be balanced against other objectives.

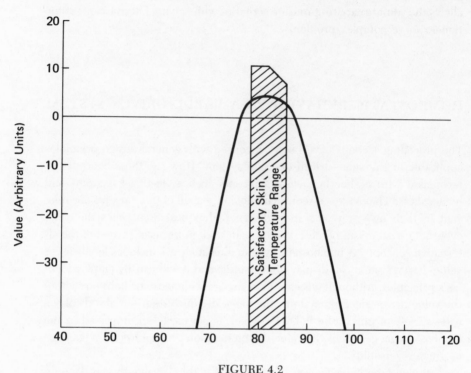

FIGURE 4.2

Illustrative "Value" versus Skin Temperature to Induce Homeostatic Behavior in a Value-Driven System

ANALOGUE VERSUS DIGITAL CYBERNETIC CONCEPTS

During the early development of electronic computers two distinct design concepts, analogue and digital, emerged. The analogue systems do not use digital representations. Instead, the magnitude of a physical parameter is represented in the computer by a voltage that varies in proportion to the physical parameter. In the familiar digital computer, numbers and other symbols are coded in digital form. For most applications the digital computer soon came to dominate, because it could easily be used to simulate an analogue computer, whereas the analogue computer could simulate a digital computer only at prohibitive cost.

On the other hand, analogue components are more efficient than digital for certain system functions. This is particularly true for data input sensing systems and for output functions that involve control of machinery. In such applications, analogue systems can usually accomplish the job with fewer components than is possible with digital equipment. For this reason, the sensing systems and the mechanical actuators for most automated systems still use analogue components.

Of course, digital systems can always be used to simulate analogue systems, and when they are used in this way, concepts such as feedback and homeostasis which were originally developed to describe analogue systems are still applicable. But when the digital systems are used in their characteristic role, as processors of numerical and symbolic information, such analogue concepts do not provide a satisfactory way of discussing their operation.

The basic building block of the biological systems (the neuron) is essentially a digital device—either it discharges or it doesn't; there is no middle ground. But at least in the input-output circuits, nature seems to have used these digital elements in such a way that they simulate an analogue system. The intensity of sensory stimuli is represented in these input channels by the frequency of discharge of the neurons. The frequency of discharge is usually a logarithmic (or power-law) function of the intensity of the stimulation. Similarly, the contraction of muscles is controlled by the frequency of discharge of motor neurons and the number of motor neurons discharging. Feedback circuits in the motor control system therefore operate very much like analogue feedback systems. The higher intellectual processes in contrast seem to be symbolic or digital in character, as is indicated by the use of language, which is obviously symbolic rather than analogue.

It seems likely that as our understanding of neurophysiology increases we will find that nature has chosen analogue or digital design concepts for specific system components in much the same way a human designer would have made the same selection. In all probability the higher control and decision processes operate primarily on symbolic or digital principles, while the more routine mechanical functions are based on analogue concepts.

REFERENCE

1. Smith, Nicholas M., and Milton C. Marney, *Foundations of the Prescriptive Sciences.* McLean, Va.: Research Analysis Corporation, March 1972.

Chapter 5

The Human Decision System

When subjective values are conceived to have objective consequences in the brain, they no longer need to be set off in a realm outside the domain of science. The old adage that science deals with facts, not with values, and that value judgments lie outside the realm of science no longer applies in the new framework.

ROGER SPERRY (8)

As a value-driven decision system, the human brain must accomplish the same basic functions as an artificial decision system, but it must also accomplish a number of other functions (such as development of its own world model) that are not usually carried out by an artificial system. The preceding chapters developed certain fundamental design principles for artificial decision systems. The design principles used in the brain are similar, but of course not identical, to the principles normally used in artificial decision systems. This chapter examines the way the principles have been applied and modified in the design of the human decision system.

During the discussion of artificial decision systems, it was sufficient to treat the problem entirely from an external functional point of view. It seemed irrelevant to ask how things might look from *inside* the system, or how it might feel to *be* such a system. But the subjective personal experience of the human decision system is the essence of our personal reality. A satisfactory model of the human mind must correspond to our own *subjective* experience as well as to the external *functional* observations and the physical *structure* of the brain.

These three points of view—the functional or system design perspective, the structural or physiological view, and the introspective or subjective view—are quite different but complementary ways of looking at the same system. The present chapter is concerned primarily with a functional or system design point of view, which parallels the previous discussion of artificial systems. Chapter 6 shows how the physical structure of the brain can be interpreted in terms of the "decision system" concept. Chapter 7 deals primarily with subjective personal experience, and shows how this introspective experience can be interpreted in terms of the same basic concepts.

To clarify the basic functional concepts of nature's design, we will begin in the present chapter with an idealized, deliberately oversimplified model of the human decision system. By adhering at first to an impersonal, idealized treatment, we hope to provide a clearer understanding of the basic design principles. These design principles of the brain tend to be obscured in our everyday experience for several reasons:

1. The human mind, like all evolutionary products, is unfinished. Because it is in the process of evolving from one set of principles to another, the actual system is a complex mixture of many design concepts. The idealized model we will discuss here probably approximates the design that the system is evolving toward.
2. In our personal introspective experience we are aware only of those parts of system operation that fall within the fluctuating hazy veil of consciousness. Some of the most important functions fall outside this veil. Our intuitive self-understanding is incomplete because subjectively we notice only part of the system.
3. The operation of the brain includes many practical imperfections because the brain is a finite system. Information recall is imperfect. Many data-processing operations take place subconsciously in such a way that the logic cannot be recovered or verified. Thus, many practical decisions are based on "intuitive" judgments which may be in error and which certainly cannot be subjected to critical review. It is unclear to what extent such judgments ought to be classified as "rational" thought. Nevertheless, they form an important, albeit imperfect, part of the human decision system.

In developing our *idealized* concept of the human decision system, we will ignore such computational limitations. All of the rational analysis capability of the system (whether conscious or subconscious, logical or intuitive) will be treated simply as part of a rational analysis subsystem. Although this simplifying perspective is appropriate for an external functional description, it is far from adequate as a basis for understanding personal subjective experience. The reconciliation of the idealized model with the available physiological data and with our subjective experience will be dealt with in the following two chapters.

In the present discussion, we will be concerned with the mental processes or decisions that determine voluntary behavior (specifically those aspects of behavior that are commonly considered to be under conscious control). For this initial discussion we will ignore various complicating factors, such as the role of habit in the routine execution of such decisions, or the fact that the decisions themselves sometimes are not consciously developed. We will be thinking of the conscious mind in the role of a master control or system monitor, whose functions include the allocation of cybernetic resources, the monitoring of results provided by subsystems, and the final choice of decision alternatives when all relevant information is at hand.

FUNCTIONS OF THE HUMAN DECISION SYSTEM

In Chapter 2 we defined five functional components normally included in an artificial decision system. These components are as follows:

1. A *data input procedure* to supply information which defines the environment as it affects action alternatives.
2. A *model* of relationships in the environment which defines action alternatives and their consequences.
3. A *procedure for searching* through action alternatives to find desirable options.
4. A *method for assigning values* to the estimated consequences of alternatives.
5. A *decision mechanism* for selecting alternatives that show the best value.

As we develop the analogy between the human mind and artificial decision systems we will be concerned with a comparison of functional properties at about the level of detail listed above. The specific procedures or mechanisms by which the functions are accomplished within the brain are undoubtedly very different from those used in a computerized decision system (although at present we have no way of knowing what procedures may actually be used in the brain). Nevertheless, it can be a useful beginning to discuss design concepts at this simple functional level.

Even at the functional level, the human brain is much more complex than a typical computerized decision system. The brain must construct its own model or models of the environment. The analysis of input data for this purpose of model building is an important function which at present distinguishes the biological systems from the typical artificial system. In most artificial decision systems, the designer provides a ready-made model of the environment, which is included as part of the basic system design. In some sophisticated artificial systems, the system is allowed to refine or improve the model, but this is more the exception than the rule.

The human mind works with many different models or theories about the environment. The different theories deal with different aspects of the environment and at different levels of detail. For the present discussion, all such models will be treated simply as different aspects of a total "world model" which is developed by the individual for the purpose of making better decisions. Although important parts of this model are simply accepted from authority, the total world model used by the individual is fundamentally a personal creation. The individual must *decide* what authorities to believe and what concepts to accept from the authorities. Thus, the creation of the world model is itself a decision process. It is one of the most remarkable and important functions of the human decision system.

There is one other important difference in the design concept. In artificial decision systems the algorithm for the assignment of primary values is provided by the designer and it is *utilized directly* to evaluate the consequences of various action alternatives. Thus, the primary value system is directly exercised for *every* alternative that is considered. In the biological designs, the primary value system is exercised *only* during the course of actual experience. As an experience takes place, the primary value system generates the sensory and emotional *value* content of the experience. These values are generated, and automatically attached to the experience, by value subsystems that lie outside the rational mind. From the point of view of the rational mind, these valuative components appear to be an intrinsic part of "reality," so the models developed by the rational mind inevitably include the valuative components. As action alternatives are considered and the outcomes are projected, the valuative content of the projected outcomes is included automatically as a part of the predicted outcome; so it is quite naturally taken into account in selecting a preferred action alternative.

We can now summarize the main functions of the human decision system, taking into account these basic differences between the human and the typical artificial decision systems. Of course, we will be focusing on the functions of the normal mature or maturing human system. Obviously such a description would not be appropriate for the early fetal stages and it may not be appropriate for the young infant. In the discussion we will speak of certain system functions as being "built-in" or genetically determined. Obviously "normal" biological maturation of the individual (in accordance with the intended genetic design) can occur only in the context of interactions with the environment. Just as normal maturation will not occur if the individual is denied food, so it will also fail to occur if the individual is denied normal psychological experience with the environment. Consequently when we speak of characteristics being "built-in" or "genetically determined" it does not mean that the characteristics would develop normally if the individual were deprived of normal psychological stimulation. It means only that within a normal childhood environment the functions seem to develop automatically, without excessive dependence on the details of the environment.

In the normal mature human decision system we can identify the following basic functions:

1. *The reception of sensory data.* The sensory data provides the information needed both to build the world model and to define the current state of the environment. This input data is extensively analyzed and interpreted by built-in system hardware, before the information is forwarded to the conscious decision system.

2. *The assignment of "values" to experience* is a vital function of the human decision system. The assignment of values is accomplished automatically by built-in system hardware, in accordance with fixed rules that are part of the evolutionary design.

   Sensory experience is normally accompanied by an intrinsic value assignment. We feel pain when we burn a finger. We do not decide to feel pain. The "pain" is automatically attached to the experience before the sensory data is received by the conscious mind. This concept is obvious with regard to most simple physical stimuli. They are intrinsically pleasant, unpleasant, or neutral: we do not decide which they will be.

   It is less obvious that the same principle applies to our social, artistic, intellectual, esthetic, and athletic experiences. Yet, with some reflection it is apparent that the same principle is at work. We do not decide what we will enjoy. The question, "Why do you like to do ____?" is unanswerable. We like it because we like it. As always, the primary value judgments are automatic, they are not rationally derived. These primary or innate values are the chief determinants of human behavior. They are the values that drive the human decision system.

3. *Data storage and retrieval (or memory).* The ability to make or use models of the environment is dependent on the ability to compare experiences. This requires the ability to store and recall past experiences. A capability for data storage and retrieval is also essential if the models themselves are to be stored for future use.

4. *Objectification and symbol use.* The ability to make models of the environment depends on the ability to classify related stimuli, and to identify them with "objects" or "events" to which symbols can be assigned. Some of the rudimentary steps in this classification of stimuli are accomplished automatically by the built-in system hardware as described in the data-input function above. However, much of it is a real learning process that is accomplished rationally by the decision system on the basis of experience.

5. *Building and refinement of the world model.* The model-building process requires the recognition of relationships among the symbols, or entities, which gradually develop into conceptual models of the external world. These models may include "self" as one of the entities. The "models" may be as unsophis-

ticated as a child's model of his mother, which he uses to predict her response, or as sophisticated as quantum mechanics or general relativity.

This model-building process is concerned not only with the development of theories or models, but also with the development of hypotheses about the present status of the world environment. A child, having heard the front door open and close, may try to decide whether mother is still in the house.

The world model may also include certain secondary or derived values that are used (together with certain rules of thumb) to help select good decision alternatives.

6. *The allocation of intellectual effort.* Cybernetic resources are limited. The available intellectual effort is a scarce resource which must be wisely allocated. The allocation of these resources is illustrated by our ability to focus attention on specific activities or problems and to ignore others that seem less urgent. Many of our innate intellectual "values" serve as criteria to manage this allocation of intellectual effort.

7. *Simulation—using the model.* This function involves the use of the model to identify possible courses of action and to predict the consequences of the possible action alternatives. In the human decision system it is not necessary to artificially attach values to the alternative consequences of action. Since experience is received with innate values already appended, the valuative aspects of experience are inevitably incorporated as a natural part of the model. When we predict the consequences of alternatives we also predict, on the basis of experience, whether we will like or enjoy the consequences.

8. *Decision—based on the outcome of simulation.* Decisions are reached by testing the alternative outcomes against a value scale. The purpose is to select the alternative which will prove "best" on the innate value scale. To permit quicker or more efficient decisions, the system may also utilize secondary values or rules of thumb that experience has shown to be an aid to good decisions.

Obviously, the resulting decision can be in error. Mistakes can be made if insufficient information is available, if the model does a poor job of predicting the consequences, or if it does a poor job of predicting the value or enjoyment of the consequences. Because the primary or innate values are orchestrated by complex rules that are not known to the rational part of the decision system, errors can be made in estimating the value of alternatives even if the physical consequences are accurately predicted.

There is one other basic problem which can degrade the accuracy of the decision process. The different elements of value which are intended to guide decisions, such as pain, hunger, or fear, are perceived as qualitatively different, and not necessarily commensurable. To make a decision, it is sometimes necessary to estimate the weight or importance of different value components. The weighting may not be done with great accuracy or consistency. This is an important problem which will be discussed in more detail later.

9. *Control of action.* This function involves the ability to implement a decision. That is, the ability to initiate the physical or mental activity required to carry out a decision.

There is nothing remarkable about the list of functions above. They represent processes with which we are all familiar. What is interesting is how neatly these familiar processes fall into place when we interpret the human mind as an automatic decision system. Admittedly, the description of the system as implied in the foregoing functions is idealized. It seems to correspond better with the way we would like to operate if we were smarter and were not subject to so many human failings. Nevertheless, it defines a basic system concept which reflects important characteristics of the human mind.

From the point of view of the conscious mind the first two processes (data input and value assignment) are processes over which little or no control can be exercised. These processes occur automatically, they are not consciously perceived, and they are not subject to rational control.* The last process (implementation) is the output process of the conscious mind. The remaining *six* are internal functions of the *conscious* mind, and they are the ones we most commonly identify with rational thought. These rational thought processes constitute the strategic decision level in the human system. It is remarkable how completely the preceding functions encompass our conscious experience, so it is tempting to conclude that the "conscious mind" is a natural and inevitable consequence of the evolution of a value-driven biological decision system.

There is, however, one glaring omission. While talking about the functions of the "conscious mind" we have said absolutely nothing about *consciousness* itself. The intuitive reaction to this omission is a very personal issue. Some will say that the functions specified above *automatically* encompass consciousness. Others will say that consciousness is fundamentally a sense of awareness, a sense of the wholeness and continuity of experience, that is not even suggested in the preceding functions. This issue about the subjective nature of consciousness must ultimately be addressed, but for the present we will adhere to our plan of discussing the human decision system initially from an external or functional perspective. The issue of consciousness and personal subjective experience will be postponed to Chapter 7.

One of the most important distinguishing characteristics of the present concept is the role that is played by the mental model as a key functional component of the control system. This sets the present theory apart from many simpler cybernetic and behavioral concepts such as feedback systems, optimal control systems, homeostasis, tropisms, and operant conditioning which do not require (and indeed could not use) such a detailed mental model. In view of the large cybernetic resources that evolution has devoted to the development and maintenance of this mental model, it seems obvious that the model must be of major importance for the survival of the species. The value-driven decision system provides a cybernetic construct in which the model is an es-

* The conscious mind does however have some ability to decrease the intensity of input data by diverting attention to other matters.

sential part of the system design. Moreover it helps to explain why it is necessary to provide such a detailed and comprehensive model.

The preceding paragraphs have sketched the outlines of a concept of the human decision system which is both simplified and idealized. Although the approach may seem to overemphasize rational and systematic intellectual processes, it defines an objective model or design concept which encompasses the higher intellectual functions. In its idealized form, it has the merit of providing a scientific model which in some ways seems better than human, in contrast to some of the older behavioral models which distort the human image because the models are so much less than human.

EVOLUTION OF RATIONAL THOUGHT

To provide a better understanding of the actual design of the human decision system it is necessary to pause briefly and consider how such a design principle might have developed during the course of biological evolution. The following discussion is necessarily speculative, but it seems generally consistent with the available evidence.

As biological species evolved, the information-processing capacity became larger and more sophisticated. In the most primitive species, behavior consisted primarily of automatic reflex responses to specific stimuli. As the species became more sophisticated, such responses evolved into rather elaborate instinctive behavioral patterns known as tropisms (such as those observed in insects) which can be released by specific stimuli. Such purely automatic responses were gradually made more versatile. With the development of a learning and association capability, the organism could be *conditioned* to respond (on the basis of experience) to a previously neutral stimulus, much as it would to the standard or built-in stimulus.

As the species became more sophisticated, the dependence of behavior on learning increased. The original neural connections were augmented to allow the automatic response to be suppressed as a result of inhibiting neural signals from higher control centers. The animal with such a structure might feel he would like to respond but because of adverse past experience he might choose not to. Gradually behavior came to be controlled less by automatic responses and more by generalized urges in which the actual response was subject to modification based on learned experience.

Automatic passive learning from past experience gradually evolved into a mode of behavior in which the learning itself became an innate goal. In such animals, curiosity is a basic motivating drive, or value, which has a profound influence on behavior. The practical function of such curiosity, of course, is the development of better mental models of the environment.

There is no sharp boundary in this development of curiosity and rational thought. There is only a gradual evolution in the capacity to develop a more elaborate and complete mental model and toward greater behavioral dependence on the resulting world model. Although the world model has immense influence on behavior in the higher animals and in man, one must not forget the continuing role of the innate values in supplying the ultimate criteria of decision. In the higher animals, the residual urges can be appropriately described as instinctive motivating values because a rational thought process intervenes between the motivating drives and the actual selection of behavior. With increased intellectual capability, the appropriate drives could be less specific but more accurately tuned to broad evolutionary goals. Old urges or drives either evolved to meet the new need or they atrophied. New motivating values specific to the needs of each species appeared.

Although mankind shares the essential features of this design concept with the higher mammals, the human capability for speech has produced a rapid widening in the intellectual gap between man and other animals. The capability for speech makes it possible for humans to engage collectively in the model-building process, so that the world model used by modern man is a product of the collective intellectual efforts of generations. The development of speech undoubtedly interacted with the evolution of the brain to produce brain structures better suited to speech and better suited to deal with more sophisticated world models.

The great evolutionary advantage of man over animals lies fundamentally in the superior world model that is used by the human decision system. The evidence, however, is quite strong that other intelligent animals also utilize some form of world model as an essential part of their decision system. Although at present we can only guess what levels of sophistication different animals may have in their world model, the existence of a world model appears to be a necessary precondition for the emergence of symbolic speech. Recent work in linguistics at MIT suggests that all human languages are based on a common underlying logical structure which may be at least in part a genetic inheritance of man (1). This logical structure seems to presuppose an underlying mental model of the environment (10). Thus, there are strong reasons for believing that speech is a recent evolutionary innovation that has been added to the basic primate decision system. It seems most unlikely that two such major evolutionary inventions as speech and the world model could occur sequentially during the brief evolution of the human species. The preexistence of the intellectual machinery required for the use of the world model helps to explain the explosive cultural impact of the emergence of speech.

THE STRUCTURE OF HUMAN VALUES

RELATIONSHIP TO MOTIVATION

From the discussion of the production scheduler, it is apparent that values can provide motivation in several ways. In an organism with little or no ability to think ahead, rewards or penalties must be almost concurrent with the associated activity. They must take the form of an immediate release of discomfort or of liking or disliking the activity. In a system that can think ahead, a specific activity can be motivated in three distinct ways: by a penalty for *not* engaging in the activity which precedes the activity, by the pleasure of the activity itself, or by a reward that will automatically follow the activity. An activity can be discouraged by a dislike for the activity itself or by a penalty that will follow the activity.* In one way or another, nature seems to have used all of these techniques. Typically, however, only a subset of the methods are used to motivate or discourage a specific form of behavior.

DISTINGUISHABILITY OF VALUES

Usually in artificial systems, the motivating values are distinguishable from each other only by their sign, magnitude, and timing. In biological systems, however, this is not the case. The motivating values are typically presented to the conscious mind in qualitatively different forms. For example, thirst is subjectively different from hunger. Pain resulting from a burn on the finger is qualitatively distinguishable from pain resulting from a burn on the elbow. Thus the motivating values are delivered to the conscious mind in the form of specific "drives" or "urges" that can be easily distinguished from one another. There are at least two reasons for this method of delivery of the primary values.

The first reason concerns evolutionary history. In an earlier evolutionary phase, when these drives or urges were directly linked to automatic re-

---

* Theoretically it might seem that an activity could also be discouraged by a prior reward for not engaging in the activity. However, such a reward would almost always be in operation, so it would amount to little more than a change in the origin of the value scale. It would serve no useful purpose and would be contrary to the cybernetic objective of parsimony in the use of motivating values.

sponses, the drives had to be neurologically different to generate different responses. Because the drives were triggered by different stimuli and were designed to produce different responses, they originally had little in common.

The second reason concerns a fundamental design consideration for the biological form of a value-driven system. To be able to make decisions, the system must learn to *predict* the value consequences of alternatives. The use of distinguishably different values makes it easier to associate specific value components with specific causal factors. One of the functions of rational thought is to classify the motivating drives or values and associate them with the specific types of activity they are intended to motivate. By delivering each drive to the conscious mind in such a way that it is separately distinguishable, the association problem is greatly simplified.

IDENTIFICATION OF SOURCES OF VALUES

The lower and more primitive drives or values appear to be built into the hardware in such a way that they require little or no learning. For example, the tactile sensations of localized pain and pleasure are delivered to the same location in the brain as other sensations from the same part of the body. Thus, a single association of physical with mental locations (whether built-in or learned) is applicable to all such sensations.

Although most of the lower drives are easily identifiable with specific types of activities, some of the higher values are not. For example, we have probably all experienced the sense of satisfaction, well-being, or euphoria that is associated with certain accomplishments. Nature seems to use this same type of "reward" for a wide range of activities. The association of the reward with the particular activities therefore may not be obvious.

This type of association ambiguity appears to be particularly important with regard to the innate "social" values. Indeed some of our problems in coping with the modern social environment may arise because of inability to recognize (in the new urban environment) which types of activities a particular value component is intended to motivate. In all cases, however, the recognition of the value content of a stimulus is built-in and does not have to be learned. There is no doubt that pain is bad and should be avoided, and there is no doubt that the sense of satisfaction or reward that follows an achievement is one of life's good experiences which one should try to repeat.

THE COMMENSURABILITY OF VALUES

Values, as they are used in artificial decision systems, are rigorously numerical or quantitative. Although there may be considerable doubt about whether the values correctly reflect the objectives of the designer, there is never any doubt about the actual magnitude of the values involved.

But this is not the case with the innate human values. The values are perceived by the rational mind as qualitatively different. When different drives are in conflict it may not be obvious which one dominates. The comparison is a little like subtracting apples from oranges, so it cannot be done easily or accurately. Of course, it is not nearly as bad as subtracting apples from oranges, for in the case of the values the relative importance (or weighting) of the values is built into the system. The weighting is just not very precise.

The different valuative components are comparable only as the conscious mind itself chooses to compare them. The final conversion from noncomparable valuative sensations to comparable values occurs only in the conscious mind. For example, if the alternatives involve conflicts between the discomfort of hunger and the discomfort of cold, then the importance of the components must be subjectively compared. Although this comparison is not accomplished easily or with great precision, it is accomplished. The comparison involves a conscious or subconscious projection of the noncomparable drives, urges, or values onto a comparable scale of values (or utility scale). It is only when this has been done that the instinctive values assume the characteristics of a unified value scale. The lack of precision in the comparison of conflicting values is a price that nature has paid to make it easier for the conscious mind to develop a good mental model of the value structure.

Although the comparison of different value components lacks precision, the capability for comparison is fundamental to the system design. The drives and values are carefully orchestrated in magnitude, so that when they are received by the conscious mind they are perceived to be of varying levels of intensity. The only practical function of this variation in magnitude is to permit and encourage quantitative comparison.

AGE DEPENDENCE OF VALUES

The orchestration of values depends not only on the external circumstances but also on the age and sex of the individual. This point is dramatically demonstrated as children of both sexes pass through adolescence. It seems obvious that the change in sexual and social behavior at this age is too sudden to be solely the result of learning or social conditioning. Although experience and learning may have laid the intellectual foundation for the new behavior, the behavior does not really occur until the innate sexual interest develops. The suddenness of the resulting change can be confidently ascribed to a change in instinctive drives or values.

Oddly enough, there seems to be great reluctance to recognize the importance of changes in innate values in connection with more gradual changes in behavior. It seems obvious that the innate drives appropriate to an infant are not appropriate to a child of school age. Moreover, the innate drives appropriate to the young adult may not be the most efficient, from an evolutionary point of view, for the middle-aged citizen. It would be very surprising if evolution had not provided for the automatic adjustment of values with age, in order to maintain behavior most efficient for group survival at all ages.

Admittedly, when changes of behavior with age are gradual, it may be difficult to determine the extent to which the change results from changes in innate values, as opposed to learning and cultural conditioning. But in much of the psychological literature, there seems to be an unstated assumption that if a behavior was not present in the child, then its presence in the adult must reflect cultural conditioning or learning. It seems very likely that many of the subtle changes in innate motivation that occur with age have been overlooked as a consequence of this unstated assumption.

In very young children, the changes in values can be very easily confused with the effects of learning. In the infant and young child, curiosity is a dominant motivating force. The child is intensely busy developing and refining his world model. In the detailed child behavior descriptions of Jean Piaget (3, 4) one can almost visualize the model-building drive in operation. The child exhibits intense interest in objects that are new enough to be interesting (i.e., which fit into his current phase of model development). He is less interested in familiar objects that he has fully explored, or in new phenomena that are too strange to fit in his current model-development activity. This intense pace of model development naturally results in a rapid improvement in the child's coordination and the appropriateness of his responses. However, underlying these changes due to learning there may also be important changes that are the result of changes in the innate drives.

SUBJECTIVE PERCEPTION OF VALUES

When we speak of the innate, or primary, human values we are usually referring to specific valuative sensations which seem to be a very concrete part of our subjective experience. It is a good rule of thumb that a value criterion is a secondary rather than a primary value unless it is linked with a recognizable valuative sensation. The valuative sensations are experienced as *intrinsically* good or bad, pleasant or unpleasant. Table 5.1 provides an illustrative list of some of the more obvious and familiar valuative sensations—indeed, so familiar that the reader's immediate reaction to the list may be that the whole concept of a value-driven decision system must be trivial. This reaction, however, is quite unjustified. The total human value structure is very complex. In many cases the relationships between the value sensations and the activities that are being motivated is very subtle. Without a careful analysis of the value structure, it is hard to believe that such sensations can explain the complexities of human motivation.

TABLE 5.1
Illustrative List of Innate Values or
Valuative Sensations

NEGATIVE (BAD OR UNPLEASANT)	POSTIVE (GOOD OR PLEASANT)
Discomfort	Comfort
Pain	Tactile pleasure
Bad taste	Good taste
Bad smell	Good smell
Sorrow	Joy
Shame	Pride
Fear	
Anger	
Hunger	
Thirst	
Itch	

To understand the operation of the value structure it is necessary to study it from an evolutionary perspective as an integrated system design. Only in this context can we begin to make sense of the very complex human value structure. Because Part II of the book is devoted entirely to such an analysis, no effort will be made to anticipate the analysis here.

NATURE'S DESIGN CONCEPT

We can now summarize the basic concepts that nature has used in the design of its value-driven decision systems. As our complex sensory experience unfolds, instinctive or innate values are automatically added to the experience and are delivered to the rational mind. From the perspective of the mind, the values themselves are an *essential* part of the experience. The specific structure of the values has no rational basis, except that it is a part of our genetic inheritance. This value structure is uniquely human, and it lies at the foundation of human behavior.

Because the innate value structure seems to be an essential part of *experience*, "rational" decisions are automatically made in terms of this value structure. When mental models are used to predict the outcome of alternatives, the predicted outcome will also include the valuative component. The desirability of future alternatives can thus be judged without any separate assignment of values, for the *values* are an *integral* part of the model that has been developed.

The value structure is somewhat unpredictable, because it changes with age and circumstance, so that what seemed rational at one time may seem irrational at another. What is rational for one person is irrational for another. Because of the variability and unpredictability of the value structure, the rational mind can never develop an entirely satisfactory model of its own value structure.

The innate values, of course, are actually experienced *only* for the specific alternative that is selected. To choose between alternatives the human decision system must try to *estimate,* on the basis of past experience, what the values would be for each alternative. This estimate involves uncertainty, both because the future is uncertain and because the value structure is imperfectly understood. If an individual has not previously had access to a full spectrum of social experiences, the projection of the valuative consequences can be defective and the resultant decisions will be inappropriate.

CORRESPONDENCE WITH TRADITIONAL VALUE CONCEPTS

In philosophical discussions of human values, it has been found useful to make a distinction between "instrumental" and "intrinsic" values. In some ways this traditional distinction is analogous to our present distinction between secondary and primary values. According to the traditional view, a thing which is valued only as a means to an end has "instrumental" value, but no "intrinsic" value. In contrast, a thing which is valued not as a means to an end, but as an end in itself is said to have "intrinsic" value. Although there is general agreement about the usefulness of this distinction, there is disagreement about exactly where the distinction should be drawn.

Almost everyone would agree that a farming tool used to plant a crop has only instrumental value; it is a means to an end, which is the harvesting of the food. On the other hand, as to the food that is harvested, some would say that it has intrinsic value, whereas others would say that the food is only a means to other ends (namely, the enjoyment of eating and the nourishment needed for survival). But the logic of the instrumental value concept demands that there must be some set of values that can be identified as the ultimate or "intrinsic" values.

When the present theory is viewed from a subjective perspective, it becomes clear the innate valuative sensations play a role in the theory that corresponds to these traditional "intrinsic" values. The value sensations are valued (positively or negatively) for themselves alone. According to theory, they are the ultimate source of human motivations. This does not mean that we eat solely for the enjoyment of eating (although that is an important factor in the motivation). Obviously we may also eat to survive. However, when we ask why we want to survive, we once again come back to the innate values. Our expectation of the enjoyment of future value sensations: joy, happiness, pride, and even the taste of good food in the future seem to play a major role in our desire to survive. Thus the theory leads naturally to a rather unambiguous version of the traditional intrinsic values concept. It identifies our innate value sensations uniquely as the "intrinsic" values of our subjective experience.

On this basis, it is possible to clarify the relationship between the present *primary* versus *secondary* value distinction, and the traditional distinction between *intrinsic* and *instrumental* values. The traditional distinction is a subjective concept based on the introspective experience of the human decision system. The distinction between primary and secondary values is an objective system-design distinction that can be applied to any value-driven decision system. However, when the present theory is interpreted subjectively from the perspective of the human decision system, then there is a correspondence between the two concepts. The primary human values correspond to the traditional intrinsic values, whereas the secondary human values correspond to the traditional instrumental (or extrinsic) values. In the material that follows we will continue to use the primary versus secondary value distinction, not only because it is more general, but also because it is less subject to ambiguity and misinterpretation.

THE CLASSIFICATION OF HUMAN VALUES

The primary values that motivate human behavior can be broadly classified in terms of the evolutionary goals they are intended to serve. The three main categories we will use to organize the discussion of values in Part II are:

(1) *Selfish Values:* values directly associated with individual welfare and survival.
(2) *Social Values:* values whose purpose is to motivate individual behavior contributing to the survival of the group—generally altruistic values.
(3) *Intellectual Values:* values whose purpose is the motivation of efficient rational thought—generally esthetic and intellectual values.

To avoid confusing these three categories of *innate,* or primary values, with the corresponding secondary values that are referred to as selfish, social, and intellectual values, we will hereafter adopt a stylistic convention which will capitalize these categories. This will serve as a reminder that we are not referring to the traditional cultural values that commonly might go by the same name.

In traditional literature the values in the first category have been classified as our animal drives or lower values. The values in categories (2) and (3) have been referred to as our spiritual or "higher" values. In some religious literature secondary values derived from the "selfish" values are treated as if they are man's values, while the corresponding "social" and "intellectual" values are treated as if they are God's values. In the present context these distinctions seem inappropriate. All the values have their origin in human evolutionary history, and they are all needed to motivate effective behavior of man as a social and intellectual animal.

The traditional social, ethical, and religious emphasis on the "higher" values (particularly the social values) undoubtedly reflects the interest of society in encouraging behavior conducive to the welfare of society, rather than just the selfish interests of the individual. Paradoxically, modern industrial society has generally done a very good job of providing for those basic "selfish" human needs that are reflected in the category (1) values. In the process, however, it has interfered with the individual's ability to find appropriate psychological satisfaction in socially beneficial activities connected with the "higher" values. Thus, modern society may be inadvertently frustrating the higher motivations that it overtly seeks to encourage.

Although the primary focus of this book is directed to the "higher" values, the discussion in Part II, nevertheless, begins with the Selfish Values. This sequence of presentation is dictated because these lower values are simpler and easier to study. Our real purpose in analyzing the simpler values is to develop a theoretical foundation for considering the Social and Intellectual Values. To provide a balanced understanding of human behavior, it is necessary to consider the rational thought processes as well as the innate or built-in value structure. Although we are postponing the discussion of the rational mental processes, they will not be ignored. The Intellectual Values can be understood only within the context of the mental processes they are intended to motivate. Therefore, the discussion of our rational thought processes will be encountered quite naturally in connection with the discussion of the Intellectual Values.

VALUE TERMINOLOGY

The preceding sections have informally introduced a number of concepts that need to be defined with more precision. This section provides some simple but somewhat idealized definitions for concepts such as "value function," "value structure," and "value system." The definitions apply most accurately to an idealized model of the human decision system. The actual biological design of course does not always adhere exactly to concepts as defined here.

INNATE HUMAN VALUES

Innate human values are the value criteria that are genetically built into the human brain to provide the human species with its fundamental criteria of decision. From a system-design perspective these values are simply quantitative measures of "desirability" that are assigned to different types of experience to provide the brain with its ultimate or fundamental criteria of decision. Within our subjective or introspective experience these innate values are recognized as more or less concrete valuative sensations such as pain, hunger, joy, or fear which seem intrinsically good or bad, pleasant or unpleasant.

VALUE SENSATION

Although from a system-design perspective the values are simply quantitative measures of desirability, the values have a concrete physical manifestation in the form of a cybernetic signal that is received by the rational mind. Presumably the signal corresponds to a specific pattern of neural discharges in a specific set of neurons (which should in principle be measurable). This neural signal is experienced subjectively within the conscious mind as a valuative sensation. When we wish to relate the innate values to our personal subjective experience we will often refer to them as "valuative sensations." The innate values include many qualitatively different sensations such as pain, hunger, or fear that can be easily distinguished from one another.

We could represent each of these valuative sensations in a mathematical model as a scalar (or one-dimensional) quantity. The numerical magnitude associated with a value sensation in such a model should, of course, depend on the intensity of the sensation. The innate value sensations respond systematically to specific types of experience. Hunger occurs when the body begins to run short of nourishment; specific types of pain occur when our skin becomes too hot.

THE VALUE FUNCTION

To provide an adequate model of the human decision system we would need to know the specific functional relationship that exists between the value sensation and the relevant aspects of experience. In mathematical terms, the numerical magnitude of each value sensation should be expressed as a function which is defined over the possible experiences for the system. In some cases, the functional relation may be quite simple. For example, the experience of pain from the prick of a pin is likely to be simple and reproducible. In other cases, the relationship may be very complex. For example, the intensity of the sensation of fear depends not only on the present experience, but also on related past experience. To provide a quantitative model, we would need to know how the fear sensation for a specific external stimulus is moderated as a consequence of past experience. The total functional relationship (whether simple or complex) between experience and the valuative sensation is what we call the "value function."

In some cases, where the value function itself is very complex, it will be convenient to consider the function as if it were made up of separate components. Thus the value function may be described as if it were composed of the sum or product of several different functional parts.

THE VALUE STRUCTURE

Each of the different valuative sensations, such as those listed in Table 5.1, has its own value function. To provide a full understanding of the human decision system we would need to know the specific value function for *each* of the value sensations. This ensemble of all the value functions constitutes the innate "value structure" for the human decision system.

THE VALUE SYSTEM

Each of the innate human value sensations is generated by some biological mechanism within the brain. These mechanisms are part of a built-in human "value system." The biological mechanism is the physical counterpart of the mathematical value function. The ensemble of all such biological mechanisms constitutes the total human "value system." The "value system," of course, is the physical counterpart of the "value structure."

The value system cannot operate in isolation. The values are defined as functions of human sensory experience; therefore, the value system cannot operate without access to sensory information. Because many of the human values are very complex functions of experience, the total biological value system involves a rather large neural or cybernetic system.

THE UTILITY CONCEPT

Almost any practical decision will involve a number of different valuative considerations which, in a biological system, may be reflected in a number of different valuative sensations. However, to make decisions, the biological decision system must place all valuative sensations on a common scale (a utility scale). Consequently, the numerical value assigned to a given valuative sensation within a model of a biological decision system should correspond to the value on such a "utility" scale. Von Neumann and Morgenstern (9) showed how such a scale can be constructed, at least in principle, simply by testing the preferences of an individual for different alternatives. To develop the scale, new alternatives are compared with a probability mix between standard "good" or "bad" alternatives. The position of the new alternative on the "utility" scale is defined by the percentage mixture between the good and bad alternatives that is required to leave the individual indifferent in a choice between the new alternative and the probability mix. In principle, the same procedure could be used to define the numerical "utility" associated with different intensity levels for each of the valuative sensations. The sum of all the values at any time on this common utility scale defines the *instantaneous* value (or utility) of any alternative. To compare any two alternatives this instantaneous value (or utility) would have to be integrated (or summed) over time for each alternative. The alternative that offers the highest integrated "utility" over the foreseeable future would be the preferred alternative.

THE SPAN OF RATIONAL CONTROL

Obviously, not all human behavior is under conscious control. We have reflex responses (such as the familiar knee-kick response) that are not under conscious control. All the complex chemical processes which maintain the supply of energy and the proper range of body temperatures occur automatically. We do not sense the operation of these systems and they are not normally under conscious control.*

* Recent experiments in biofeedback indicate that many of these processes can be influenced by the conscious will. The processes themselves must first be brought to conscious awareness. This can be done with instruments that allow a subject to monitor certain normally automatic internal processes. Once this has been done the subject may be able to learn, probably by trial and error, to influence the operation of the processes. However, the processes are not normally subject to conscious control.

The fact that such functions are not under rational control is simply a matter of good system design. The required operations can be routinely controlled in a satisfactory way without burdening the strategic control level with unnecessary information and unnecessary decisions to make. Moreover, from an evolutionary design view, the rational thought process is somewhat unpredictable. Because it can reach erroneous decisions, it is better not to trust it with vital system functions, especially those that will not benefit from a very flexible behavioral repertoire. For this reason, the rational control of behavior is concerned mainly with those aspects of behavior that involve interactions with the unpredictable external environment. Those aspects of behavior that are concerned mainly with interactions with the more predictable internal environment need not be under rational control.

Indeed, most of the routine aspects, even of external behavior, are not under direct conscious control. These activities, when once learned, are delegated to "habit." In another sense, however, these activities remain indirectly under rational control. When we choose, we can bring them to the level of awareness and place them under conscious control. The ability to delegate such responses to habit is a matter of system efficiency. It frees central cybernetic resources to deal with problems that cannot be routinely treated. Of course, occasionally, habits can lead us into mistakes, but on balance they result in an improvement in system efficiency.

The ability of the peripheral nervous system to develop conditioned responses that are not under conscious or rational control is also well established. Most of the experiments on conditioned responses and many of the experiments dealing with operant conditioning have dealt with this type of learning. Such subconscious learning in the neural system, beyond the range of the conscious mind, is an important contributor to the efficiency of the overall design. It is important, not only in the acquisition of habits, but also in acquisition of manual dexterity and in the development of athletic coordination. Because this type of learning is relatively reproducible (and easier to analyze) there has been a tendency in psychological experiments and literature to overemphasize its importance.

MECHANISMS OF LEARNING

Extensive evidence showing permanent storage of detailed memory traces from one-time experiences shows conclusively that very efficient neurological mechanisms exist for data storage which can permit efficient computer-like learning in the central control system. While the conditioned response mechanism may be an appropriate learning process for the peripheral parts of the nervous system, it is not an efficient mechanism for the central control system. It would be very surprising if evolution had chosen to use it, where much better mechanisms seem to be available.

Of course, it is quite probable that something like simple operant conditioning plays a significant role in conscious behavior. Some of the instinctive values or drives may themselves be subject to operant conditioning. Indeed, it can be argued that such a relationship is to be expected because of the evolutionary history of the motivating drives. Such an operant conditioning mechanism would introduce a more complex relationship between the motivating values and past experience. It would not, however, require any fundamental modification in the basic concepts developed here concerning the relationship between motivating values, rational thought, and behavior.

While such a role for operant conditioning in conscious behavior undoubtedly exists in some degree, it is almost certainly overestimated in the present behaviorist literature. Almost all the experiments which purport to show behavioral changes as a result of operant conditioning can also be explained as a purely rational response to learning experience which changes the animal's world model. Within the present theory, a change in the world model will produce changes in the behavior pattern even when there is no change in the basic value system. Of course, in the extreme behaviorist view, it would be meaningless to try to analyze such experiments to distinguish between changes in values or motivation on the one hand, and changes in the animal's concept or theory about the external world on the other. For this reason the results traditionally have not been analyzed to distinguish between the different possible causes.

At present, little is known about the actual neurological mechanisms for any of the learning processes. The information code for the storage of information is unknown. It is not even known whether data elements are locally stored in one or more neurons, or whether they may be distributed in some complex way in the correlated states of a large network of neurons. Some experts think that information may be somehow coded in the structure of large molecules within the neurons, or even in the cells between neurons. Others think that learning takes place as a result of changes in the sensitivity of the junctions between neurons. Any or all of the above hypotheses could be correct, or they could all be wrong. Obviously, much needs to be learned concerning the neurological mechanisms of information storage. But for the present, we will simply accept the fact that mechanisms of learning and data storage exist, and we will assume that evolution would normally use the different mechanisms in the roles for which they are best suited.

Fortunately it is not necessary to know the specific physical mechanisms. The logic of a computer system can be completely described without knowing whether it is to be used in a modern solid-state or old-fashioned vacuum tube computer. Just as the science of electronics seems irrelevant to the study of artificial intelligence, so also the science of neurophysiology may be irrelevant to understanding the natural intelligence (6). This view is consistent with our overall approach, which deals with the cybernetic system design from a broad functional perspective without much concern for the detailed physical structure in which the design concepts may be actually implemented.

RELATION TO OTHER MODELS OF BEHAVIOR

It is now possible to summarize the distinguishing features of our model of human behavior. The "decision system" model corresponds very closely with widely accepted views concerning biological drives. However, it provides a more rigorous logical and philosophical framework in which those concepts can be developed and quantitatively analyzed. The analogy with artificial decision systems emphasizes the importance of the internal world model as an intrinsic part of biological behavior, to a degree that was not obvious in most "drive" formulations. It clarifies the role of rational thought, interacting with a nonrational value system, to produce a versatile repertoire of adaptive behavior.

Most studies of animal behavior (as well as the behavioral approach to human behavior) have tended to ignore the role of rational thought or a world model as a factor in behavior. Conversely, most of the classical philosophers have tended to overemphasize the role of reason and underestimate the importance of instinctive drives in human behavior. It seems clear that an adequate science of behavior must provide a balanced treatment of both the rational and irrational factors in the behavior of both animals and man.

It is obvious that the world model used by most humans is far more sophisticated than that used by animals, but this does not mean that the use of a world model can be safely ignored in animal behavior. Emil W. Menzel, a psychologist at State University of New York, has recently reported on experiments which show conclusively that the behavior of chimpanzees involves a mental picture or model of their environment (5). Chimps that were shown food being hidden in different locations within a large outdoor enclosure were able to go efficiently to the sequence of hiding places, regardless of the route they were carried over when the food was hidden. Moreover, regardless of the order in which the food was hidden, the sequence the chimps used to find the food was very nearly optimal to minimize the walking distance required. In all probability the same experiment would yield similar results with animals much less intelligent than chimps.

Although it is easy to overestimate the importance of theory in experimental research, it is obvious that theories influence both the choice of experiments and the description of experimental results. Recent studies in comparative ethology have done a great deal to clarify the evolution of some of the major biological drives, but the observed behavior is described almost entirely in terms of ritualized motor patterns and operant conditioning. Undoubtedly this is a good explanation for *some* of the observed behavior. I find it hard to believe that it is the best explanation for all the behavior.

Even a human scientist has ritualized motor patterns. He smiles when he is pleased; he stands straight when he is confident; he slumps when he is dejected. Yet an explanation of his day in terms of ritualized motor patterns omits some very important elements. It seems likely that such explanations are similarly incomplete with regard to a large fraction of animal behavior.

The woods behind my bedroom is populated with a large number of common squirrels. I am impressed by the apparent importance of a world model in the behavior of these animals. As they leap from limb to limb among the trees they do an excellent job of judging the geometry and distances involved. They compensate for the flexibility of the limb from which they leap, and they avoid jumping to limbs that are too small to bear their weight. It seems obvious we are missing something in our descriptions of animal behavior. We are not likely to find out how much until we can approach both human and animal behavior with a theoretical model that is rich enough to encompass both innate motivations and rational thought.

REFERENCES

1. Chomsky, Noam. *Syntactic Structures*. Atlantic Highlands, N.J.: Mouton, 1957.
2. Chomsky, Noam. *Aspects of the Theory of Syntax*. Cambridge, Mass.: MIT Press, 1965.
3. Piaget, Jean. *The Origins of Intelligence in Children*. New York: International Universities Press, 1952.
4. Piaget, Jean. *The Construction of Reality in the Child*. New York: Basic Books, 1954.
5. Menzel, Emil W. "Chimpanzee Spatial Memory Organization." *Science*, Nov. 30, 1973, p. 943.
6. Newell, Allen. "Artificial Intelligence and the Concept of Mind." In *Computer Models of*

Thought and Language. Edited by Roger C. Schank and Kenneth Mark Colby. San Francisco: W. H. Freeman, 1973.

7. Smith, Nicholas M., and Milton C. Marney. *Foundations of the Prescriptive Sciences.* McLean, Va.: Research Analysis Corporation, March 1972.

8. Sperry, Roger. "Messages from the Laboratory." *Engineering and Science,* January 1974. (Special issue on behavioral biology at Caltech.) Published by California Institute of Technology.

9. Von Neumann, John, and Oskar Morgenstern. *Theory of Games and Economic Behavior.* Princeton, N.J.: Princeton University Press, 1944.

10. Winograd, Terry. *Understanding Natural Language.* New York: Academic Press, 1972.

Chapter 6

Nature's System Design for the Brain

> The new interpretation, or reformulation, involves a direct break with long-established materialistic and behavioristic thinking that has dominated neuroscience for many decades. Instead of renouncing or ignoring consciousness, the new interpretation gives full recognition to inner conscious awareness as an important high-level directive force or property in the brain mechanism.
>
> ROGER SPERRY (15)

THE PRECEDING CHAPTER developed an idealized functional view of the human decision system. The present chapter relates this decision system concept to what is known of the structure and function of the brain. The match to the known physiology seems remarkably good. It provides a consistent interpretation of the brain as an integrated cybernetic system and suggests some previously unrecognized functions for parts of the brain that have previously been a mystery. Of course, the theory of values is not dependent on where or how specific functions are performed in the brain. On the other hand, if either the structure or evolutionary history of the brain proved to be incompatible with the decision system concept, it would cast serious doubt on the concept as a functional model.

The main purpose of the present chapter is to show that such a problem of incompatibility does not arise. For readers with some prior knowledge of the brain, the chapter should provide an interesting new functional and structural interpretation of the brain as a "decision system." Since this interpretation goes somewhat beyond presently confirmed factual knowledge of the brain, this system-design "concept" can also be viewed as a "prediction" of the present theory. Other nontechnical readers may find that this overview of the physiology of the brain will add realism to our interpretation of the brain as a value-driven decision system.* Those who find the material too detailed may skip the chapter without missing information that is essential to the theory of values.

Before describing how the decision system concept fits into the physical structure of the brain, it is necessary to summarize some of the physiology of the brain.

THE NEURON: NATURE'S CYBERNETIC BUILDING BLOCK

Nerve tissue, both inside and outside of the brain, is composed primarily of cells called neurons. These cells accomplish most of the data-communication and data-processing functions of the brain. Figure 6.1 shows a schematic diagram of a neuron. The enlarged portion is the body of the nerve cell. The twisted rootlike projections from the body are known as dendrites.

Input signals from other neurons are received through the surface of the dendrites at junction points called synapses. If appropriate signals are received from enough input synapses, the neuron will trigger and a signal will be sent down the output channel which is called the axon. Signals received at an input synapse may either encourage or inhibit the firing of the neuron, depending on the characteristics of the neuron from which the signal is received. The response (firing or nonfiring) of the neuron depends on the balance of favorable versus inhibiting signals received.

The interior of the axon is normally maintained at a voltage of about −70 millivolts relative to the exterior walls of the axon. When the neuron fires, an electrical "short circuit" occurs which cancels out this normal voltage. The short circuit propagates down the axon at a speed which depends on the diameter of the axon. The larger the diameter of the axon, the higher the transmission speed. Typical transmission speeds range from 2 to 200 miles per hour.

The axon branches as it nears its destination, so that the pulse is carried to

* For readers interested in more complete information about the physiology of the brain, the following are a few references that I found to be particularly useful: Wooldridge (17), Pines (12), Rose (14), Milner (6), Neisser (9), and Thompson (16). The last three are relatively technical texts.

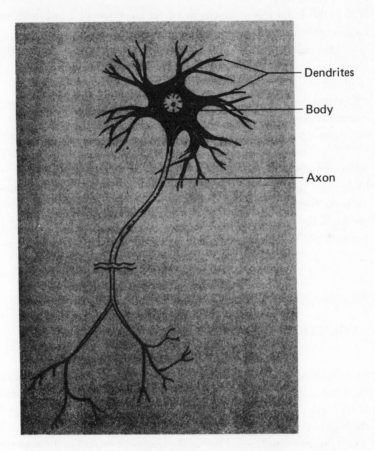

FIGURE 6.1

Schematic drawing of a Neuron, or Nerve Cell

From *Machinery of the Brain* by Dean E. Wooldridge, copyright © 1963 by McGraw-Hill Book Co.; used by permission of the publishers.

numerous other neurons through their input synapse junctions. Although neurons differ greatly in the number of synapses, the length and branching of the axon, and in the location of the cell body relative to the input (or dendrite) zone, the basic functional design is very similar for all neurons.

Data-processing logic occurs in the dendrite (input) zone which determines when a neuron will fire. The resulting information (a signal or nonsignal) is transmitted via the axon. In nature's design, groups of hundreds or even thousands of axons may be bundled together to form information transmission cables. When they are used in a transmission cable to link to the extremities of the body, the axons from individual neurons may be as long as two or three feet, even though the cell body is less than 1/100th of an inch in diameter.

Within the brain the distances involved are typically much less, so that the total length of the axon may be less than 1/1000 of an inch.

Although the neuron is much more complex in its operation than typical electronic components, it has nevertheless all the essential properties required for data processing and data communication. It seems clear that with a little practice computer designers could learn to use the neurons as circuit elements in place of transistors, assuming that methods could be found for hooking them up in accordance with desired specifications.

It is estimated that the human nervous system includes about 10^{10} or 10 billion neurons, and that about 90 percent of these are located within the brain. The total number of synapses in the brain is about 100 times larger. Thus on the average, a neuron will have about 100 input synapses and the axon will link to synapses on about 100 other neurons. Obviously the nerve network within the brain is exceedingly complex.

EVOLUTIONARY DEVELOPMENT

From an evolutionary point of view the brain developed as an enlargement of the front end of the spinal nervous system. In lower animals the enlargement is little more than a group of swellings at the front end of the spinal cord. In the higher vertebrates there are three swellings that are classified as forebrain, midbrain, and hindbrain. During the course of evolution these swellings have become larger and have differentiated into what appear to be separate functional components. Figure 6.2 illustrates the early phases of this evolution.

Figure 6.2a, at the top, shows a section of a simple neural tube such as might be encountered at a very early evolutionary stage. Two rows of sensory nerves servicing the right and left sides of the organism enter from the back of the neural tube. Two rows of motor nerves servicing the right and left sides of the organism exit through the front face of the tube. Figure 6.2b, at the lower left, shows an early enlargment near the head of the neural tube to form the three main divisions of the brain (forebrain, midbrain, and hindbrain). Figure 6.2c, the middle figure, shows further expansion and lateral growth of the forebrain. The final figure, Figure 6.2d, illustrates a stage of evolution at which the main features of the mammalian brain are apparent. The forebrain has expanded and folded back on itself to produce the main components of the modern forebrain. The outside layer, known as the cerebral hemisphere (one hemisphere for each side of the brain), is the most obvious external feature of

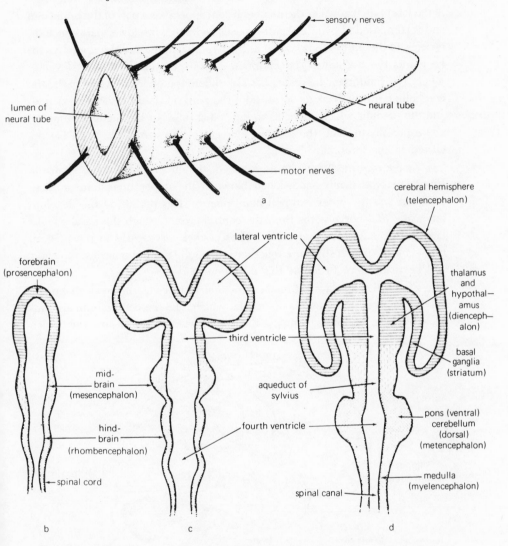

FIGURE 6.2

Early Evolution of the Mammalian Brain (after E. Gardner, 1968)

From *Physiological Psychology* (1970), by Peter Milner; with permission of Holt, Rinehart and Winston, Publishers.

the modern brain. The under fold of this hemisphere, which is pressed down toward the brainstem, includes the basal ganglia and striatum. The inner part of the forebrain (known as the diencephalon) appears as a part of the brainstem and is composed of two differently specialized parts; the lower part which interfaces to the midbrain is known as the hypothalamus, and the upper part is known as the thalamus. The midbrain shows relatively little new development. The hindbrain, however, has also differentiated into several parts: the cerebellum on the back or dorsal side, the pons on the front or ventral side, and the medulla which connects to the spinal tube. In the higher mammals, and especially in man, the greatest increase in the size of the brain has occurred in the forebrain.

As the cerebral hemispheres enlarged, the under fold which is shown in Figure 6.2 was firmly sandwiched between the upper brainstem or diencephalon and the outer cerebral hemispheres. This permitted the development of *radial* nerve paths from the central area, through the basal ganglia layer to the outer cerebral cortex. These radial nerve paths in the modern brain knit the whole structure together into a solid mass of neurons. No effort will be made here to follow this evolution in detail.

Figure 6.3 displays the main structural features of the modern human brain. The pons and medulla oblongata, which (like the cerebellum) are outgrowths of the hindbrain, compose the lower part of a physical structure which

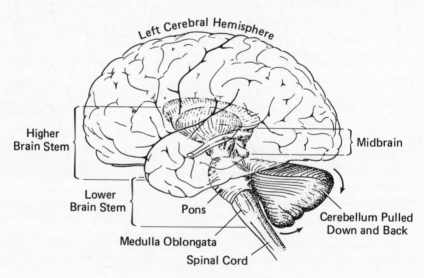

FIGURE 6.3
The Three Major Parts of the Brain (left side)

From *Speech and Brain-Mechanisms* by Wilder Penfield and Lamar Roberts (copyright © 1959 by Princeton University Press), p. 15. Reprinted by permission of Princeton University Press.

is known as the brainstem. The midbrain and the lower part of the forebrain (diencephalon) compose the upper part of the brainstem. The two cerebral hemispheres spread out from the top of the brainstem, and completely encircle the rest of the brain, like two halves of a football helmet. The two hemispheres are physically almost completely separated, except that they are linked (at the base of a separating crevice) by a dense cable of nerve fibers (called the corpus callosum) which provides for communication between the right and left cerebral hemispheres.

AN OVERVIEW OF "NATURE'S DESIGN"

With this brief review of the basic physiology of the brain it is possible to outline the way the decision system concept fits into the physical structure of the brain. The nontechnical reader may find it helpful to refer to Figure 6.2d as an aid in following the discussion. The simpler evolutionary version of the brain shown in this figure includes all the main features needed for an overview of the basic "system design." The following paragraphs provide an introduction to our basic structural hypothesis. The ideas are expanded and the rationale for the hypothesis is developed in the later part of the chapter.

LOCATION OF THE DECISION SYSTEM

As is apparent from Figure 6.2d, the brain is composed of a number of distinguishable components. Each of the components is specialized for certain specific functions. According to our structural hypothesis, the value-driven decision system is the largest of the major components, and it is located in the forebrain.

In this early evolutionary version, the source of values for the decision system is located at the base of the forebrain in the hypothalamus. The hypothalamus operates as a special-purpose computer which provides values for the decision system. As was mentioned earlier, the operation of this "value system" lies outside the realm of consciousness.

The conscious mind (which is the "rational" part of the decision system) lies above the hypothalamus and includes almost all of the rest of the forebrain. It is helpful to think of this part of the decision system as a large general-purpose computation center. This conscious computation system corresponds to what has traditionally been called the "mind."

From this structural perspective, the "mind" is simply a particular computational subsystem of the brain. This computation system operates on a decentralized basis so that data-processing and data-storage functions are distributed over the entire surface of the cerebral hemispheres. Such a large decentralized computation system needs a central control that can allocate computational resources and coordinate the information flow. This central control and coordination function seems to be the responsibility of the thalamus or upper part of the diencephalon.

STRUCTURE OF THE MIND

To link the central control center of the mind to the peripheral computation and memory areas, evolution has provided radial nerve pathways (not shown on the figure) which link the thalamus, through the intermediate layer, to the cerebral hemisphere, providing a sort of three-layer sandwich. The intermediate layer, which is located where we would expect it to serve a switching and a temporary information storage role, interfacing the control center to the peripheral computation centers, has differentiated into a series of very specialized subsystems which may serve such interface functions.

The sensation of consciousness which seems to arise within the decision system is one of the most profound mysteries of the human mind. This "mystery of consciousness" will be discussed at some length in the next chapter when the decision system is discussed from a subjective point of view. In this chapter we will continue to think of this remarkable computation system primarily from a functional and structural point of view.

Nevertheless, it is worth commenting on the relationship between control and computation subsystem and the boundaries of consciousness. Evidently the sensation of consciousness arises primarily out of the interactions between the central control system (the thalamus) and the peripheral processing components in the cerebral cortex. By shifting its interactions to different parts of the cerebral hemisphere, the control center can focus attention on different types of problems. Although the control center can interrogate and interact with almost any part of the cerebral cortex, it apparently cannot interrogate any other part of the brain. The boundaries of consciousness seem to be limited by the scope of interactions of the control center. The control center routinely receives value signals supplied by the hypothalamus. The reception of these value signals (anger, fear, joy, sorrow, etc.) at the control center, combined with the "rational" interactions of the control center with the cerebral hemispheres, apparently produces (within the control system) a sensation of consciousness.

INPUT AND OUTPUT INFORMATION CHANNELS

To make the decision system operate usefully it is necessary to provide for input and output information channels. Experience with large-scale artificial computers has demonstrated the need for "buffering" data input and data output operations so that they will not interrupt the main computer. The ongoing logical operations (thinking) of the main computation system must not be interrupted every time a trivial bit of new information happens to come in. On the other hand the new information should be available when the system is ready to use it. This isolation of the computer from the routine input/output operations is accomplished through the use of "buffer" storage areas.

In large artificial computation systems, certain memory cells in the data storage area are set aside to be used as input and output storage areas. These areas are known as buffer storage areas. Information from data-input sensors is transmitted to a buffer storage area where it can be interrogated as needed by the main computer system. Output information generated by the computation system is stored in an output buffer area where it can be interrogated by the output data processing subsystems.

It appears that nature has used some very similar concepts in the design of the human decision system. Certain specific areas on the cerebral hemisphere are specialized as input and output areas. Certain areas are set aside for data from specific sensory organs, and this sensory information apparently enters the realm of consciousness only to the extent that the control center of the mind interacts with that area on the cerebral cortex. In the human body a large amount of preprocessing of sensory input information is carried out before it reaches the cerebral cortex. Such preprocessing takes place subconsciously and is inaccessible to conscious experience.

Other areas on the cerebral cortex are set aside for storage of outgoing muscle control information directed to specific muscle groups. Information placed in these output buffer areas enables the mind to command "voluntary" muscle activity. When such information is placed in the output area it becomes available for processing by the body's output processing unit. In the human body, the processing required to transform the action decisions of the mind into detailed muscle commands normally occurs outside the "mind." It is accomplished primarily in an output processor called the cerebellum which is an outgrowth of the hindbrain. All of the processing which occurs after the information leaves the output area of the cerebral hemisphere occurs subconsciously.

DATA INPUT CHANNELS FOR THE VALUE SYSTEM

Just as sensory input data is needed for the operation of the "mind," so it is also needed to operate the value system in the hypothalamus. To provide necessary sensory data for the value system (as well as the rational mind) the sensory pathways in this primitive brain divide so that information is carried to the midbrain as well as the forebrain. The upper branch carries information to the "mind." The lower branch, which goes to the midbrain, evidently carries information to the midbrain below the hypothalamus. It appears that in the primitive brain the midbrain may serve as the preprocessor for the sensory information needed by the value system.

In the past, the midbrain has been viewed primarily as an information conduit and as an arousal mechanism to alert the "mind" when potentially dangerous information is sensed. The sensory branches leading to the midbrain were therefore classified only as reflex and arousal information links. The present perspective suggests that these sensory linkages to the midbrain may be much more fundamental to the design of a value-driven decision system. They provide the essential sensory data that is needed to generate the value structure.

Of course the same sensory information is undoubtedly also used by the midbrain to trigger the arousal mechanisms. In the design of a large-scale computer system it is important to provide for interruption of normal processing when critical events occur in the environment. Thus modern computers include a component known as an interrupt system. Nature seems to have incorporated the interrupt as one of the functions of the midbrain.

EVOLUTION'S BASIC STRUCTURAL DESIGN

This overview defines, in a very simple form, evolution's original division of functions among the main parts of the brain. Evidently, the forebrain houses a "rational" decision system; the midbrain houses a sensory processor and special-purpose computer that drives the value system; and the hindbrain houses an output processor which converts general decisions into specific muscle commands. If this perspective is correct, it suggests that the rudimentary value-driven decision system is a surprisingly old evolutionary invention. The three main parts of the brain are apparent, even in most fish, although some fish (for example the shark) have a very limited forebrain. This suggests that the basic concept for the value-driven decision system predates the evolution of mammals and probably occurred prior to the emergence of the reptiles.

RECENT EVOLUTIONARY DEVELOPMENTS

In the more recent evolution of the mammalian brain something unex-
pected has happened. The midbrain has stopped enlarging at a pace commen-
surate with the rest of the system. Indeed, there is evidence of regression in
this system. Evidently something else is taking over many of the functions that
used to reside in the midbrain.

It appears that these functions may be in the process of migrating to the
much larger frontal lobes of the cerebral hemispheres. Thus the extreme
complexity and refinement of the innate human value system may reflect the
operation of these large frontal lobes. Although these frontal lobes are physi-
cally part of the cerebral cortex they do not seem to be part of the "conscious
mind." They are linked with the hypothalamus and midbrain in such a way
that they appear to be part of the value system rather than of the rational
mind.

THE TWO CEREBRAL HEMISPHERES

The modern human brain involves one other very important departure
from the basic system concept just discussed. In the previous discussion we
spoke of the cerebral cortex and its control system as if it were a single unified
decision system. Actually the human forebrain is divided into two separate cer-
ebral hemispheres, and each seems to be coordinated by its own half of the
thalamus. The two halves of the thalamus are almost completely disconnected.
Thus it is almost as if there are two separate and independent control centers,
and therefore two decision systems.

Of course, the two decision systems do not really operate as if they are in-
dependent. The two cerebral hemispheres share a great deal of information
via the large neural linkage cable called the corpus callosum. Moreover, the
hypothalamus, which provides the values, is a more unified structure than the
thalamus, and it probably sends almost the same value signals to both sides of
the brain. Thus the two control centers are intimately linked, much like Sia-
mese twins, so that they are unlikely to get much out of step with each other.
Although the reasons for this duality in the human decision system are not yet
really clear, some possible advantages of the arrangement will be discussed in
a later section.

Having completed a brief overview of the design concept, we will now
discuss the system components in somewhat greater detail. One of the best
ways to study a cybernetic system is to begin with the input/output functions.
We will therefore begin with a brief discussion of the hindbrain (which serves
as an input and output subsystem) and then proceed to our real interest, the
human decision system.

INPUT/OUTPUT FUNCTIONS IN THE HINDBRAIN

In addition to its role as an input/output processor for the decision system, the hindbrain (particularly the lower brainstem) plays an important role in the control of basic functions such as body metabolism. But in the present treatment we are concerned with only those functions of the hindbrain that are important in the operation of the decision system itself. As sketched in Figure 6.3 the hindbrain includes three major structural features: the cerebellum, the pons, and the medulla oblongata. The cerebellum is the most prominent part of the hindbrain. Its function is only vaguely understood, but it plays an important role in the coordination of muscle movements. Damage to the cerebellum almost always results in a degradation of muscle coordination. From a system design perspective there is an obvious function for such a system. Once action decisions have been made, the generalized decision must be converted into specific commands to individual muscles. This requires detailed information that is irrelevant to the decision processor. According to the hierarchical design principle discussed in Chapter 4, this processing function is best separated from the decision process.

The left to right roles in the cerebellum are reversed relative to the cerebral hemispheres. Whereas the left cerebral hemisphere controls the right side of the body, the right half of the cerebellum seems to control the right side of the body. Consequently, there is a need for a crossover network between the cerebellum and the cerebral hemispheres. The information transmitted from the cerebral hemispheres to the cerebellum comes by way of the pons, which provides the necessary crossover network.*

The lower brainstem or medulla oblongata obviously serves as an information conduit for communication with the rest of the body. Sensory messages from the body travel up the rear half of the spinal cord, while muscle control messages travel down the front of the spinal cord. This region also serves an important role as a communication monitor. In the center of the medulla there is an interconnecting set of nerve fibers known as the reticular formation which serves to monitor incoming messages. When the brain is concentrating on other activities, the amplitude of irrelevant incoming messages can be reduced by the reticular formation, but if the incoming messages seem to be particularly significant the reticular formation can sound a general alarm (4). Even if the person is asleep, certain signals passing through this area can

---

* The left to right reversal of the forebrain probably reflects the importance of vision in evolution of this system. The roles in the forebrain had to be reversed to compensate for the reversal of the optical image.

result in an awakening of the entire upper brain. The response is quite selective. For example, a baby's cry may waken the mother but not the father. The growl of a watchdog may waken the father but not the mother.

THE FOREBRAIN: A GENERAL-PURPOSE COMPUTER

Most of the higher intellectual functions of the brain are centered in the forebrain. The basic structural plan of the forebrain is such that memory storage and the rational thought processes use the general-purpose computational capability of the large cerebral cortex area. The control and orchestration of the value system is centered in a special computation area in the vicinity of the hypothalamus. In all probability the center of control for rational thought is accomplished in the thalamic region above the hypothalamus.

Computer scientists are familiar with both special-purpose and general-purpose computers. The special-purpose computer is custom-made for a specific task. It can be highly efficient at that task but may be almost useless for any other application. The general-purpose computer is designed for versatility. It is not as efficient as a special-purpose computer at any specific task, but it can be applied with moderate efficiency to a wide range of tasks. Whereas the value system that operates the hypothalamus seems to resemble a special-purpose computer, the cerebral cortex resembles a general-purpose computer. This cerebral cortex area is the part of the brain in which the greatest growth has occurred during evolution of the higher mammals.

One should not expect a structural examination to tell us much about the functional use of a general-purpose computer. By definition, a general-purpose system is designed so that the function can be changed without changing the structure. If a general-purpose electronic computer were examined, the physical components would be exactly the same whether it was being used to schedule a chemical plant or to assign children to schools. Probably for very similar reasons, little is known about the function of the cerebral cortex in those areas that are used for general-purpose applications. The best information is available for those areas that are committed to specific input or output activities.

MAPPING THE MIND

Figure 6.4 is a sketch of the left cerebral hemisphere showing some of the input and output areas. The sensory nerve signals that are carried up the back half of the spinal cord continue through the brainstem and terminate generally in the light shaded data input area labeled "somatic sensory." The signals from the right side of the body go to the left side of the brain, while signals from the left side of the body go to a mirror image area (not shown) in the right cerebral hemisphere.

Once the signals have reached this area of the cortex they are available to be interrogated by the conscious mind. The output signals from the brain that control voluntary muscle movement are placed in the parallel (darker shaded) output strip, which supplies the motor signals that travel down the front of the spinal cord.

A casual inspection of the human cerebral cortex shows that it is highly convoluted, as if nature had gone to great lengths to fit a maximum number of neurons into the cortical surface. The cortex area in lower animals is both smaller and less convoluted. In humans the size of the skull has increased, the thickness of the skull has decreased, and the cortex seems to have expanded to fill all the available space. Nature has been willing to pay a considerable

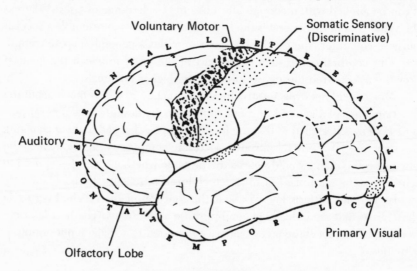

FIGURE 6.4
The Primary Sensory and Motor Areas of the Cortex (viewed from the left)

From *Speech and Brain-Mechanisms* by Wilder Penfield and Lamar Roberts, copyright ©1959 by Princeton University Press, p. 25. Reprinted by permission of Princeton University Press.

penalty in increased risks in childbirth and the reduced protection of the brain to achieve the present size of the cortex. It seems unlikely that evolution would have accepted these penalties unless the increased brain size offered very important compensating advantages in system performance.*

Recognizing that the cerebral area is a valuable and scarce resource, we must expect it to be carefully allocated on the basis of priorities and needs. The allocation of the area within the sensory and motor strips that were shown in Figure 6.4 is particularly revealing. This allocation in the motor strip is illus-

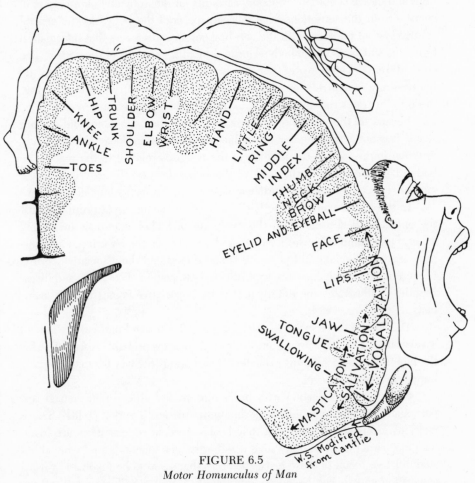

FIGURE 6.5
Motor Homunculus of Man

After Penfield and Rasmussen, *Cerebral Cortex of Man*, copyright © The Macmillan Co., 1959. In Buchanan, A. R., *Functional Neuroanatomy*, Philadelphia: Lea and Febiger, 1961. Reprinted by permission of the publishers.

* Obviously the size of the cortex is only one of the many factors affecting system performance. The efficiency with which the design concept is implemented in the structure of the brain is probably of even greater importance.

trated in Figure 6.5. (The allocation in the sensory strip is almost identical.) The gross distortion of the imaginative "little man" shows the high priority that nature has placed on dexterity in the fingers and in the control of speech muscles. These two functional areas account for almost half the total area of the motor control strip! A similar analysis in other mammals shows quite a different allocation of these areas, reflecting a different set of behavioral priorities.

The very systematic layout of the cortical areas dealing with motor functions and tactile sensations probably accounts for our subjective impression of continuity in the sensations from adjacent parts of the body. As the central control system interacts with the cerebral cortex it discovers that information from adjacent parts of the body is located in adjacent areas of the cortex. The cortical areas devoted to other senses (smell, hearing, and vision), however, do not bear any similar relation to positions in the basic sensory motor strip. Figure 6.6 illustrates the approximate location of the other sensory areas.

Because of the separation of these areas from the basic motor and sensory strips, it is not surprising that our subjective perception does not convey any intuitive impression of the location of these senses relative to the rest of the body. The large area at the back of the brain devoted to analysis of visual images reflects the great importance of vision in conscious human behavior.

In the lower mammals the basic sensory and motor areas occupy almost the entire cortical area. With the evolution of higher mammals increasing areas of uncommitted cortex have opened up between the basic input and output areas. This uncommitted area is sometimes called the associative cortex, because it is available for more sophisticated association of concepts and ideas. Figure 6.7 illustrates the emergence of the associative cortex in four mammals.

The area of the associative cortex in the primates and particularly in man is very large. This large area of uncommitted cortex is probably responsible for the high intelligence of most primates. It also opens the way for that uniquely human invention, speech.

Neurosurgeons dealing with brain tumors and other brain damage are particularly careful to avoid interfering with a patient's verbal ability. Before incisions are made in areas that may be involved in speech, tests are made with electrical probes to determine whether speech functions are involved. As a cumulative result of these tests, the areas that seem to be involved in word association and speech have been identified. Figure 6.8 illustrates the three main speech areas.

In very small children, speech areas usually begin to develop in both the left and right hemispheres of the brain. But as the child matures, the speech function on the right hemisphere seems to fade away, leaving only the highly

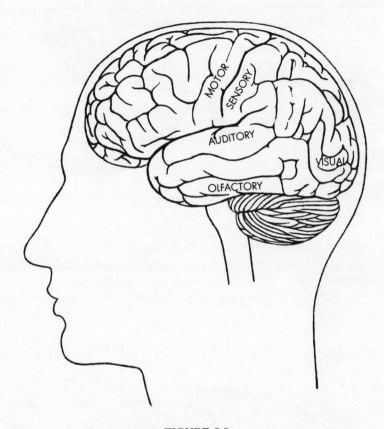

FIGURE 6.6
Illustrating the Cortical Location of Sensory Areas

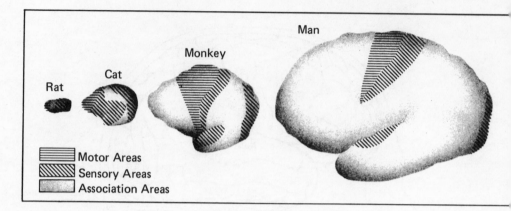

FIGURE 6.7
The Emergence of the Associative Cortex

Approximate scale drawings of the cerebral hemispheres of four mammals. Note both the absolute and relative increase in the size of the associative cortex.

From *The Conscious Brain*, by Stephen P. Rose, copyright © 1973 by Stephen P. Rose. Reprinted by permission of Alfred A. Knopf, Inc. and Weidenfeld and Nicolson, Ltd., Publishers.

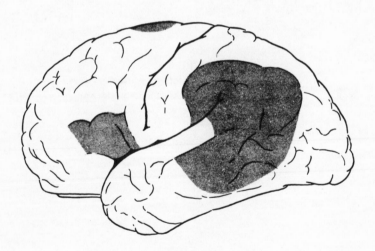

FIGURE 6.8
The Three Speech Areas of the Cortex

From *Speech and Brain-Mechanisms*, by Wilder Penfield and Lamar Roberts, copyright © 1959 by Princeton University Press, p. 135. Reprinted by permission of Princeton University Press.

developed speech areas as shown on the left side of the brain. Presumably, the corresponding cortex area on the other side becomes available for other applications.

The results of brain stimulation in all three speech areas are very similar. Word association processes are blocked in unpredictable ways, but no specialization among the areas is apparent. The most obvious observation is that the three speech areas seem to be separate and distinct. They have filled in around other areas but do not overlap them. The location of the speech areas relative to the previously identified brain areas is very striking and may be indicative of the way these areas became involved in human speech.

The largest and most important speech area behind the left templar lobe apparently developed as an outgrowth of the auditory region on top of the left temple. It interfaces with the main sensory strip in the region that provides feedback from the lips, tongue, and other vocalization components, and it interfaces with the visual area in a region that may be suitable for symbolic representation of visual images. The central location of the main speech area relative to these three sensory areas probably reflects the evolutionary origin of speech as a symbolic elaboration of sensory data. The portion of the speech area near the upper boundary of the shaded region and closest to the visual area seems to be involved in reading, as might logically be expected.

The location of the secondary speech area in front of the motor strip suggests that this area may have developed as an outgrowth of the adjacent motor area, which is involved in control of vocalization. The least important speech area is located near the top of the head. At present, there is no obvious reason for this specific location. The uncommitted cortex area in Figure 6.8 between the two frontal speech zones (and adjacent to the motor control area for the hand) has been shown (as might also be expected) to be involved in writing. The uncommitted areas on the templar lobes seem to be involved in the detailed recall of past events.

These functional areas do not seem to be rigidly fixed. If damage to some of these areas occurs at an early age, it is often possible to transfer the function to some other part of the cortex. On the other hand, if the damage occurs late in life, it may be very difficult, if not impossible, to recover normal speech. Because of the flexibility of the cortex as a general-purpose system, the role of areas associated with less routine activity have been much more difficult to identify. There is, however, considerable evidence that the area on the right cerebral hemisphere which is not involved in speech may be used instead for functions such as geometric thinking, recollection of music, and spatial perception and judgment.

THE CENTRAL CONTROL SYSTEM

As mentioned earlier, the sensation of consciousness seems to arise from the interactions of a central control system with the data-storage and data-processing units of the cerebral cortex. The central control system can interact with any of the relevant areas just identified in the cerebral cortex. Our subjective experience suggests that the control unit can interact simultaneously with several of the input areas and that the intensity of interaction is adjusted as attention shifts.

Almost all large-scale electronic computers include a central monitor or master control subsystem which has the responsibility for allocating the cybernetic resources of the system. The control system assigns tasks to be accomplished by other components, and it may control the flow of information between components. It seems likely that the central control system serves an analogous role in the human cybernetic system.

Because the content of consciousness seems to shift as we shift our attention, we have the feeling that the boundaries of consciousness are vaguely defined. This is quite consistent, however, with the definition of a finite realm of consciousness which includes only those parts of the cortex with which the control system can interact. Since the control system is always involved in consciousness, it might be thought of as a sort of "center of consciousness." In all probability the sensation of consciousness cannot be achieved with the control system alone, for to achieve the continuity and wholeness of experience that we think of as consciousness, the control center must almost certainly have information sources with which to interact.

There is as yet no general agreement concerning the physical location of such a center of consciousness. Penfield (11) has speculated that it lies somewhere within the diencephalon. The thalamic area is very centrally located; so it is a logical candidate for the center of consciousness.

The circumstantial evidence for this choice is considerable. Because of its central location, this area is almost ideally located for a master control system. It is almost completely encircled by the large cerebral hemispheres and it serves as a massive crossroad for neural pathways between the upper brainstem and the cerebral hemisphere. Pathways to other important brain components from this central location are as short as possible. Whereas damage to most other areas of the brain can occur without loss of consciousness, damage in this area seems inevitably to result in the loss of consciousness. Efficient system design seems to dictate that major neural pathways leading from the control center should branch radially outward toward the peripheral data-processing and data-storage components. The general radial information flow within the brain is confirmed both by the existence of the radial pathways and by surgical evidence which shows that the normal operation of specific parts of the cortex is not impaired by removal of adjacent parts of the cortex. For instance, Penfield notes his experience in which (on different occasions) it has been necessary to remove sections of the cortex on all sides of the speech areas, but the speech function in all cases has remained essentially unimpaired. Thus it seems probable that the most important interactions of the peripheral data-processing and data-storage units in the cortex are indeed those with the central control system.

THE MIND AS AN ASSOCIATIVE PROCESSOR

The previous sections have discussed the interaction of the control center with the peripheral processing units of the cortex only in a very general way. Although little is known about the actual interaction mechanism, our subjective experience suggests that the mind may operate much like an "associative processor."

This method of information retrieval is quite different from what is used in most modern computers. In typical digital computers each piece of information is stored in a specified memory location. If we do not know where the information is stored we cannot recover the information. In an associative memory, information is stored together with certain retrieval keys. To recover the information we must know the retrieval keys, but we do not have to know where the information is stored. To recover the information we simply broadcast an information request to all memory locations. The request has a form something like, "Who has information relevant to the following (retrieval keys)?"

To make the difference clear it is useful to consider an analogy. Suppose a classroom teacher wanted to find out when her student, Mary, was born. If she did not know where Mary sat in the classroom (the address of the information) she might proceed as follows. She could ask each student in turn, "Is your name Mary?" This method corresponds to the technique now used in most digital computers.

However, a few electronic computers are designed on what is called an associative memory principle. To follow this principle the teacher would simply call "Mary?" and the proper student would respond "Yes." The teacher could then ask, "When is your birthday?" In this example the name "Mary" serves as a retrieval key. When a retrieval key is broadcast, a response is obtained only from memory cells with the proper retrieval key. Thereafter any associated information such as the student's birthday or address can be obtained.

Associative memories can be designed so that there may be several alternative retrieval keys, or so that more than one key must be used to retrieve the information (for example, first, last, and middle name might be required). Similarly, they can be designed so that retrieval will occur even when the match to the keys is not perfect. For example, Johnny Jones might serve to retrieve Jon Jones.

It is important not to push the analogy with existing computers too far. Efficiency considerations in the design of a biological cybernetic system are quite different from those that apply in existing electronic computers. The fact that neurons can serve either as memory devices or logical processors makes it practical for the biological system to achieve a better balance between memory and data-processing capability than has been practical in the past with artificial computers.

It seems likely that local data-storage areas in the brain are also provided with *local* data-processing resources. Whereas in electronic computers the associative principle is usually used only in a simple memory, the brain may use the same principle to provide communication between central control and peripheral units that have their own self-contained memory *and* data-processing resources. Such peripheral units could be used to search for similarities between new and old experience. Moreover, they could proceed independently with separate analysis tasks and provide information to master control only when "significant" results are obtained.

The experiments in brain surgery that were mentioned in Chapter 2 seem consistent with this design concept. When electrical probes are inserted in the temporal lobes (in areas that seem to be responsible for the memory of past events) the patients may experience either of two alternative responses: they may experience a vivid recall (almost a reliving of a past experience), but alternatively they may experience an intense sensation of "familiarity," as if their current status on the operating table duplicated a familiar past experience. These two alternative responses occur almost interchangeably as a result of electrical stimulation of the *same* general areas of the cerebral cortex. It is natural to guess that both the information-storage function and the association logic are located in the same physical area. Thus, the electrical probe can stimulate a retrieval of stored information resulting in the reliving of an old experience, or alternatively, it can falsely stimulate a "familiarity" response which is intended to notify central control of the presence of "relevant" past experience. The ability of a peripheral unit to send a "familiarity" message may serve the same purpose as the schoolchild's response "yes" when her name is called. After that initial response, central control can request whatever more specific information it may need. To minimize data overload in the central processor, no signal would be returned from any area where a match was not found.

The convention of a nonresponse from all areas which do not find a match is probably essential to efficiency in the operation of a really large associative memory. It also provides a natural explanation of why brain surgery patients seem unaware of the absence of excised parts of the cortex. The conscious mind expects to hear from peripheral areas *only* when they volunteer that they have something relevant to say. Removal of a portion of the cerebral cortex of course eliminates such volunteering of information, but since a response is not ordinarily expected, the absence of a response seems perfectly normal to central control.

THE HUMAN ASYMMETRY

Earlier in the chapter we observed that the cerebral cortex is really divided into two separate hemispheres which seem to be able to function almost independently as separate associative processing systems. What practical function is served by this arrangement?

In lower animals this division of the brain into two hemispheres may be of relatively little significance. Both halves are in the same skull and except for mirror image responsibilities they seem to accomplish the same functions. The interconnecting nerve fibers of the corpus callosum ensure that essential information is available to both halves. Experiments in which this connecting link is cut, either before or after some learning experience, make it clear that dual memory traces are laid down in both hemispheres. Since both sides of the brain share the same learning experiences the two sides may serve only to provide some redundancy and increased reliability in system performance.

However, this is not the case in humans. In providing for human speech, nature has abandoned the traditional plan of bilateral symmetry. One side carries the speech function and specializes in logical relations; the other side seems to specialize in spatial and geometric relations. One can think of several considerations that might have encouraged such specialization.

The need for speed and efficiency in processing the abstract symbolic information of speech may have required short communication lines that could only be achieved if the entire speech function was localized on one side of the brain. One can envision an instability such that when one side became a little more proficient in speech, it would tend to dominate and the capability on the other side would atrophy. Moreover, human speech evidently requires a large cortical area, so duplication of the function on both sides could crowd out other intellectual functions.

Although these considerations suggest some advantages in localizing the speech function in one hemisphere, they do not explain why it almost always seems to be the *same* hemisphere. But as experience with left-handed children shows, it can be a disadvantage to be unable to decide which hemisphere should be verbally dominant. Given such efficiency considerations during the early evolution of speech, almost any mutation that would make the two hemispheres functionally asymmetrical would tend to be favored.

It is unclear what form such a mutation may have taken. Although superficially the two hemispheres appear almost completely symmetrical, careful measurements seem to show statistically significant differences in the size and shape of corresponding cortical areas in the two hemispheres (2). Whatever the cause, the speech function is almost always in the left hemisphere. The only exception seems to be those cases where there was early physical damage to the left side, in which case the speech function may develop in the right hemisphere. Even in left-handed children, speech is almost always in the left hemisphere, with the exception of about 15 percent of the left-handed people who seem to retain some speech functions on both sides.

It is tempting to speculate that some of the superior intellectual capability of modern man may have been obtained simply by abandoning the duplication of higher intellectual processes in both hemispheres. The concentration of speech on one side made extra cortical area available for other intellectual tasks.

While this asymmetry substantially increases the intellectual potential of humans, it may also make communication between the two cerebral hemispheres more difficult. To what extent will the two sides understand each other's language? Certainly the right side cannot communicate using the verbal symbols of the dominant left hemisphere. Thus, the available communication must be nonverbal. Experiments of many kinds have shown that the nonverbal side of the brain is more proficient with spatial and geometric problems. It dominates in recognizing faces, and it dominates in music and art.

When Einstein was asked how he arrived at some of his most original ideas he explained that he rarely thought in words at all. "A thought comes and I may try to express it in words afterwards," he said. His concepts seemed to take the form of physical entities that he could perceive and work with in the form of more or less clear images. The words had to be searched for laboriously later. This description seems consistent with the hypothesis that these ideas originated largely in the right side of his brain.

Each of these observations seems consistent with a specialization of function between the two sides of the brain. The dominant verbal hemisphere specializes in symbolic logic. It is concerned with verbal and other symbols, and its association processes are essentially digital. The right hemisphere, on the other hand, seems to deal with concepts in a less symbolic but more concrete or representational way.* Whereas the dominant hemisphere operates much like a digital computer, the other hemisphere may operate more like an analogue computer.

THE HYPOTHALAMUS: SOURCE OF BASIC VALUES

Numerous experiments using both electrode stimulation and localized surgical excision have demonstrated that the hypothalamus plays a central role both in control of emotions and in the associated regulation of related body functions. The hypothalamus is important in the operation of the pituitary gland, the control of body temperatures, and secretion of hormones. Through the use of electrical or surgical interference with the hypothalamus it is possible to produce behavior in animals corresponding to thirst, hunger, anger, fear, sex, or even a sense of generalized pleasure or well-being.

Looking at these results in overview, the hypothalamus emerges as a highly differentiated and efficiently organized special-purpose system. From an evolutionary point of view, the hypothalamus is the oldest part of the forebrain. It is responsible for the modulation of a wide variety of body functions. Over its extremely long evolutionary history this lowest part of the forebrain has had time to become extremely compact, efficient, and specialized. The highly specialized nature of this old area of the forebrain undoubtedly explains the great similarity in the dominant emotions that are observed in almost all the higher mammals.

From an evolutionary point of view, it seems natural that the basic value system would be located in the old part of the forebrain. At an earlier stage of evolution these values, in the form of instinctive responses, may have been the highest control mechanism of the organism. Output from this area was

* For a more comprehensive discussion of hemispheric specialization see Dimond and Beaumont (2).

necessary both to control the instinctive responses and to provide proper conditioning of learned responses in the lower nervous system. Moreover, the existence of such a rudimentary response system was probably a necessary precondition for the evolution of a more sophisticated rational decision system.

The orchestration of the value system requires access to sensory information. To know when to be afraid, the value system needs access to the senses of vision, hearing, and probably smell. To identify some tastes as "good" and others as "bad," it needs access to the taste information. To provide an appropriate emotional response to sexual signals, the value system needs to be able to identify the sexual signals whether they come to the eyes, ears, or nose.

When a man sees a beautiful woman he can rationally observe and analyze the contours of her body, but such rational analysis cannot explain the emotional response that says "Wow!" The "Wow!" is not a result of rational analysis, but is a value judgment supplied *to* the rational mind, through the value system. How does the value system observe and analyze the female contours to deliver the message "Wow!"?

THE ROLE OF THE MIDBRAIN AND FRONTAL LOBES

Structurally, and from an evolutionary perspective, the midbrain is the natural source of the sensory information needed by the hypothalamus. The midbrain interfaces directly to the hypothalamus. Moreover like the forebrain, the midbrain has a highly developed sensory processing system. Almost all sensory input channels branch—with one branch going to the forebrain and one to the midbrain. Our understanding of the information requirements for a value-driven decision system explains the need for such elaborate sensory inputs to the midbrain.

The earlier observation concerning the tradeoff between complexity in the value system and the rational analysis capacity helps to explain why in its early evolutionary form the midbrain had a size comparable to the forebrain. But in the recent evolution of the mammalian brain, the growth of the midbrain has failed to keep pace with the rest of the brain. Indeed there is evidence of regression in both the midbrain itself and the sensory input to the midbrain. What explains this unexpected change in the evolutionary trends?

The function of the frontal lobes of the cerebral hemispheres has long been a mystery. They do not appear to be directly involved in any measurable aspects of rational thought. Nevertheless, damage to these lobes can produce very distinct behavioral changes. Individuals with damage in these frontal lobes typically show decreased sensitivity to the more subtle aspects both of human emotion and of social motivation. Moreover, strange changes occur in their intellectual motivation. They may care less about making mistakes, and they are less likely to change an established pattern of behavior—even when it is obvious that the behavior is not longer producing the desired results. Apparently these frontal lobes are somehow involved in human *intellectual* and *social* motivation.

The probable involvement of the frontal lobes in human motivation is also supported by observations on the limbic system underlying this frontal part of the cortex. This part of the limbic system is located where one would expect it to serve as an interface between the frontal lobes on the one hand and the hypothalamus and midbrain on the other. Indeed stimulation and surgical procedures in this area often produce emotional and behavioral responses comparable to stimulation of the hypothalamus itself. Thus evidence is accumulating that the frontal lobes are an important part of the innate human value system.

Generally, during the evolution of the brain, there has been a tendency for functions of the lower brain to migrate to the larger, newer parts. The confinement of the skull has caused the frontal lobes to fold under so that they are in close proximity to the midbrain. It would not be surprising therefore if these frontal lobes have been taken over by the value system, to serve as a general-purpose associative learning system for the value system. Thus it seems likely that the frontal lobes and the midbrain may now operate together, to provide the hypothalamus with a more sophisticated and finely tuned innate value system.

Recent physiological evidence provides strong support for this hypothesis (7, 13). Walle Nauta of MIT has developed a much improved method for tracing neural connections in the brain and has used the technique to study the neural linkages of the frontal lobes. As a result of these studies, Nauta has concluded that the frontal lobes, the hypothalamus, and the midbrain are so intimately linked that they need to be considered as an integrated system (8).

This interpretation of the role of the frontal lobes as part of the value system helps to explain the recent evolutionary regression in the role of the midbrain. But how can we explain the regression in the sensory channels to the midbrain? How do the frontal lobes gain access to the sophisticated sensory data needed to operate the human value system? Again, Nauta's studies provide a remarkable answer. His analysis of neural pathways leading to the frontal lobes (7) shows that many of them originate near the sensory input areas of the cerebral cortex. Indeed there seems to be a systematic pattern of connections from *all* the sensory input areas of the cerebral cortex leading to the frontal lobes! These connections do not seem to be connected directly to the sensory input areas, but rather to neighboring areas which might be expected to contain a more refined and meaningful representation of the sensory data.

This physiological information provides a critical missing link which rounds out our understanding of the evolutionary development of the frontal lobes. Really satisfactory orchestration of the value system could not be achieved without access to sophisticated and highly refined sensory data. The old midbrain design concept was inherently inefficient. It required dual systems (one in the midbrain and one in the forebrain) for processing and analyzing the sensory data. This inefficiency was removed in the frontal-lobe design concept. The frontal lobes tap into highly refined sensory data, as it is developed in the cerebral cortex. This information is then processed further in the frontal lobes to provide the essential information needed by the hypothalamus to define the human value structure. This remarkable design concept makes it possible for the human value system to use the most highly refined and meaningful representation of sensory data that is available to the rational mind. It eliminates the need for duplicate systems for processing the sensory data. This remarkable change in the design of the value system may be one of the greatest evolutionary inventions of the primate species. Following their introduction in this role, the frontal lobes have expanded rapidly during the evolution of the primates. The frontal lobes are exceptionally prominent in the human species. As we proceed in Part II with our study of the human value system, we will be trying to understand the role of this remarkable value system in defining the human personality.

SUMMARY

In this chapter we have tried to provide a general overview of the brain and to relate the known physiology to the concept of a value-driven decision system. Any such brief review is necessarily oversimplified. The brain is a very complex system, and it does not adhere precisely to any simple system-design concept. Nevertheless, the concept of the value-driven decision system seems to be a useful guide in understanding and interpreting the operation of the brain as a cybernetic system.

In the preceding discussion we have tried to focus on what is known about the brain and on what seems probable on the basis of the available evidence. We have not dwelled on our vast areas of ignorance about the brain. Consequently, there is a risk that the reader may get the false impression that the human brain is well understood. It is easy to confuse the initial steps of naming and classifying with real understanding. While there has been some success in identifying the functions of certain areas, there is as yet almost no understanding of the real physical principles underlying the system's operation. Thus we have a long way to go before we can expect physiological theory to yield a real understanding of human behavior or human values.

REFERENCES

1. Buchanan, A. R. *Functional Neuro-Anatomy.* Philadelphia: Lea & Febiger, 1961.
2. Dimond, Stuart J., and J. Graham Beaumont. *Hemisphere Function in the Human Brain.* New York: Halsted, 1974.
3. Eccles, John C. "The Psychology of Imagination." *Scientific American,* September 1958.
4. French, J. D. "The Reticular Formation." *Scientific American,* May 1957.
5. Gardner, Ernest. *Fundamentals of Neurology,* 5th ed. Philadelphia: Saunders, 1968.
6. Milner, Peter M. *Physiological Psychology.* New York: Holt, Rinehart & Winston, 1970.
7. Nauta, Walle J. H. "The Problem of the Frontal Lobe: A Reinterpretation." *Journal of Psychiatric Research* 8 (1971): 167–87.
8. Nauta, Walle J. H. "Connections of the Frontal Lobe with the Limbic System." *Surgical Approaches in Psychiatry,* 1972.
9. Neisser, Ulric. *Cognitive Psychology.* Englewood Cliffs, N.J.: Appleton Century Crofts, 1967.
10. Penfield, Wilder, and T. Rasmussen. *Cerebral Cortex of Man.* Macmillan, 1950.
11. Penfield, Wilder. "Speech, Perception, and the Uncommitted Cortex." Conference on Brain and Conscious Experience, Sept. 28 to Oct. 4, 1964, reported in *Brain and Conscious Experience.* New York: Springer-Verlag, 1966.
12. Pines, Maya. *The Brain Changers.* New York: Harcourt Brace Jovanovich, 1973.
13. Powell, T. P. S. "Sensory Convergence in the Cerebral Cortex." *Surgical Approaches in Psychiatry,* 1972.
14. Rose, Steven P. R. *The Conscious Brain.* Weidenfeld and Nicolson, 1973.
15. Sperry, Roger. "Messages from the Laboratory." *Engineering and Science,* January 1974. (Special issue on behavioral biology at Caltech.) Published by California Institute of Technology.
16. Thompson, Richard F. *Foundations of Physiological Psychology.* New York: Harper and Row, 1967.
17. Wooldridge, Dean E. *The Machinery of the Brain.* New York: McGraw-Hill, 1963.

Chapter 7

Mysteries of the Mind

If the human brain were so simple
That we could understand it,
We would be so simple
That we couldn't.

EMERSON M. PUGH *

TO PROVIDE a satisfactory model of the brain, we must be able to reconcile our introspective mental experience with the previous functional and structural theory. Although the previous chapter discussed some of the mechanisms that may underlie our subjective mental experience, other issues remain to be addressed. What is the nature of consciousness? How can we explain the role of the "subconscious" as we perceive it in our subjective mental experience? How does the interpretation of the brain as a biological decision system influence our understanding of "free will"? Although these issues are not essential to the theory of values, they need to be addressed; otherwise the unanswered questions may operate as mental blocks to preclude any further consideration of the theory.

THE CONSCIOUS MIND

In humans, the cognitive decision process is intimately linked with our sense of awareness or consciousness. Indeed, it seems likely that consciousness itself may have evolved as a natural (perhaps inevitable) by-product of the value-

* Author's note: Quote from my father around 1938.

driven decision system. In our everyday speech we speak of consciousness as if it were a well-defined object. Yet there are reasons for believing that consciousness is a physically distributed and diffuse property of the brain. The content of consciousness at any single moment is quite limited, but as we shift our attention, the content of consciousness changes. We can bring into our consciousness a wide variety of alternative elements, varying from specific parts of present sensual experience to a preoccupation with conceptual or theoretical problems.

This subjective experience of consciousness is consistent with our comments about the way the thalamus, acting as a central control system, can adjust its interactions with different parts of the cerebral cortex. The sensation of continuity of experience is facilitated both by the ability to retrieve old memories and by the logical way sensory data (stored in the input buffer areas of the cortex) parallels the evolution of external events.

But many readers will claim that this functional discussion does not really address the riddle of consciousness. They will say that in postulating the thalamus as a central control system we have really postulated a little decision system inside our main decision system, and have thereby only moved the basic dilemma of consciousness from the main decision system to the little decision system. And then, they will ask, what about a micro-decision-system inside the little decision system?

I believe this concern is unfounded. Electronic computer systems also have control systems. Moreover, they operate autonomously, without an infinite sequence of control systems. The central control is usually quite simple, and it does not require a control for the control, etc. On this basis it seems reasonable to assume that the sensation of consciousness may indeed arise from the interaction of the control center with the peripheral processing units and that it does not require an infinite regression of control systems to provide an explanation.

On the other hand, the true physical basis of this sensation of consciousness remains a mystery.* It is even unclear whether the mystery is philosophical, metaphysical, or scientific. This basic theoretical dilemma about the origin of consciousness will be addressed in more detail in a later section.

---

* For a summary of some recent speculations see Eccles (1).

THE SUBCONSCIOUS MIND

In our subjective experience things are always happening which we did not plan. Ideas come to us which we cannot explain. Feelings arise which we cannot rationally explain. These experiences convince us that the "subconscious" plays an important role in our intellectual lives. How can we reconcile this subjective experience with our understanding of the brain as a system?

Much of our subjective experience with subconscious mental processes can be attributed to the presence of many almost autonomous information-processing centers. In a large cybernetic system where the basic components (or neurons) are quite slow (relative to electronic components), efficient operation demands extensive use of peripheral processing subsystems. These processing centers communicate with the conscious mind (and probably with each other) only when they have something relevant to say. The unexpected messages from these processing centers are experienced subjectively as a profound mystery.

The peripheral processors are not, however, the only source of subconsciously generated information. A number of qualitatively different system functions can be identified which contribute in different ways to our awareness of a "subconscious mind":

1. *The operations of instinctive drives and values.* The selection of primary goals and objectives is not under conscious control. Instinctive drives can produce almost uncontrollable urges that may be inconsistent with our rational objectives. They can also generate an intuitive sense of "right" or "wrong." This intuitive conscience may be perceived to be of either subconscious or supernatural origin, depending on the religious persuasion of the individual.

2. *Automatic analysis of sensory data.* The "sensory data" received by the conscious mind has already been subjected to much logical analysis by other system components. This is particularly true of visual and auditory information. Basic two-dimensional visual images are interpreted as three-dimensional structures. The perceived colors of objects are automatically compensated for the redness or blueness of illumination. These automatic interpretations of the input data can sometimes be in error, as illustrated in some illusions earlier in the book. Nevertheless, the information is presented to the conscious mind as if the interpretation were an essential part of the original sensory data. This prior processing of sensory data is not accessible to conscious thought, and it provides another subjective manifestation of a "subconscious mind."

3. *Operation of the associative memory.* The recall of past experiences cannot be voluntarily commanded. An associative memory requires a certain correspondence between the interrogation and the stored retrieval keys in order to recover the stored information. Since the precise retrieval keys are not usually known by the conscious mind, the success of retrieval (or recall) is somewhat

unpredictable. Either success or failure of recall can come as a surprise. The recall keys for little-used information can be encountered almost accidentally. As we retrieve related information, we encounter additional retrieval keys. Thus, by a chain of association, we may ultimately recall the desired information.

4. *Operation of peripheral processors.* In all probability the brain includes a complex hierarchy of peripheral processors that carry out analysis and comparisons at different levels of abstraction. The processors that operate at low levels in the hierarchy may never be directly accessible to the conscious mind. Others that operate at a higher level may report to the conscious mind only when they have relevant results to report. The results generated by these processors can be unexpectedly delivered to the conscious mind long after the original stimulus for the activity. Scientists frequently report that critical ideas have "come to them" suddenly when they were relaxed or engaged in quite unrelated activities. The sense of suddenly "knowing the answer" can be very strong, even before there is conscious awareness of the answer itself. In such cases, the individual may feel extremely confident that the answer is valid, despite the fact that much work may remain before the conscious mind really "understands" the result.

All of the above phenomena, which once seemed to be impenetrable mysteries, now appear to be a natural consequence of the semiautonomous peripheral processors that are needed to provide parallel processing in a large cybernetic system. Although the relationships just described remain speculative (because we do not have a detailed physiological understanding of the neural structures involved), the circumstantial evidence for this type of design concept seems very strong.

THE PARADOX OF "FREE WILL"

The interpretation of the human brain as an automatic decision system provides a rather simple resolution (at least in scientific terms) of the long-standing controversy concerning "free will" versus "determinism." The traditional discussions of free will have been concerned with reconciling two extreme points of view. On one side, there are those who wish to demonstrate the existence of an almost metaphysical entity known as the "will." The will is presumed to exercise ultimate and unpredictable control over human behavior without regard for (and perhaps even in violation of) physical laws. On the other side there are those who view the mind as an essentially predictable physical system which fatalistically follows deterministic laws. The decision-system concept offers a well-defined scientific model which occupies a position intermediate between these two traditional extremes. Whether this model confirms or denies free will depends, of course, on how we define "free will." Rather than become involved in an unproductive discussion of the definition, we will try to focus attention on those traditional attributes of free will which seem to be scientifically observable.

From a scientific perspective there seem to be three key issues. First, does the system make decisions or choices? Second, are the resulting decisions "free" choices? Third, are the choices predictable? The decision-system model provides rather direct answers to each of the above questions.

First, a decision system does in fact make "choices." It considers alternatives and decides which alternative to select. Evidently at one time it was assumed that the process of making choices could not be "explained" in terms of physical laws. Thus the mere existence of such a process was accepted as evidence for a metaphysical entity called the will. Experience with artificial decision systems shows decisively that "choice" is not inconsistent with physical laws. It is a form of natural behavior which man shares both with other animals and with artificial decision systems.

The argument that freedom of choice does not exist was reiterated recently by B. F. Skinner (4). The underlying reason for Skinner's belief that freedom of choice is an illusion seems rather obvious. His "operant conditioning" model of human behavior does not include decisions or the concept of "choice," so obviously it cannot include "freedom of choice."

But Skinner makes his formal argument on an entirely different basis. He claims that free choice cannot exist because everything man does follows automatically as a consequence of the interaction between his genetic inheritance and his experience with the environment. If we were to accept this logic and apply it to the decision-system model, we would have to conclude that free choice does not exist because the individual will always select the decision alternatives that seem best in terms of his world model (which is a consequence of experience with the environment) and his innate value system (which is a genetic inheritance). Obviously the resulting choices in the decision theory model are "natural" consequences of experience and inheritance. But does this mean that the choices are not "free"?

The answer to this second key question depends on what we mean when we describe a choice as free. According to our decision-theory model, the brain will always try to select a choice that is most "desirable" in terms of a *built-in* genetically inherited value system. Because the brain is not free to select its own primary "value system" one might claim that the brain does not really have "freedom of choice." From the decision theory perspective, however, such an argument makes little sense. Deliberate free choices can *only* be made *relative* to a scale of values or some other primary decision criterion. In the absence of a criterion of decision there could be no real "choices," but only random "chance" decisions. When we speak of freedom of choice we surely are not referring to random mindless decisions. If decisions are to be based on "choice" rather than "chance" then there must exist some fundamental criterion of decision. It appears that if "free choice" is possible at all, it is possible only in the context of an a priori value system.

This difficulty cannot be avoided by allowing a decision system to select its own value system, for such a selection itself could be only a random chance decision, unless it was based on an a priori set of values. Thus we are led inevitably to the conclusion that "choice" can be meaningful *only* in the context of a preexisting set of values and that the fundamental values can never be supplied by the system itself. From a decision-science perspective real "choice" is not possible except in the context of preexisting values. If our definition of "free choice" requires that the system must rationally select its own primary values, then "free choice" is a logical impossibility, and we must concede that we do not exercise "free choice."

But in everyday conversation what we usually mean by "free choice" is that we are free to consider alternatives and to select what seems best. Obviously, by this definition we do exercise freedom of choice. To claim that we do not have "freedom of choice" because we will always choose what seems "best" in terms of our own innate values is a commonsense absurdity. It is equivalent to saying that we do not have freedom of choice because we will always choose to do what we like! Although we did not choose our innate values we certainly do make choices in terms of those values. The choices are as "free" as it is theoretically possible for a "choice" to be. Apparently, in terms of any reasonable commonsense definition, we must admit that freedom of choice exists.

But, if we focus our attention on the "deterministic" extreme of the traditional "free will" debate, the issue appears to center neither on the *existence* of "choice" nor on the freedom of the choices, but rather on whether the choices are "inevitable" deterministic consequences of natural laws. Obviously, in any scientific theory of behavior, "choices" would have to be "explained" in terms of natural laws. But it is an open question whether the choices are inevitable, predictable, or deterministic consequences of the natural laws.

We must therefore address the question of inevitability or predictability. In this respect biological systems differ fundamentally from present artificial decision systems. Of course, adherence to physical laws no longer requires a strict determinism. According to the laws of quantum mechanics, systems at the atomic and molecular level are predictable only in a statistical sense, so microscopic phenomena are subject to an irreducible uncertainty.

Modern electronic computers are engineered so that the predictability of their results will not be impaired by random thermal noise or by quantum uncertainties. However, biological cybernetic systems are designed very differently. The all-or-nothing response of a single neuron in the brain depends on input from hundreds of other neurons, some of which will tend to facilitate a response while others will tend to inhibit a response. The response of the neuron thus depends on input from numerous sources. When the input signal is very close to the level required for a response, minor fluctuations in temperature or chemical composition and possibly even quantum fluctuations can determine the response of the neuron. The *unavoidable* uncertainties are thus *amplified* by the all-or-nothing response of neurons so that they can have important effects in the subsequent system behavior. Thus, the brain is a mechanism capable of amplifying microscopic thermal and quantum fluctuations to the macroscopic level of human behavior. Even if two physically identical brains could be placed in exactly the same environment, the results produced by the brains would be different. Evidently the operation of the brain is far from predictable, even in principle.

But the practical difficulties in predicting the behavior of a complex decision system are acute even without the fundamental uncertainties. This can be seen by considering the behavior of an artificial system which is completely deterministic and therefore predictable, at least in principle. But how predictable is such a system in practice? The answer is that its behavior is really not very predictable, even to the designer of the system. The purpose of the system is to make or recommend decisions. If the resulting decisions could be easily predicted, there would be no need for the system. Of course, if two identical systems were prepared and provided with identical input data, the subsequent behavior would be identical. In a sense, each system could predict the behavior of the other. But this is almost an academic observation, for the slightest difference in the two systems can completely destroy the "predictability" of the results.

Almost any decision system will encounter numerous points where there is approximate indifference between alternatives. The final selection of one of the alternatives then depends on trivial details, such as the way numbers are rounded in the machine, the sequence in which the alternatives are encountered, or the number of significant figures used to represent values. Once a different decision has been made at such an indifference point, the subsequent behavior of the two systems will diverge, for they are no longer facing the same problems.

Most computer scientists are very familiar with this type of unpredictability. A typical example occurred when the production scheduler was transferred to another computer with a slightly different level of accuracy in the representation of numbers. Despite the fact that no change had been made in the system design, it was impossible to duplicate the schedules produced on the previous computer. The schedules would usually start out identically and would often remain so for a few weeks of the schedule. But sooner or later a decision of near indifference would be taken differently. Thereafter there was little resemblance between the two schedules.

Fundamentally, the difficulty in predicting the behavior of a decision system arises because there are thousands of alternative courses of behavior that are almost equally optimum. The choice between such alternatives is largely a matter of chance. It depends in great detail both on the structure of the world model and on the order in which decision alternatives are examined (which of course determines when the first "adequate" alternative will be found). As a practical matter, the accuracy of knowledge required to predict how an organism will resolve such choices of near indifference can probably never be attained.

Thus, a science of human behavior can never do more than to identify a plausible range of action alternatives. It is unlikely that it will ever be possible to really predict behavior. Human beings operating in an environment of other human beings are continuously making unpredictable decisions in an environment that is intrinsically unpredictable. The state of any individual's mind, as well as the state of his values, is the result of his experience during a long complex chain of past decisions. The level of predictability that can be expected without an unattainable precision of knowledge is necessarily very low.

From the preceding discussion we can conclude, first, that human beings do make choices; second, that the choices are as "free" as it is theoretically possible for choices to be; and third, that the choices are intrinsically unpredictable. At best, human decisions can be only imperfectly predicted. Obviously, one can continue to argue about the definition of free will—and in the context of some of the definitions, one can deny its existence. However, if we accept a definition based on the concepts of "choice" and "unpredictability" then it appears that the affirmation of "free will" corresponds better with reality than does its denial.

THE MYSTERY OF CONSCIOUSNESS

On page 104 we listed nine major functions of the human decision system. Six of these could just as well have been described as functions of the "conscious mind" because, from a purely functional point of view, they constitute the main activities of the conscious mind. Yet, paradoxically, they do not even include the concept of consciousness! There is at present no consensus with regard to this paradox, but there are at least three conflicting points of view. Probably the best way to clarify the nature of the problem is to outline briefly these three alternatives.

1.   *The functional view.*   According to this view, a theory of consciousness needs only to encompass the functional activities of consciousness. The six basic functions of the human mind previously listed do, in fact, encompass both the objective and subjective functions, and thus they provide an adequate explanation of consciousness. After all, if a system can receive information about the environment, recall and compare past experiences, sense the quality of experience, make conceptual models of the environment, project the consequences of alternative courses of action, and then select and implement preferred courses of action, then what else does consciousness entail?

However, this purely functional view does not seem satisfactory to most people. According to the functional view any system which performed these basic functions would be conscious. A computer programming expert could easily design computer programs which, at least in a rudimentary way, would have all of these characteristics. But even if a program were designed to have all of the characteristics at a very sophisticated level, most experts would see no reason to believe that the program would acquire consciousness. Indeed, it is difficult to see how a computer program could acquire any subjective sensations at all.* It is even more difficult to see how it could acquire the sensation of wholeness and continuity of experience that we call consciousness.

Supporters of the functional view, however, would reject these objections, on the good scientific grounds that we have no way of knowing whether the computer system is conscious. Indeed, there is no reason to believe that the system would behave differently if it were conscious.

Obviously, a large part of the difficulty lies in the fact that there is no known way for an external observer to determine whether a system is conscious or not. This functional perspective, however, is probably the most simple and internally consistent theoretical view, so we will return to it after considering the other points of view.

2.   *The metaphysical view.*   According to this concept, consciousness does not occur and cannot occur in any machine. It is a unique characteristic of complex biological systems, and it may be unique to the human brain. It is a metaphysical concept, perhaps related to the "soul," and it cannot be explained by any physical laws. The obvious objection to the metaphysical view

---

* The word "subjective" in this context refers to experience as perceived *within* the system, as opposed to externally observable experience which can be shared with others and which is therefore called "objective."

is that it is a scientific cop-out. It provides no explanation, and if scientific investigation were founded on this premise, it would almost guarantee that no theory would be provided in the future.

3. *The biological view.* According to this view, consciousness does not occur, and cannot occur, in computers or computer programs as we know them today. The programs might behave as if they are conscious, and might even assert that they are conscious when asked; but they would not "really" experience the subjective sensations of consciousness that are common to the higher animals.

The reason these computer systems cannot experience consciousness is that they are fundamentally different from biological systems in their cybernetic structure. Present digital computer systems are essentially sequential processing systems. While they have large high-speed memories (where results can be filed and later retrieved), and they may have several peripheral processing units, their actual processing in any unit is done sequentially one step at a time. The results of each processing step are dependent only on the status of a very narrowly defined set of system elements. This narrow, disciplined approach to data processing makes it much easier for human programmers to control the computer and to obtain predictable results.

The logical operations in the brain, however, are probably much more complex. Information, images, and concepts may flow through the neural network in a completely parallel wavelike form that is unlike anything that can occur in present digital computers. The phenomena of consciousness therefore could be a consequence of this informal parallel processing approach. According to the biological view, therefore, the subjective sensation of consciousness is a natural but as yet not understood consequence of the foregoing functions of consciousness in a cybernetic system like the brain; but such consciousness would not be expected to occur in the narrowly disciplined operation of present-day computers.

The obvious objection to Concept 3 is that it is not really an explanation. It offers the hope of an explanation, but unlike Concept 1 it does not purport to provide one.

THE MISSING TEST

It was mentioned earlier that one of the most critical problems facing a theory of consciousness lies in the lack of an objective test for the existence of consciousness. A useful scientific theory must make predictions that can be tested. The functional view does in fact make predictions. It predicts that any cybernetic system with certain specified properties will be conscious. But the prediction is not subject to experimental verification, because there is no known test for the existence of consciousness.

The seriousness of this difficulty can best be understood by imagining circumstances where such a test would be needed. Suppose we were to construct a complex robot which encompassed all the previously listed functions of consciousness. With appropriate choice of design parameters we would expect that the robot might behave much like an intelligent animal. With appropriate input/output routines, the system could probably be designed to communicate in English. Terry Winograd (7) in fact developed a computer program capable of limited communication in natural English language. Thus, one would be able to carry on a conversation with such a robot. One might ask the robot if it is conscious. In response, the robot might say yes.

Supporters of the "functional view" might believe the robot, but most other scientists would not. They would reject consciousness in the robot on the grounds that they know the system design and know of no mechanism by which any component (or group of components) in the system could acquire the sense of whole experience that we call consciousness. But supporters of the functional view would claim that consciousness is a distributed sensation of the whole system, which cannot be localized to any single component or any single location in the system logic.

Although most people will reject the robot's claim of consciousness, they will readily accept the same claim by another person or even by another intelligent animal. Basically, we are willing to attribute consciousness to people (and perhaps to animals) because they seem similar to ourselves—and we "know" we are conscious.

The logical strength of the "functional view" can best be perceived by reversing the encounter with the robot and allowing the robot to ask if the man is conscious. When the man answers yes, the robot can proceed to ask what he means by consciousness, and what are the attributes of his experience that he identifies as consciousness. As the man responds, the robot murmurs, "Just like me!"

The issue of consciousness raises profound questions about how much one can or should ask of a theory. In dealing with the paradox of "free will" it became apparent that the "paradox" is largely a result of problems in the *definition* of "free will." Is it possible that similar problems are involved in the definition of consciousness? Perhaps if we could obtain an adequate definition we could also provide an adequate theory.

THE MISSING DEFINITION

To address the problem of the definition of consciousness we can imagine a scenario involving a brilliant computer designer and his skeptical supervisor. The computer expert claims he can make a computer system that will be conscious. The supervisor does not believe him, so they agree to a test. The supervisor writes down a complete list of all the characteristics the system must have if it is to meet *his* definition of consciousness. Some months later the designer returns and reports success in the project. To prove his point, he puts on a demonstration which shows that all the specifications have been met. He then challenges the supervisor.

"Now," he asks, "do you believe it is conscious?"

"No!"

"Why not? It meets all of your requirements!"

"Well, I'm not sure, but I think I must have omitted some important factors in my definition of consciousness. Let me think for a while."

A few days later the supervisor has collected some additional critical specifications and he gives them to the designer. The designer goes back to work and returns a few months later again to report success. Again, he puts on a demonstration, and again the supervisor is unsatisfied. The process is repeated several times until the supervisor can no longer think of any new requirements for his definition. The designer claims success, but the supervisor still is not satisfied. The designer wants to know why.

"It met all your criteria; you can't even think of any others! Why do you stubbornly refuse to believe it is conscious!"

"Well, I just can't believe that that pile of wires and circuit elements is conscious."

"In what way would it behave differently if it were conscious?"

"I don't know. I admit it acts as if it were conscious! It shows emotion; it reasons; it remembers; it seems to have a highly developed ego! It even asks the right questions when we don't give it complete information. But I still don't think it's conscious."

"Then for heavens sake, why don't you believe it's conscious?"

"I don't really know. But I do know how you designed it. We talked about all the components. You never showed me any part of the design that would give it consciousness; I don't believe it is conscious. I believe it is just a complex computer program, and an assembly of electronic components. I don't know what I omitted in my definition of consciousness, but I am sure I forgot something."

The dialogue is imaginary, but it could easily be real. Why is the supervisor still skeptical? Basically he wants to know whether the system is "really" conscious. He wants to know how it "really" feels, not how it *says* it feels. There is no way he can find out. All his complicated specifications are really efforts to specify how the system must *feel*. In the end he can only observe how it acts. We can never know how it "really" feels—if indeed it *feels* at all.

There is one critical experiment that he would have to do, to be sure. He would have to get *inside* the system and *be* the computer. That he can never do. The ultimate test of consciousness can be applied only to ourselves. We can never be sure that anyone else is "really" conscious. We are willing to accept that they probably feel as we do, on the circumstantial evidence that they look and act much as we do.

If a theory is supposed to make predictions that can be verified, then perhaps the *most* we can ask of a theory of consciousness is that it be able to explain why a system acts *as if* it is conscious. Perhaps it is asking too much to require a theory to tell us when a system "really" is conscious. The existence or nonexistence of an internal consciousness is not experimentally observable outside the system. It is only observable in an introspective sense. This does not mean that it is unreal; it is probably the most "real" of all our personal experience.

Although we are dealing with the cybernetic functions of consciousness, both the essence of consciousness and the mechanisms of consciousness remain a profound mystery. Any or none of the three concepts discussed earlier could be correct. It is possible that in the future a theoretical perspective will evolve that will provide a more satisfying explanation of our personal sensation of consciousness. But such an explanation does not seem to be needed for the present development. Although we have not been able to provide either an explanation or a definition of consciousness, it appears that we may be able to examine the logic of human decisions and the structure of human values without providing any answer to this most basic of human mysteries. In the pages that follow, we will proceed in that way.

THE POSSIBILITY OF A DUAL CONSCIOUSNESS

In the preceding chapter we identified the thalamus as a probable center of consciousness of the human mind. However, the forebrain includes two cerebral hemispheres, each serviced by its own half of the thalamus. Since the two halves of the thalamus appear to be structurally almost completely separate, this raises the strange possibility of a dual consciousness within the human brain. Is the consciousness really divided? If so, does one side of the brain serve as a sort of peripheral processor for the other? In terms of our discussion of values, the issue is not critical. It does not really matter whether we are dealing with a unified system (in which the central control is really unified but has two halves), or two cooperative independent systems in which each serves to assist the other, or two cooperative independent systems in which one side leads and the other serves as a peripheral processor. Since we have already allowed for peripheral processors, the problem of the two halves of the brain can be treated simply as a special case of that very general concept.

Nevertheless, the question is very significant both in terms of our understanding of the human consciousness and in terms of our perception of the subconscious. By now a wide variety of experiments have confirmed that the two halves of the brain are in fact capable of operating independently. When the information transfer between the two cerebral hemispheres is stopped by surgically cutting the connecting fibers of the corpus callosum, the two sides of the brain appear to operate like two completely independent centers of consciousness. This is consistent with what we would expect based on the physical structure of the thalamus, and it adds support to our structural interpretation of the system.

It is still not conclusive whether the two halves of the brain *normally* operate independently or as a single unit. The weight of evidence, however, seems to favor the hypothesis that the two sides operate rather independently. If there really is a hidden, independent intelligence in the nonverbal side of our brains, we know (from our earlier discussion) that we will never be sure whether it is "really" conscious. The most we will ever learn, is whether it acts *as if* it is conscious. From the point of view of our conscious verbal mind we would have to think of the nonverbal side as an imperfectly controlled peripheral processor. Intellectual activities and action decisions of the nonverbal half would appear to the verbal side to be subconscious or intuitive.

If this point of view is correct, then the scope of the "conscious mind" may actually be limited to only one hemisphere of the brain. The nonverbal hemisphere should then be considered to be a major subsystem of the brain that lies outside the "mind." The possibility of such a division within our own brains seems very difficult to believe. The only thing that makes it credible is the actual results from the split-brain experiments (2, 5, 6).

RELATION TO IDEALIZED MODEL

The purpose of the present chapter has been to provide a better understanding of the relationship between the idealized model and the imperfect realization of that model in the actual human decision system. The human system suffers from many limitations because of design compromises necessary to realize a versatile decision system despite the limitations of the size of the skull and the size and processing speed of the neuron.

The human brain suffers from imperfect memory storage, but more importantly from imperfect ability to recall stored information. It suffers from imperfect control over peripheral processors and almost total inability to monitor the activities of peripheral processors. It may also suffer from difficulties of internal communication between the left half, which specializes in digital or symbolic logic, and the right half, which specializes in analogue or representational logic. It suffers from limitations in the scope of activities that can be encompassed within conscious awareness at one time.

Like any finite decision system, it suffers from an inability to think usefully very far ahead in an environment of uncertainty. For this reason, it has been supplied by evolution with a time-dependent innate value system. These values are designed to motivate satisfactory behavior without excessive dependence on uncertain projections of the more distant future. Like any finite decision system, it must operate with imperfect and incomplete world models. It must make decisions on the basis of an exploration of a small number of action alternatives. Within these practical limitations, however, it remains true to its basic concept as an optimizing value-driven decision system. Many readers, of course, will object to the use of the word "optimizing" to characterize a system that so readily accepts action alternatives that are so far from optimum.

Herbert A. Simon in his challenging book *Models of Man* (3) observes that man does not operate on principles of optimization. Human behavior, he points out, is better described as a policy of "sufficing"—that is, of finding solutions that are adequate, but almost never optimum. If an adequate solution cannot be found the individual will continue to "worry." He will explore and reexplore alternatives until a satisfactory solution is found, or until the urgency of time forces acceptance of an unsatisfactory alternative. The present perspective makes it clear that in a finite system, just such a policy of "sufficing" is an inevitable consequence of a broader policy of "optimization" when finite cybernetic resources must be conserved and allocated.

One of the most important characteristics of the human decision system is the dependence of behavior on the quality of available world models. This dependence was clearly illustrated in some experiments designed to measure actual human behavior in game situations. Considerable uniformity of behavior was observed among most of the subjects. However, the possibility of any universal conclusions was destroyed by a few subjects who had studied game theory. These subjects played quite consistently in accordance with that theory. Obviously, successful prediction of behavior requires an understanding of *both* the goals or values *and* the underlying world model. A better world model will usually facilitate better or more effective behavior.

It is worth emphasizing that a better model is not necessarily a more accurate or more complete representation of reality. It only needs to be more useful. The simpler the model, the easier it will be to examine alternatives. The quality of decisions involves a tradeoff between accuracy, which permits a better evaluation of individual alternatives, and simplicity, which allows more alternatives to be examined. The dependence of system behavior on the available world model of course explains the importance of education and the importance of simplifying theories.

One of the most important aspects of a world model is the self model. A better model of self should permit a better focusing of one's efforts through a better understanding of personal objectives. It is hoped that, for some readers, the model developed here will serve that purpose. One of the wisest pieces of advice ever given was contained in two words, "Know thyself."

The model of the human brain as an automatic decision system provides a suitable model only at one level in our intellectual hierarchy. We can hope and expect that the decision-system model will someday be "explained" in terms of more fundamental physiological knowledge. Such a theory might "explain" both the *decisions* and the *orchestration of values* in terms of changes in synapse sensitivities and the detailed structure of the neural network. Such a model could be more *complete* and more *accurate*, but it would not necessarily be more *useful*. The decision-system "model" of human behavior is offered in the hope that it will provide a model that has practical value in human decision making.

The finite cybernetic limits of the human decision system explain a number of phenomena that seem superficially inconsistent with the behavior of an optimizing system. Human beings are suggestible, and they can often be misled by authority. In a mob environment, they may collectively commit atrocities that they would not consider as individuals. All of these weaknesses are consequences of limitations either in the value structure or the analysis capacity of a finite decision system.

A human being can consider no more than a small number of action alternatives at any time, so he is inherently susceptible to suggestions. If a suggestion is offered which provides an adequate or "sufficing" alternative, the suggestion is likely to be accepted. The suggestion changes the state of the system by making available an alternative that otherwise might not have been perceived. The "suggestion" inevitably increases the probability that the suggested course of action will be chosen. By accepting a satisfactory suggestion, the individual avoids wasting intellectual effort to devise his own alternative.

Individuals are sometimes misled by authority, because reliance on authority is often a good way to conserve cybernetic resources. If prior experience has shown that an authority is usually right, it will be more efficient to accept ideas from the authority than to develop independent ideas. After all, even independently generated ideas can be in error.

The foregoing considerations about cybernetic efficiency are often interwoven with value considerations. It may seem desirable to accept a suggestion in order to please, or to accept the ideas of an authority to avoid adverse consequences. The time dependence of values and the innate emotional response of the Social value system to a group environment accounts for much of the apparent irrationality of mob action.

The emphasis in this chapter on the limitations of the human mind should not obscure the achievement of evolution in the design of the human brain. Any real cybernetic system must be finite, and any finite cybernetic system would suffer in some degree from the human limitations. In such a design, compromises are inevitable, and the quality of the system will reflect the quality of the compromises.

We have now completed the theoretical development of our model of the human decision system. The next step is to apply the model to obtain a better understanding of human values and human behavior. As was shown earlier, the behavior of an automatic decision system tends to be dominated by its innate values. Therefore, the first step in the application of the model will be to study the structure of the innate human values.

For the purpose of this development, which is the subject of Part II of the book, we will think of the human mind as a unified but imperfect decision system. The reader, if he chooses, can think of the nonverbal side of the brain as an imperfectly controlled peripheral processor. During this phase of the development, we will not be concerned with detailed capabilities or limitations of the rational mind. We will simply assume the existence of the cybernetic resources necessary to compare alternatives and make decisions based on the innate value structure.

REFERENCES

1. Eccles, John C. *Brain and Concious Experience*. Pontificia Acadamia Scientiarum. New York: Springer-Verlag, 1966.
2. Gazzaniga, Michael S. "The Split Brain in Man." *Scientific American*, August 1967.
3. Simon, Herbert A. *Models of Man*. New York: Wiley, 1957.
4. Skinner, B. F. *Beyond Freedom and Dignity*. New York: Knopf, 1971.
5. Sperry, R. W. "The Great Cerebral Commissure." *Scientific American*, January 1964.
6. Sperry, R. W., M. S. Gazzaniga, and J. E. Bogen. "Interhemisphere Relationships; The Neocortical Commissures; Syndromes of Hemispheric Disconnection." *Handbook of Clinical Neurology*, vol. 4. Edited by P. J. Vinken and G. W. Bruyn. New York: North-Holland, 1969.
7. Winograd, Terry. *Understanding Natural Language*. New York: Academic Press, 1972.

PART II

STRUCTURE OF HUMAN VALUES

Introduction to Part II

PART I developed a theoretical interpretation of the human brain as a "value-driven" decision system and showed how principles of efficiency appropriate to such a decision system might have molded the present design of the brain. The evolutionary evidence also suggests that evolution's basic design concept for a value-driven decision system probably predates the emergence of the reptiles. It remains to be shown whether this simple design concept can provide a practical basis for understanding intelligent human behavior. Our objectives in Part II are to show:

1. That the decision-system model is generally compatible with what is known about human behavior and human motivation.
2. That there is a substantial body of behavioral data that supports the theory.
3. That the main features of the innate human value structure can be "explained" in terms of the evolutionary experience of the species.
4. That the theory shows promise of providing a very realistic model of conscious human behavior.

To accomplish these objectives, Part II develops a preliminary theory concerning the specific value structure that evolution has built into the "human decision system." The purpose is not to provide a definitive value structure but only to show, through the use of an illustrative value structure, that the value-theory approach can provide a generally satisfactory representation of human motivations.

The human mind is motivated by a very complex structure of built-in "values." These "innate values" include both the *emotions* and the traditional *biological drives*. The values also include a number of other important elements that are less frequently identified as part of an innate motivation system. The analysis in Part II attempts to understand the evolutionary origin of this complex system of innate values.

To understand the innate human values we must study their evolution during the primitive past. Biological evolution is a slow process. It is estimated that the primate order is about 75 million years old. The evolutionary branch that led to man separated from other primates some 5 to 10 million years ago. Small-brained apelike ancestors of man may have begun to use tools as much as 2 to 3 million years ago. About 1 million years ago the first big-brained upright ancestors of man appeared.* Truly modern man, *Homo sapiens sapiens* (or Cro-Magnon man), is estimated to be only 40,000 to 60,000 years old. Agriculture was introduced only about 10,000 years ago.† Thus 99 percent of the evolutionary process that separated man from the apes occurred before the advent of modern man. About 99.8 percent of the history of man occurred prior to the introduction of crop agriculture! For all practical purposes the evolution of modern man was complete before the introduction of agriculture.

From a broad evolutionary perspective modern man is a remarkably new species. The tremendously accelerated pace of *cultural evolution* in the 40,000 to 60,000 years following the appearance of modern man suggests that the really important distinction between modern man and his predecessors involved an improved *intellectual* capacity. Perhaps the most likely change of this type that may have occurred with modern man was a refinement of the brain to support a more elaborate modern language, as opposed to more primitive forms of verbal communication.

This history of human evolution suggests that most of man's innate motivating values were designed to operate in a primitive, preagricultural, hunting society. It seems probable that most of the values evolved in a society that had only rudimentary verbal communication. It is a remarkable testimony to the adaptability of the human "decision system" that the same basic behavioral repertoire could operate in an agricultural society and then in an urban society. But the ancient human values may not be really appropriate for a modern urban society. There is reason to believe that much of the conflict and frustration experienced by modern man may reflect basic incompatibilities between the ancient human motivations and the modern social environment.

Why should we be interested in the innate values that motivate human behavior? One obvious reason is to gain an improved understanding of human behavior. But from the point of view of decision science there is a deeper reason. The primary values of any decision system are its ultimate criteria of decision. They are the fundamental source of all values used by the system. According to this view, human beings can find fulfillment only in terms of the innate *human* values. In the final analysis, according to this decision-science

* For a summary of the evolution of man see Pfeiffer (9).
† For a more detailed analysis of human genetics and the evolution of agriculture see Cavalli-Sforza (1).

perspective, all human decisions, whether they involve personal or public issues, must ultimately be measured against the primary human values.

EMPHASIS ON PRIMARY VALUES

The discussion in Part II is concerned almost exclusively with the "primary" rather than with "secondary" human values. But we must not lose sight of the vitally important role of "secondary" criteria in motivating human behavior. The human mind is a sophisticated decision system and it makes extensive use of secondary criteria. Indeed, most of the day-to-day activities of modern man can be explained more directly in terms of secondary values, rather than primary values. For example, a great deal of activity can be explained as the pursuit of money. Obviously, the acquisition of money is a secondary not a primary goal. Similarly as members of a society we follow many cultural traditions. These traditions reflect "secondary" decision criteria that have been developed during the cultural experience of the society. They do not correspond to specific primary values.

In modern man it is often hard to distinguish primary from secondary values, because the behavior attributable to the two types of values is so intertwined. For these reasons, our one-sided concentration on the "primary" values in Part II may seem, at first, to be rather strange. But there are some valid reasons for the concentration on "primary" values in Part II, before moving on to deal with "secondary" values in Part III:

1. If we wish to assess the "validity" of secondary decision criteria, we can do so *only* by evaluating them against the *primary* human values. To make such value assessments it is essential to know which values are primary and which are secondary.
2. Because the primary values are genetically built into the human decision system, they constitute an *essential* part of the system description. If we are to understand human behavior, the most basic requirement is to understand the *innate* value structure that underlies that behavior.
3. Regardless of what changes we may consider in society, we can expect that individuals will continue to be motivated by the *same* structure of *innate* values. People may learn to change their secondary values, but they cannot change their innate values. Thus the innate human value structure places important limits on man's ability to adapt to change. If we are to propose realistic concepts for social reform, we must be sure that the reform concepts are compatible with the innate values.

For these reasons Part II is concerned almost exclusively with the primary or "built-in" human value structure. In selecting evidence and examples for Part II we have deliberately concentrated on material which highlights the *innate* values as opposed to secondary decision criteria.

RELATION TO EXISTING INTERPRETATIONS OF BEHAVIOR

The interpretation of human behavior developed in the following chapters is different from current orthodox interpretations of psychology and behavioral science. Obviously, the observations and experiments are the same, but the interpretation of the observations is quite different. Before proceeding it is appropriate to discuss briefly the rationale for this revised interpretation.

Following the discovery of what seemed to be preprogrammed or instinctive behavior in lower animals, behavioral scientists began to look for similar well-defined or "instinctive" responses in human behavior. Careful research failed to reveal anything except a few reflex responses that could be correctly described as "instinctive." This dilemma gave rise to two quite different schools of thought.

The behaviorists concluded that human behavior must have almost no innate component, and that it had to be attributed almost entirely to conditioning experience of the individual within the environment. But there were also some drive theorists who contended that behavior could be explained as a purposeful response to underlying *innate* drives. Although the specific forms of behavior were obviously not innate, the drive theorists believed that the basic drives were innate. In keeping with the traditional scientific principle of parsimony in assumptions, the drive theorists tried to explain behavior with a very small number of simple innate drives.

Thus both orthodox schools of thought were in agreement that the innate component of human behavior is neither very complex nor very important. Although the present theory is obviously closest to that of the drive theorists, our analysis of the "design principles" for a value-driven decision system puts the problem in a very different perspective. The decision theory suggests:

1. That the *innate* values provide the primary "decision criteria" underlying all conscious behavior. They are therefore *very* important in understanding behavior.
2. That there is no a priori reason to expect the innate value structure to be

simple. Indeed there is every reason to expect that it will be quite complex. Experience with artificial decision systems shows that a complex value structure can often compensate for the intellectual weaknesses of a "finite" decision system. On a priori grounds, therefore, we expect that the "innate values" will be very complex, involving *many* components, and that the relationships between sensory experience and the resulting values will be quite complicated.

The following chapters represent an effort to reinterpret the experimental data in the context of this new decision-system perspective. The structure of "innate values" developed in this analysis includes a large number of quite specific and detailed value components, but the mere existence of a "value component" does not imply that it will dominate the resulting decisions. According to the theory many of the "innate value" components must be quite *weak* so that they can easily be *overruled* when other values (either primary or secondary) are more important.

The present interpretation differs strikingly from the orthodox interpretations because of the increased emphasis on *detailed* innate values. However, the increased emphasis on innate behavioral traits is not unique to this theory. A growing body of opinion is placing a renewed emphasis on innate motivations to explain the behavioral differences between species. This changed perspective is most clearly represented in the work of ethologists and other scientists who have been concerned with behavioral comparisons between species (4, 5, 6, 7, 8, 11, 12).

ANALOGY WITH NONHUMAN PRIMATES

Adult human behavior is influenced by so many factors that the specific role of the innate values is difficult to decipher. One cannot be sure to what extent a specific behavior is innate or derived from "socializing" influences. When we see devotion, compassion, and self-sacrifice in human behavior, we cannot be sure whether it reflects innate values or whether it is the result of early teaching at home, in school, or in the church. But when we see the same basic behavior patterns in the social life of nonhuman primates, then it seems very likely that the motivations are innate, and not the result of religious or moral teaching. The fact that apes and other nonhumans do not use a symbolic language limits the complexity of concepts that can be assumed to influence their behavior, and this simplifies the interpretation of their behavior.

In previous chapters when we were developing the theory of decision systems, it proved useful to discuss the concepts first in the context of artificial systems. After the principles were developed they could be applied to our understanding of the human brain. In the next few chapters we will employ a similar strategy. The innate motivations will be developed first in the context of nonhuman primate societies. The ideas will then be applied to improve our understanding of human motivations. Of course, modern man is not an ape. Some of the innate human values are uniquely human. To provide insight concerning such uniquely human values, we will look for basic behavioral differences that distinguish primitive human society from the societies of the nonhuman primates.

In the past, much attention has been concentrated on the "unique" characteristics that distinguish humans from the "lower" forms of life. There has been a tendency to assume that those characteristics which we admire most are uniquely human and that the rest are part of our "animal" inheritance. Our approach here will be quite different. We will begin with the hypothesis that innate human motivations are similar to the corresponding motivations in other primates. We will abandon that hypothesis only in specific cases as required by specific conflicting evidence. This approach is consistent with our belief that the main source of values lies in the hypothalamus, which is the oldest and most specialized part of the forebrain. Thus, it is reasonable to expect that the broad design of human values will be much the same as in neighboring species.

Table 11.1 shows a summary classification of some of the most familiar of approximately 240 known species of living primates. Although detailed behavioral studies are available for only a small number (perhaps 20 to 30) of the species, sufficient information is available to permit certain broad generalizations. Within the primates one observes a very broad range of behavioral patterns; but when we try to interpret the behavior in terms of innate motivational factors, we find a great deal of similarity among the species. The major behavioral differences seem to result more from differences in the relative strength of the innate values than from qualitatively different types of motivation. It is as if nature had a fixed list of innate value components to use in the design of all primate species. Each species (and each individual within a species) seems to have a characteristic motivational profile, just as it has a characteristic facial profile.

The situation is analogous to the genetic inheritance of skeletal structure. All primates have essentially the same skeletal structure. The large differences in the appearance among the species can be traced to differences in the size and shape of the different parts. This same basic principle seems to apply to the motivational structure. Similar motivational factors are common to most primates, but the relative importance and shape of the factors varies widely among the species. Of course, no nonhuman species will have a motivational profile which corresponds to the human profile, but we will expect that most of the important factors in the human profile will also appear in the profile of *some* other primates. As we study the role of these factors in nonhuman species, we will gain a better understanding of their role in human behavior.

VARIATIONS IN THE HUMAN GENETIC INHERITANCE

For many years the reason for the gradualism in the effects of genetic mutations remained a mystery. Intuitively one would expect that mutations should produce discontinuous changes—adding (or subtracting) new bones and new organs and new value components to (or from) the basic design. Recent studies, however, have shown that the amount of genetic diversity within the normal inheritance of a species is surprisingly high. The evidence from other species suggests that approximately 12 percent of all genes carried by the typical individual may be unmatched in the sense the maternally and paternally inherited genes in a single gene pair are different.* This high percentage of mixed or unmatched genes can only be explained on the basis that the mixtures *themselves* have survival value and that evolution is in fact actively maintaining in the population of each species an equilibrium mixture of genes which is most advantageous for survival. A number of specific cases have now been documented where the most advantageous inheritance involves an unmatched gene pair and where a pure inheritance of either gene is less advantageous.

In retrospect, it is not at all surprising that this should be the case. A new mutation cannot begin to spread in a population *unless* it is advantageous as one member of an unmatched gene pair. It is only after a mutation has become successful in this unmatched form that it even becomes relevant whether it might be more advantageous in a pure form. This important new finding obviously helps to explain how equilibrium genetic mixtures can arise and be maintained. It also helps to explain both the speed with which populations can respond genetically to changes in environmental pressure and the gradualism in the design modifications that usually result from genetic mutations.

This new information on the great genetic diversity within an equilibrium population helps to put into proper perspective the genetic differences be-

* Lewontin and Hubby (3), Clarke (2), Wilson (12), pp. 70–72.

TABLE II.1 *
Summary of Primate Classification

SUBORDER	SUPERFAMILY	FAMILY	SUBFAMILY	GENUS	VERNACULAR NAME(S)
	Tupaioidea	Tupaiidae	Tupaiinae	*Tupaia*	Tree shrew
				Dendrogale	Smooth-tailed tree shrew
				Urogale	Philippine tree shrew
			Ptilocercinae	*Ptilocercus*	Pen-tailed tree shrew
		Lemuridae	Lemurinae (greater lemurs)	*Lemur*	Lemur
				Hapalemur	Gentle lemur
				Lepilemur	Sportive lemur
			Cheirogaleinae (lesser lemurs)	*Cheirogaleus*	Dwarf lemur, mouse lemur
				Microcebus	
Prosimiae (prosimians)	Lemuroidea	Indriidae	Indriinae	*Avahi*	Woolly lemur
				Propithecus	Sifaka
				Indri	Indri
		Daubentoniidae		*Daubentonia*	Aye-aye
	Lorisoidea	Lorisidae		*Loris*	Slender loris
				Nycticebus	Slow loris
				Arctocebus	Angwantibo
				Perodicticus	Potto
		Galagidae		*Galago*	Galago (bush-baby)
	Tarsioidea	Tarsiidae		*Tarsius*	Tarsier
		Callithricidae		*Callithrix*	Marmoset
				Leontocebus	Tamarin, pinche
			Callimiconinae	*Callimico*	Goeldi's "marmoset"
				Aotes	Douroucouli (night monkey, owl monkey)
				Callicebus	Titi

Suborder	Superfamily	Family	Subfamily	Genus	Common name
Simiae	Ceboidea (New World monkeys, platyrrhine monkeys)	Cebidae	Cebinae	*Pithecia*	Saki
				Chiropotes	Saki
				Cacajao	Uakari
				Alouatta	Howler monkey
				Saimiri	Squirrel monkey
				Cebus	Capuchin
				Ateles	Spider monkey
				Lagothrix	Woolly monkey
	Cercopithecoidea (Old World monkeys, catarrhine monkeys)	Cercopithecidae	Cercopithecinae	*Macaca*	Macaque
				Cynopithecus	Black ape
				Papio	Baboon, drill, mandrill
				Theropithecus	Gelada
				Cerceocebus	Mangabey
				Cercopithecus	Guenon
				Erythrocebus	Patas monkey (hussar monkey, red monkey)
			Colobinae	*Presbytis*	Langur, leaf-monkey
				Pygathrix	Doue
				Rhinopithecus	Snub-nosed monkey
				Simias	Pig-tailed langur (Mentawi Islands langur)
				Nasalis	Proboscis monkey
				Colobus	Guereza
	Hominoidea (apes and man)	Hylobatidae (lesser apes)		*Hylobates*	Gibbon
				Symphalangus	Siamang
		Pongidae (great apes)	Ponginae	*Pongo*	Orangutan
				Pan	Chimpanzee
				Gorilla	Gorilla
		Hominidae		*Homo*	Man

NOTE: Names in parentheses are synonyms for the names they immediately precede. Names separated by commas but not in parentheses are not synonyms.

* From *Behavior of Non-Human Primates*, edited by A. M. Schrier and F. Stollnitz (1971). Reproduced by permission of Academic Press, New York.

tween racial and ethnic groups. Although there are real differences in the average composition of the genetic mix between racial groups, the statistics show that these differences are almost trivial compared to the differences between typical individuals *within* any racial group (1).

Since the motivational profile is largely an inherited characteristic we can also expect significant differences between individuals in the motivational profile. Characteristic differences also occur because of differences in the age and sex of individuals. Finally, the motivational profile of individuals can be strongly affected by early experience, just as skeletal size can be affected by early nutrition.

ORGANIZATION OF PART II

It is helpful to classify the innate human values into three major categories:*

1. "Selfish Values," which are concerned with individual survival.
2. "Social Values," which are concerned with the welfare and survival of the social group.
3. "Intellectual Values," which serve to motivate intellectual activities.

Two chapters are devoted to *each* of these categories of innate values. In each case, the first chapter develops background concepts, and the second chapter develops a specific hypothesis about the actual structure of the motivating values.

Chapters 8 and 9 are concerned with the Selfish Values. The selfish values, which are the easiest to understand, serve as a vehicle to develop certain principles that are needed to understand the Social and Intellectual Values. In Chapter 8 the infant primate is used as a vehicle to introduce evolution's value concepts, in the context of a simple biological system. The chapter shows how the value theory can be applied to a biological system and demonstrates that the theory has predictive power. Chapter 9 develops the concepts into more general theory and proposes a specific structure for the Selfish Values in the human adult.

Chapter 10 and 11 are concerned with the Social Values. Chapter 10 summarizes evidence that demonstrates the evolutionary need for and existence of innate Social Values. Chapter 11 proposes a specific theory concerning the

---

* As noted previously on p. 116, we have adopted a stylistic convention of capitalizing these categories of innate, or primary, values to distinguish them from the secondary cultural values that go by the same name.

structure of these innate social values. These values are the ones of the greatest current interest, because they appear to be the ones most in conflict with the structure of our modern society.

Chapters 12 and 13 are concerned with Intellectual Values. The purpose of these values is to motivate efficient operation of the human mind, particularly in the development and improvement of a mental model of the environment.

In its broad perspective, the theory seems surprisingly successful in explaining previously mysterious characteristics of human nature. It appears to explain the role of humor, joy, sorrow, pride, shame, and anger in "nature's design" for the human species. Because this "decision-system" approach to the analysis of human motivation is new, the proposed details of the value structure are necessarily conjectural. Some of the motivations that are treated as innate may later prove to be culturally conditioned, and vice versa. The illustrative innate value structure developed here has been chosen because it seems, on the basis of system design experience, to be generally consistent with available behavioral evidence; but it is really intended only as a starting point for research that should ultimately confirm or disconfirm the specific details of the illustrative value structure.

REFERENCES

1. Cavalli-Sforza, L. L. "The Genetics of Human Populations." *Scientific American*, September 1974.

2. Clarke, Brian. "The Causes of Biological Diversity." *Scientific American*, August 1975.

3. Lewontin, R. C., and J. L. Hubby. "A Molecular Approach to the Study of Genic Heterozygosity in Natural Populations." *Genetics* 54, no. 2 (1966): 595–609.

4. Lorenz, Konrad. *On Aggression*. New York: Harcourt Brace & World, 1966.

5. Lorenz, Konrad. *Studies in Animal and Human Behavior*. Cambridge, Mass.: Harvard University Press, 1971.

6. Morris, Desmond, ed. *Primate Ethology*. Chicago: Aldine, 1967.

7. Morris, Desmond. *The Naked Ape*. London: Jonathan Cape, 1967.

8. Morris, Desmond. *Intimate Behavior*. New York: Random House, 1971.

9. Pfeiffer, John E. *The Emergence of Man*, 2nd ed. New York: Harper & Row, 1972.

10. Schrier, Allan M., and Fred Stollnitz, eds. *Behavior of Non-Human Primates*, vols. 3 and 4. New York: Academic Press, 1971.

11. Tiger, Lionel, and Robin Fox. *The Imperial Animal*. New York: Holt Rinehart & Winston, 1971.

12. Wilson, Edward O. *Sociobiology: The New Synthesis*. Cambridge, Mass.: Belnap Press of Harvard University Press, 1975.

Chapter 8

The Infant: A Case Study in Values

"... the cradle's but a relic of the former foolish
  days,
When mothers reared their children in unscien-
  tific ways;
When they jounced them and they bounced
  them, those poor dwarfs of long ago—
The Washingtons and Jeffersons and Adamses,
  you know."

"A Complaint that for Hygienic Reasons He Was
Not Allowed to Play with His Grandchildren in
the Old-fashioned Way." Attributed to BISHOP
DOAN

THE PRESENT CHAPTER uses the primate infant to illustrate the value
concept in the context of a relatively simple biological system. The infant pro-
vides a particularly simple case study because past experience and learning
is minimized. The first part of the chapter develops a detailed hypothesis
about the "value structure" that motivates the primate infant. Obviously
this "value structure" is constructed so that it "explains" the normal behav-
ior of the infant; but the *same* value structure is used later in the chapter to
"predict" the *abnormal* behavior of infants that have been "socially de-
prived." The close correspondence between the "predicted" and observed
abnormal behavior suggests the potential usefulness of the "value theory" as a
predictive behavioral model.

The "innate" value structure was treated in Part I as if it is rigidly defined
by the genetic inheritance. The final sections of the chapter develop evi-

dence which suggests that the biological "value system" may be responsive to a simple learning process somewhat like "operant conditioning." The value concepts that are developed informally in this chapter provide the foundation for a more formal development of principles in the later chapters.

BEHAVIOR OF THE INFANT PRIMATE

The nonhuman infant primate is completely dependent on his mother for protection and survival. As a baby he spends most of his time clinging to his mother's body. In most species, the baby is able to cling successfully almost immediately after birth. In chimpanzees and gorillas (the species most similar to man) the baby is heavier and less mature at birth, so that it initially requires support from the mother. But as the baby matures it clings successfully without assistance. Like all mammals the infant primate suckles for nourishment.

As the infant matures, the changes in behavior follow a rather standard pattern. Soon the baby will leave the mother for brief periods; but any unexpected noise or disturbance will send him quickly back to the mother. Gradually the tendency to explore increases and he is willing to venture further from the mother. Finally, leaving the baby stage he begins to take solid food, but complete weaning usually occurs much later. Throughout his infancy and childhood, the young primate remains dependent on the mother and continues to turn to her in times of stress.

The chronological time involved in this process varies widely between species. Although many monkeys begin to use solid food when they are between two and four weeks of age, the corresponding stage in the great apes is reached only when the baby is almost three months old. Moreover, there are rather wide variations between species in the sequence with which the various stages occur (14). Nevertheless, the broad outlines sketched above seem typical of infant development for most nonhuman primates.

As described above, the infant's behavior is clearly functional and necessary for his survival. What innate motivations are responsible for the behavior? We will discuss briefly some key experiments and then interpret the results in terms of our theoretical concept.

MATERNAL ATTACHMENT IN THE INFANT

In 1959, Harry Harlow and Robert Zimmerman (12) reported on their classic series of experiments concerning the emotional attachment of the infant to the mother. The experiments were designed to distinguish between two alternative possibilities:

1. That the bond is conditioned and reinforced by the reward of milk provided by the mother.
2. That the attachment reflects an innate response of the infant to contact comfort provided by the mother.

The results were remarkably conclusive, and they strongly support the second hypothesis.

The experiments, using infant rhesus monkeys, were carried out at the University of Wisconsin Regional Primate Center. Infants were separated from their mother within six to twelve hours after birth and placed in an environment where only artificial dummies were available to act as mother substitutes. Two types of mother substitutes, one wire dummy and one soft cloth dummy, were provided. Either type could be arranged to provide nourishment by means of a nipple which protruded from the chest. In some cases feeding was provided only by the wire "mother"; in other cases it was provided only by the cloth "mother." The experimenters recorded the amount of time the infants spent with either of the two "mothers." The most impressive results were shown when the infants were given free access to both the wire and the cloth-covered artificial mothers. Regardless of which "mother" did the feeding, the infant monkeys clearly preferred the cloth "mother" and spent the bulk of their time with her. The wire mother was visited little more than was necessary for actual feeding. When the wire mother did the feeding, there was initially some confusion (up to about fifteen days) before the preference for the cloth mother was firmly established. Thereafter, the attachment to the cloth mother seemed almost as firm, regardless of where feeding was provided.

The experiments showed conclusively that contact comfort is the most decisive element in generating the attachment of the infant to the mother. Confirming tests were made to show that the preference was not just a preference for a warm "nest." Infant monkeys would leave a warm heating pad and spend most of their time on the unheated cloth "mother." Other tests showed that the emotional dependence of the infants on their artificial mothers was remarkably similar to the attachment of a human infant to his mother. When presented with a strange fear-producing object (such as a teddy bear) the infants quickly sought comfort in the cloth mother.

This interpretation of the statistical data is supported by direct observations of the infant's response. In spite of their abject terror of the strange object, the monkeys seemed comforted after clinging to the cloth "mother." Within a minute or two the comforted infants would begin visually exploring the frightening object from the "safety" of the mother. A few of the bravest infants would actually leave the "mother" to explore (under her psychological protection) the previously threatening object.

When only the wire mother was present, the infants would sometimes retreat to her when frightened, but they did not seem to be comforted. They continued to cringe in terror. Perhaps a weak bond develops to the wire mother but it is very weak compared to the normal bond.*

AN ILLUSTRATIVE VALUE STRUCTURE

Although the literature of behavioral science contains many discussions about biological drives, the "drive" concept is rarely defined in the unambiguous terms needed to specify a quantitative behavioral model. In the following sections an effort is made to develop a motivation model for infant primates which is quantifiable in terms of specifically defined innate drives.

The treatment is detailed, not because of any misplaced confidence in the details of the model, but rather to provide a concrete example of how the value concept can be used to define quantitative behavioral theories. In all probability, future experiments will show that the specific value structure used in the example is incorrect in many ways. But that is the purpose of a quantitative theory—to allow us to systematically improve our understanding. In developing this illustrative value structure, no effort has been made to be complete. Instead we have limited ourselves to a few key motivations that are necessary to explain the infant behavior just discussed.

Infant behavior usually begins with what appear to be simple reflex responses. The transition to a more systematic or rational response occurs gradually as the infant matures. To provide a realistic discussion of infant motiva-

---

* Subsequent experiments have shown that the ease and degree of attachment to the mother can be influenced by other factors such as warmth, food, and motion, but no change has been required in the basic conclusions.

tions we must consider the early reflex behavior as well as the more rational later behavior. We will deal successively with four separate components of the infant's behavior: clinging, suckling, curiosity, and fear.

THE REFLEX CLINGING RESPONSE

When it begins to fall, the newborn infant responds reflexively, bringing its hands together as if to clasp the mother's body. It also responds to tactile stimulation, grasping an object that contacts its hands. These and other early responses are appropriately described as simple reflexes, but as the infant matures the reflex is brought under more coordinated control.

The value concept is most useful after the behavior becomes more efficiently purposeful, but there is no fundamental reason why it cannot be used to describe even a reflex behavior. A reflex response differs from a conscious or voluntary response in two important ways. First, the response decision is made outside the conscious mind. Second, the response is completely automatic in the sense that no learning is involved.

To apply the "value" concept to reflex responses, we would have to postulate a simple value-driven decision system located wherever the reflex originates. We could then postulate that the stimulus generates a large negative value which the decision system can eliminate by activating the reflex response. Of course, to make the model consistent, we must assume that the system is designed so that it "knows" and does not have to "learn" that the negative value can be eliminated by the reflex response. Obviously, in this simple case the decision-system language is only a more complicated way of describing a simple reflex response, but the fact that the concept can be used in this way illustrates the versatility of the principle and suggests that mechanisms similar to the value-driven decision system may occur outside the conscious mind.

In the remainder of this discussion we will be concerned primarily with the conscious mind as a decision system. In this context, the decision-system interpretation becomes really applicable as the infant matures and the higher control centers of the brain gain more control over the response.

VALUE MOTIVATION FOR CLINGING

A value-driven system needs a well defined value structure that can be used to evaluate alternatives. What are the specific factors that determine the "value" of clinging? How does the infant decide what to cling to? The experiments show that a soft furry or fuzzy surface is particularly important. Other factors characteristic of the mother such as shape, rocking motion, angle of slant, warmth, and perhaps the maternal heartbeat also play a role.

How many separate value sensations are involved in the clinging behavior? At one extreme, each of the above characteristics (that help to identify the mother) could be associated with a separate, or independent, value. At the other extreme we might deal with only a single value which is achieved in proportion to the accuracy with which some combination of the different attributes (shape, warmth, texture, motion, etc.) match those expected for the mother.

Although the rule of simplicity suggests that we should avoid assuming more separate values than is really necessary, an experiment by William Mason with rocking mother substitutes (which will be described later) strongly suggests that the desire to be rocked is a separate innate drive. Individuals who fail to satisfy this drive in infancy often carry it as a residual drive into adolescence and adulthood. For the present, therefore, we will assume that the clinging behavior is motivated by two separate innate values:

1. An enjoyment of contact comfort, for which the value per unit time is in proportion to the body area that can be brought into contact with a soft furry object.
2. An enjoyment of rocking that is achieved by passive body motion. The value achieved per unit time is assumed to be a function of the amplitude and frequency of the rocking motion, so that the normal walking motion of the mother is the preferred form of motion.

We assume that the strength of the valuative motivation for each of these behaviors follows an independent appetitive curve. As with the taste of good food, the enjoyment of the activity is greatest initially and gradually fades (or is satiated) as the activity continues. For the very young infant the pleasure of contact comfort is almost insatiable. It dominates any conflicting curiosity drive, and the infant remains clinging to his mother. As the infant matures the desire for contact comfort gradually decreases and becomes more easily satiated. When this drive has been largely satiated, it may fall below the curiosity or the exploratory drive, and the infant will get off the mother for brief periods to explore.

The enjoyment of rocking motion, however, is not so easily satiated. Consequently, as long as the mother continues to move the child will choose to remain on her back. He will only dismount when she has stopped to rest. Since the enjoyment of rocking remains unsatiated, the child will learn to climb aboard voluntarily when the mother signals she is ready to move on.

THE SUCKLING REFLEXES

The suckling behavior, like the clinging behavior, begins as a simple instinctive behavior. The newborn human infant makes frequent sucking motions even when not in contact with any object. When the infant's lips make contact with an appropriate skin surface, for example the back of a hand or side of a breast, a distinct sucking response occurs. The response is more vigorous and prolonged if the object (which may be an adult finger or knuckle) has a shape which (like the maternal nipple) can slip into the infant's mouth. The infant has an instinctive response such that contact of the breast with one of the infant's cheeks will cause his head to turn so that the lips contact the breast.

The infant sucking response discriminates on texture. He will quickly reject a cloth or furry surface but will accept bare skinlike surfaces. The newborn human infant is not capable of coordinated search or even systematic groping for the nipple. His motions are almost random and essentially uncoordinated; so assistance from the mother is required to bring his lips into contact with the maternal nipple. When the infant is successful in coming into contact with the nipple the instinctive sucking response fills his mouth with a fluid, which in turn triggers an automatic swallowing response.

These early responses are traditionally described as reflex or instinctive behavior patterns. They resemble the built-in behavior patterns called tropisms in lower animals. But the relatively inept suckling behavior of the human infant and the higher primates (such as the great apes) is in marked contrast to the more efficient behavior of less advanced species. In part, this may be simply a consequence of less maturity in the higher primates at birth, but it also seems to reflect a general loosening of preprogrammed patterns, so that a more intelligent adaptive control over behavior can be exercised.

VALUE MOTIVATION FOR SUCKLING

Although the early suckling of the infant primate can be explained as a purely reflexive response, it gradually evolves into a coordinated behavior that can be better explained in terms of a value-driven decision system. What are the separate components of value which motivate this suckling behavior? Present information suggests that at least two and probably four separate values are involved. The values are associated with the following activities:

1. Sucking
2. Swallowing
3. Taste
4. Hunger

There is strong evidence that the sucking process itself is intrinsically enjoyable. The innate reward associated with the sucking may be related to the ability to obtain real suction (i.e., a reduced atmospheric pressure in the mouth) as well as the physical contact of the lips and inside of the mouth with the nipple. The assumption that the act of sucking itself achieves an innate value is supported by a number of observations.

Both human and nonhuman infant primates become attached to pacifiers that provide no nourishment. Human infants that are physically deformed so that they never swallow will nevertheless maintain a persistent sucking response. Such infants of course have to be nourished by other means, but the lack of nourishment associated with sucking does not appear to diminish the sucking response. On the other hand, sucking behavior can apparently be motivated by hunger as well as just the reward of sucking itself. A hungry infant will stop and cry if the sucking fails to produce the expected nourishment. He will also cry if he cannot find an appropriate place to suck.

Although there is little objective evidence about the way the values associated with swallowing and hunger interact in the hungry infant, the following hypothesis seems plausible. The hungry infant experiences a discomfort (a negative value) associated with his hunger. The act of swallowing produces an immediate but short-lived reduction in this discomfort. The swallowing itself may also be intrinsically pleasurable. Thus the infant quickly learns that swallowing is a cure for the pangs of hunger, but the reduction in hunger pangs resulting from swallowing is short-lived so the motivation quickly returns for the next swallow.

It seems virtually certain that taste plays an important role in the sucking behavior of a small infant. Indeed, the principles of system design suggest that infant taste preference should favor milk over other flavors that might be more attractive to an adult.

To fully describe the suckling value system we need to say something about how the various drives vary with time. The sucking drive seems to be an independent drive which is not directly related to hunger. Apparently the values associated with this drive can be achieved or satiated simply by sucking. Conversely, in a child that has already been fed, the sucking drive can be satiated only when the infant has achieved a sufficiency of sucking.

The hunger drive seems to depend on both the state of nourishment of the infant and the fullness of his stomach. A full stomach seems to turn off the discomfort of hunger. Moreover, if the stomach becomes too full, a new discomfort appears, so that the infant will be motivated to stop sucking whether or not the sucking drive itself is fully satisfied.

These considerations suggest that infants fed from bottles where the milk flows too easily may be motivated to overeat simply to satisfy the sucking drive. It also suggests that infants fed on an easy bottle will be more likely to develop thumb-sucking habits than infants fed on mother's breast or a slower bottle.

THE MOTIVATION OF CURIOSITY

The curiosity drive provides the motivation to find regularity and structure in the world environment. The drive is elaborately programmed to motivate efficient mental model building. The structure of this drive will be discussed in some detail in the chapters on Intellectual Values. For the present, we will consider only those aspects of curiosity that show clearly in the external behavior of the infant. The young infant is curious about his external environment and about the control of his own movements. As he matures he must learn about both.

His muscular system is designed so that it is potentially subject to rational control. The infant decision system has at its command the neural control levers necessary to initiate body movements, but the infant decision system does not know how to use the controls. Arms and legs move in a random uncoordinated way. When the infant is agitated, they move randomly but in a frenzy. Through experience, as the infant comes to learn more about the control system, a greater degree of purposefulness and coordination is observed in the body movements.* Motor control decisions may first become important to the infant as they influence his position on the maternal breast. This experience produces mental associations leading to a gradual improvement in motor control.

The infant is also learning about the environment. The initial learning involves the recognition of regularities in sensory input—for example, the gradual recognition of the mother as a single entity responsible for a wide array of sensory information. The recognition of regularities is essential to developing an understanding of the world.

The efficient construction of a world model requires a somewhat orderly approach to learning. Things that can be fitted neatly into an existing world model are most easily learned. The infant's model of the world begins to expand much like a jigsaw puzzle as new pieces of information fall into place. The infant primate begins his model of the external world at his mother's breast. The model gradually expands to include more of her body. Only later do outside objects take on meaning. For this reason, the focus of the infant's curiosity shifts with time. His interest is concentrated at the *current* boundaries of the jigsaw puzzle that is his world model. Items that are either well understood or too strange to fit into the model are of less current interest.

This expansion of the sphere of curiosity has a direct bearing on the infant's physical behavior. Initially while the infant's curiosity is concerned primarily with the mother, it does not motivate him to leave her protective grasp. But as the circle of curiosity enlarges, it motivates the infant to leave the mother to investigate interactions with the larger world. This expansion of the sphere of curiosity coincides with and complements a decline in (or easier satiation of) the "pleasure" of clinging. Thus both factors contribute to the rise in the exploratory behavior of the maturing infant.

THE MOTIVATION OF FEAR

If we were designing a motivation system to guide animal behavior, one of the most fundamental requirements would be an avoidance drive. The purpose of the avoidance drive is to motivate the animal to stay away from dangerous situations. Nature's value system includes such an avoidance drive. We call it

---

* This should not be interpreted to mean that all improvements in body control are just a consequence of learning. During this time the nervous system is continuing to mature. Large sections of the brain are as yet immature and nonfunctional. Consequently, the gradually improving coordination undoubtedly reflects both learning and maturation.

"fear." Fear is a negative or unpleasant component of the value system, so the animal will try to avoid fearful situations.

Traditionally we think of fear in connection with "real" danger. Fear under circumstances that are not really dangerous is often considered irrational. Although this traditional view may be valid with regard to adult behavior, it is very misleading when we are concerned with the behavior of an animal or an infant.

It is sometimes asserted that fear operates like a conditioned reflex, so that we fear situations similar to those in which we have been hurt in the past. But, an avoidance behavior which did not operate until *after* an animal had been hurt would be an inefficient survival mechanism. An efficient avoidance response must operate in threatening situations regardless of whether the animal has previously been hurt in a similar situation. Almost all animal species exhibit an automatic fear response to certain specific genetically determined danger clues.

Experiments with human infants in the range up to three years have confirmed that this also seems to be true of humans. Experiments have shown that at least the following specific stimuli seem to be intrinsically fearful: *

1. Falling
2. A sudden noise
3. An object which seems suddenly to loom large near the infant
4. Animals
5. A large moving object
6. A strange or unfamiliar object or event
7. Being alone
8. Darkness

Fear in the infant begins as an innate response to specific genetically inherited stimuli. Nature has provided a remedy for the infant's fear response. Evidently the remedy is provided by the same "contact comfort" sensations that motivate the infant's clinging response. The infant finds relief from the unpleasant fear stimulus when he is able to cling to the mother's body. The degree and speed of relief from fear may again be proportional to contact area. It seems that the relief of fear may be even more rapid and complete if he is also able to suckle or be rocked. Thus, in a frightening situation the infant urgently seeks his mother and clings more tightly than usual. The mother may seek to reassure him by rocking or offering him her breast.

Thus the relief of fear provides a very important *independent* motivation for clinging. At least in theory, it appears that fear *alone* could have provided an adequate motivation for clinging. But if fear had been used as the sole motivation, then separation from mother contact, in and of itself, would have had

* See Bowlby (5), Chapters 5–10, for a review of relevant research.

to be *intrinsically* fearful. Although this might be true for very small infants, it seems unlikely that "fear" is the sole motivation for clinging. A behavior as fundamental to survival of the species as clinging will probably be associated with several independent value components, to provide greater assurance and redundancy in motivating the necessary behavior.

The existence of a "fear" response seems to be universal in vertebrate animals. It is usually accompanied by a more rapid heartbeat, which prepares the animal for intense activity. But the fear reaction is generated by a wide variety of different physical stimuli, which may be different for different species and may vary with the age of the individual. In the primate infant fear automatically generates an instinctive vocal distress signal (or crying) which notifies the mother of a problem. As the infant acquires physical coordination, fear motivates a "rational" response in the infant which causes it to seek the comfort and protection of the mother. In other species the response is somewhat different, birds fly away, turtles retreat into their shell, other animals freeze in a motionless state.

How does the primate infant learn that contact with the mother will relieve the unpleasant symptoms of fear? The answer seems obvious. The newborn infant inevitably experiences many frightening situations. Birth itself is an unfamiliar and therefore fearful experience. The primate infant experiences many occasions when his grip on the mother begins to slip. On each such occasion the infant is fearful of falling and cries. The mother responds by clasping the infant to her body. Each time, with renewed body contact, the fear symptoms fade. Thus, the infant quickly learns to associate mother contact with relief from fear.

The interaction between the two drives, fear and curiosity, is particularly noteworthy. A curiosity drive unrestrained by fear could produce a particularly dangerous kind of behavior. But curiosity, counterbalanced by fear of the unfamiliar, produces a cautious and conservative form of exploration. The survival benefits of this combination are obvious.

The level of fear induced apparently rises rapidly as a strange object is approached, whereas the attractiveness of an object (either because of curiosity or hunger) seems to be relatively independent of distance. The exploring animal thus reaches a distance where there is an equilibrium between the two drives. At this distance he may circle the object, gradually approaching as familiarity results in a fading of the fear.

The interaction between fear and curiosity often produces an oscillatory behavior. Too close an approach causes fear to rise rapidly. The animal then retreats while the fear subsides. Numerous familiar examples of such behavior are seen in the human infant—for example, the timid child approaching a high dive platform. This basic relationship between the drives of curiosity and fear is very widespread among the higher animals. Caroline Loizos reports a typical example in the play behavior of a three-year-old chimpanzee:

> I bounce a tennis ball in front of the cage several times so that she hears as well as sees it and I place it inside on the floor. She backs away, watching the ball fixedly—approaches with pouted lips, pats it—it rolls. She backs hurriedly to the wall. Hair erection . . . J. pokes at it from a distance, arm maximally extended, watching intently; looks at me; pokes ball and immediately sniffs finger . . . She dabs at ball and misses, sniffs finger; she backs away and circles ball from a distance of several feet, watching it intently. Sits and watches ball . . . (pause of several minutes) . . . walks around ball. J. walks past the ball again even closer but quite hurriedly. She lifts some of the woodwool in the cage to peer at the ball from a new angle, approaches ball by sliding forward on stomach with arms and legs tucked underneath her, so that protruded lips are very close to ball without actually touching it. Withdraws. (20)

As familiarity is gained the chimp becomes less cautious and more playful, until finally after about thirty or forty minutes all fear of the ball seems to have vanished.

It would be intresting to combine such behavioral observations with electronic monitoring of heartbeat and other physiological functions that might be objective indicators of "fear." Such experiments should provide a more reliable understanding of the relationships between the environmental stimuli, the innate values, and the conscious behavior.

In the primate infant the interaction between fear and curiosity causes the maturing infant to leave his mother for cautious exploration and then return to be comforted. The reassuring presence of the mother allows the exploration to proceed without risk of prolonged exposure to excessive fear, because the child knows that the remedy for fear is at hand.

Near the beginning of this discussion of fear, we listed eight (there are undoubtedly others) intrinsically fearful circumstances. However, some of these elements contribute to fear in quite different ways. The first five (falling, sudden noise, looming objects, animals, or large moving objects) are probably just simple instinctive fear inducers that play a role similar to the instinctive "releasers" described by ethologists. The last two elements, darkness and being alone, in and of themselves are apparently only slightly fearful, but when they occur in combination with any of the others, they seem to magnify or multiply the level of fear.

The sixth element, fear of the unfamiliar, is a very special interaction. The definition of what is familiar constitutes the beginning of a learning process. Thus, when we specify that one of the primary drives depends on what is familiar, we have specified an interaction such that the instinctive drive may depend on prior rational thought. In this respect, fear differs from the previously discussed drives. As the animal matures, the innate or instinctive causes of fear become less important and the causes of fear become more closely linked with experience and understanding.

At present, it is unclear how this transition occurs. Perhaps the nerve pathways to the frontal lobes tap directly into areas of the conscious mind where the unfamiliar is recognized. Alternatively, the instinctive fear response may be subject to associative learning or operant conditioning which takes place independently within the frontal lobes of the human motivation system, so that if familiarity with a strange object is obtained without reinforcing the fear response, the fear response may fade. Conversely, if the instinctive fear of the unknown is confirmed by loud noises, looming objects, or painful experience, the strange object may become permanently "fearful" by association. At present it is not even clear whether *real* emotional "fear" of a situation can be engendered as a result of purely rational understanding, without any prior conditioning experiences. Of course the traditional "commonsense" view is that such purely "rational" fear is very common.

One of my own experiences suggests that the commonsense view is probably correct. On one occasion I was alone in the experimental area of the MIT synchrotron. The synchrotron produces high-energy radiation called gamma rays. The gamma-ray beam cannot be detected without special instruments, but like any high-energy radiation, it can do serious biological damage. Suddenly I realized that I was directly in the path of the beam, and I was not sure whether the beam was on. As I moved quickly away I developed an emotional response which subjectively seemed like acute fear. Of course, I might have become afraid because I ran, rather than vice versa. But it would be interesting to try to resolve the issue with appropriately controlled experiments.

THE ATTACHMENT SENTIMENT

The sensation of "contact comfort" plays a central role in infant behavior. It is important not only in directing the clinging drive, but also in the control of fear. As the infant experience unfolds, these instinctive valuative responses, which originally were applicable to any source of "contact comfort," become specialized so that they are specifically attached to particular individuals or objects. Normally, of course, the primary attachment is to the mother, for she is the only figure who provides satisfaction for all of the drives. However, secondary attachment to the father and other relatives or neighbors who are frequently present are also common. The secondary attachments do not seem to detract from, and may actually enhance, the primary attachment to the mother.

Once the attachment has been specialized to specific individuals, substitutes will not do. Separation of the infant from the mother can be a source of acute distress. If an infant has become attached to a specific blanket or fuzzy toy, no substitute will do. The drive which was originally general or abstract becomes both more intense and specific as it becomes attached to a particular object or individual. The act of touching, as well as being touched, obviously plays a key role in the development of such attachments.

Once attachment to a specific individual has developed, the symptoms of separation will be much the same as if *all* sources of contact comfort had been withdrawn. Such separations are emotionally so intense that they can have permanent consequences for the development of personality. During laboratory experiments in which a rhesus monkey is separated from the mother the distress of both the mother and the infant is obvious. But the seriousness with which such separations are viewed is perhaps most clearly demonstrated by the fact that unaffected mothers and infants in neighboring cages also display a high level of distress.

GENERAL COMPLEXITY OF HUMAN VALUES

By now the careful reader will have noticed an obvious redundancy or overlap in the motivating drives. For example, the clinging behavior of the primate infant is explained in part as a positive enjoyment of "contact comfort." But it also is explained as a mechanism for avoiding the unpleasantness of fear. Similarly, suckling is motivated by many separate drives—hunger, a sucking drive, taste, and even relief from fear. Why has nature provided such complexity and redundancy in the motivating drives? Wouldn't a simpler drive structure be adequate? For example, why not motivate the suckling behavior using only the hunger drive?

There are several possible answers to these questions. First, in terms of our theoretical development, we found that it is usually impossible to provide satisfactory behavior in a decision system without using a complex set of value components. Even when the main objective of the decision system seems simple, it is usually necessary to provide a number of subsidiary value components to avoid certain forms of ineffective or even silly behavior. It is clear that evolution has repeatedly encountered this problem of inappropriate behavior and has gradually developed a complex set of interacting value components which motivate a relatively successful pattern of behavior.

In the theoretical section we made the point that there is a design trade-off between the complexity of the value structure and the required rational analysis capacity. It is often possible to compensate for a limited intellectual capacity by providing a more complex and detailed value system. A complex time-dependent value system can be used to lead a simple organism through a complicated series of steps which would require great intelligence and knowledge if only the *final* objectives were represented in the value system.

The human value system is probably more complex than is really needed for an intelligent human adult. But this is not surprising since the basic structure of the human value system evolved in animals of lower intelligence. Although such a detailed value system may not be really necessary for a creature of human intelligence, it does not seem to interfere with efficient behavior and it may actually help; so the main features of the complex primate value system are retained in man. It seems likely that the only uniquely human elements of the value structure (i.e., those not shared with any other primates) may be those that developed during primitive man's unique experience as a ground-living band of hunters.

Of course the infant motivation system must be relatively detailed even for humans because the infant has not yet acquired the knowledge necessary for good rational decisions. To carry the infant through this awkward period (until the decision system has the knowledge needed for good decisions) nature has provided a combination of "reflex" responses and a very detailed infant value system. For example, the sucking reflex, which later transforms into an innate value of sucking, gets the suckling behavior under way before it could be rationally motivated by a "hunger" drive.

A complex value system makes it possible to produce more refined behavior patterns. For example, the inclusion of a value associated with rocking or passive motion helps to synchronize the exploratory activities of the infant with those occasions when the mother is resting and not otherwise engaged. The use of clinging as a method of alleviating "fear" helps to ensure that the infant will quickly seek the safety of the mother in the face of a frightening situation. Thus, the apparently "redundant" drives actually produce a more elaborate and effective pattern of behavior than would be possible with the "contact comfort" drive alone. As we proceed with the exploration of primary human values we will find repeated examples of such redundancy.

We must, however, avoid any assumption that the present value structure is either unique or perfect. From an evolutionary perspective the only requirement is that it be sufficiently good to avoid extinction of the race. Almost all species exhibit certain forms of behavior which contribute little or nothing to survival. For these reasons, one obviously could not deduce the innate human value structure on an a priori basis. It can only be determined from observation and experiment.

EXPERIMENTS IN EMOTIONAL DEPRIVATION

The preceding sections make it clear that the concept of innate values can be used to provide an "explanation" of infant behavior. But because the assumed value structure was developed for the purpose of explaining the behavior, it is still unclear whether the value interpretation of behavior has any real predictive power. To demonstrate effectiveness as a predictive theory, we need to consider the operation of the values in an environment different from the one in which our hypothesis about the specific value structure was developed. To provide such a comparison we will consider the behavior of infants that are deprived of normal mothering and contact comfort.

In September 1958 at a meeting of the American Psychological Association, William A. Mason reported on a series of experiments showing that animals raised in laboratory cages develop abnormal behavior patterns. Two groups of rhesus monkeys, each about 2½ years of age, were used in the experiments. The restricted monkeys were separated from their mothers shortly after birth and were raised in cages where they could see and hear other monkeys but could not engage in physical contact. The comparison group was captured in the wild and brought to the laboratory at an estimated age of 20 months. One month before the experiments began the wild (or feral) monkeys were housed in cages like those of the restricted group. The results of the experiments were reported in a series of papers during the early 1960s (17).

The restricted monkeys showed symptoms similar to disturbed mental patients. In unfamiliar surroundings the restricted monkeys crouched, sucked their thumbs or toes, clasped themselves, and engaged in rocking or other stereotyped behaviors. The feral monkeys responded to the same situation with more movement and a more organized exploratory behavior. The stereotyped behavior patterns such as thumb sucking, self-clasping, and rocking were absent or at least extremely rare in the feral monkeys. In human terms the restricted monkeys seemed insecure, timid, and easily frightened.

Quantitative comparisons showed a wide range of behavioral inadequacies. The restricted group, particularly the males, were sexually inadequate. Although they were responsive to sexually active females, their attempts at copulation were uncoordinated and generally ineffective. They participated in little social grooming and engaged in more fighting. In the case of monkeys raised in complete social isolation where they could not even see other monkeys, then the abnormal responses became even more severe. Their behavior in any unusual situation involved cringing or crouching; they also engaged in head banging and self-clasping.

Mason's results were confirmed by experience at the University of Wisconsin. The rhesus monkeys that had been raised by artificial wire and cloth mothers (or entirely without mothers) showed a similar range of abnormal behavior. It is apparent that the isolation of monkeys in the laboratory constitutes an artificial environment which was not encountered during the evolutionary design of the species. In the context of this artificial environment the innate values seem to motivate "abnormal" and nonadaptive behavior patterns.

In this connection, one specific experiment was particularly revealing. Mason observes that under natural conditions the mother provides "a great deal of passive rocking movement for her infant in the course of her routine activities—we might conjecture that an infant monkey or ape deprived of such passive stimulation may supply it for himself through self-rocking or similar repetitive activities" (18).

To test this hypothesis he compared two groups of monkeys, both separated from their mothers at birth. One group was reared on a moving cloth-covered dummy and the other on a similar but stationary dummy. All of those reared on the immovable dummy developed persistent stereotyped rocking, while *none* of the monkeys reared on moving surrogates rocked! In addition, the monkeys reared on moving or robot dummies were less fearful and more active than those reared with stationary surrogates: "Robot-reared monkeys spend less time in contact with the social surrogate; they more often move about the cage and they are quicker to approach and to interact with people. . . . We can be sure that the differences were produced by adding movement to one of the dummies but beyond that it is difficult to be more precise" (18).

Mason described what he called the "primate deprivation syndrome," the symptoms of which are (1) abnormal postures and movements, for example, rocking, (2) motivational disturbances, for example, excessive fearfulness or arousal, (3) poor integration of motor patterns, for example, inadequate sexual behavior, and (4) deficiencies in social communication, for example, threat by an aggressive animal does not produce withdrawal by a subordinate animal. He noted that the addition of movement in early life did away with at least the first two symptoms of the syndrome. But what about integration of motor patterns and deficiencies in communication? *

* Summarized from an excellent review article on abnormal primate behavior by G. Mitchell (19).

CONSEQUENCES OF DRIVE FRUSTRATION

Having developed a rather detailed hypothesis concerning the instinctive drives of the primate infant, it is appropriate to ask what explanation the theory can provide for Mason's "primate deprivation syndrome." What insight does the theory provide concerning the abnormal behavior of animals reared in isolation?

We recall that the infant is motivated by certain instinctive "appetitive" drives: clinging or contact comfort, sucking, and rocking. It is characteristic of an appetitive drive that the drive becomes progressively more intense unless it is satisfied. The laboratory infant raised in isolation is unable to satisfy his instinctive drives for contact comfort and rocking. If (as is likely) he is raised on relatively "easy" nipples he may not even be able to adequately satisfy the sucking drive. Thus the laboratory infant is left in a state of intense frustration. In all probability, each of the unsatisfied drives reaches its maximum possible level of intensity; and there it remains. When any "fearsome" events occur, the isolated infant lacks the source of comfort that is essential to reduce the fear. Thus the infant is likely to spend a large fraction of its time in a state of intense terror. Lacking a mother to whom he can cling, he grasps himself. Lacking the rocking provided by a mother he rocks himself. Lacking a maternal nipple he sucks his thumbs, or his toes. With so many drives continuously at almost maximum intensity the infant's whole intellectual and physical attention is devoted to drives which are impossible to fulfill in his artificial environment.

What consequences can be expected from prolonged existence in a state of intense drive frustration? Of course, the infant cannot verbalize his mental distress, but we may guess at the consequences from our experience when adult drives are frustrated. When adults suffer prolonged thirst or hunger, hallucinations are common and the drives seem to grow into obsessions. In the case of hunger and thirst there are physical limits to drive deprivation. If the drive is frustrated too long, physical destruction of the body will follow, so the experiment in drive frustration is automatically terminated. Except for suckling, the infant's drives are not essential to his survival in the laboratory cage. Thus the deprivation can be continued indefinitely. The damage is psychological rather than physical.

When the infant is obsessed with intense unsatisfied appetitive drives his curiosity gets little chance to operate. The development of his world model is retarded, and the circle of the "familiar" fails to expand at the normal rate. More things remain strange, unfamiliar, and fearsome. To the extent that certain features of his "world model" may develop best when the intellectual apparatus is at a particularly "receptive" phase of development, the quality of his world model may be irretrievably damaged. Thus the infant raised in the deprived environment may show a permanent loss in intellectual capacity. In addition, if he fails to acquire practical social experience, he may fail to learn how his own innate value system will respond to social stimulation. The motivational effectiveness of the social value system can thus be negated.

In cases of extreme deprivation, however, the damage can be much more severe. Nature has decreed that when other drives are frustrated, a new drive comes into play. We call it "anger." So far we have not discussed anger since it was not essential to a discussion of the normal young infant. But, as the infant matures, "anger" begins to play a more important role. In the normal evolutionary environment anger serves a useful motivational role. Especially in simpler animals, it intensifies the physical effort needed to overcome obstacles. Much like "fear," it provides an automatic mobilization of body resources. Normally, anger is a directed drive. It is focused on the offending objects that seem to be frustrating the achievement of other goals or values.

As anger becomes intense it develops large negative values that can be most effectively turned off by attacking the offending object. As the deprived animal matures and the "anger" (or "frustration") mounts, the desire to attack and destroy becomes intense. In the absence of other offending objects to attack, the isolated animal alternately attacks himself and the offending cage. He bites at his flesh and bangs his head against the cage. Thus in the artificial environment normal instinctive drives lead to maladaptive and even self-destructive behavior.

The purpose of this discussion has been to show that the "value theory" has real predictive power. Within the present theory, the maladaptive behavior of the deprived animals is "predicted" as a direct consequence of the value structure that was needed to "explain" normal behavior.

So far we have discussed only the obvious automatic consequences of intense drive frustration. These general consequences would be expected even if the instinctive drives continued to develop in accordance with the normal pattern of biological maturation. But there is evidence that the prolonged frustration of instinctive drives also interferes with the normal maturation of the drive structure itself. It is not surprising that it should do so, since the emotional deprivation falls far outside the range anticipated in the evolutionary design. Of course, the specific way the development is affected could not be predicted a priori. But the evidence suggests that when instinctive drives are chronically frustrated the drives try to adapt by becoming even more intense and finally solidify into a permanent obsession. Whereas the instincts for clinging, rocking, and sucking normally wane as the animal matures, if the drives are not appropriately satisfied these infantile drives can remain as a permanent part of the personality. It is possible that other drives which should emerge later in the maturing individual may be retarded or distorted. Thus future behavior can be influenced in unpredictable ways that may not even be susceptible to psychological treatment.

RELATIONSHIP TO HUMAN INFANT MOTIVATION

The preceding analysis of infant motivation was based entirely on experiments with nonhuman primates. Because of obvious problems in conducting controlled experiments with human subjects, it is much more difficult to provide convincing experimental evidence concerning the value structure of the human infant. Nevertheless, extensive research has been done on the normal behavior of the human infant, and the results appear to be generally consistent with a value structure analogous to the one described for the nonhuman primates.

Margaret Mahler and her collaborators have reported recently on a very extensive study of this kind (16). Mahler's observations on the normal development of the infant-mother relationship were carried out over a period of more than fifteen years in a children's center where the mothers came with their children several times a week. To facilitate observations of normal behavior, the center was organized to provide a natural social environment in which the children could play under the mother's supervision.

Although Mahler and co-workers have described their results using the traditional language of Freudian psychoanalysis, the actual experimental findings seem to be compatible with an innate motivational structure similar to the one discussed for the nonhuman primates. The results show clearly the evolution of interest as the infant gradually expands his mental model of the world. In the beginning the infant's attention seems to be focused internally on purely physical sensations such as hunger and the new experience of bodily elimination. Gradually his attention moves outward and the infant begins to distinguish between himself and his mother. As his mental model develops and the range of his curiosity expands, the infant's attention moves beyond the mother to the outside environment. Like the nonhuman primates, the human infant returns repeatedly to the mother, and he turns to her most strongly in times of stress. In the value-theory formulation we explained this behavior in terms of an innate appetitive value associated with contact comfort and the relief of "fear." Mahler and co-workers describe this response to the mother as an "emotional refueling"—a description which clearly reflects the appetitive nature of the underlying motivational drives. Mahler's observations suggest that the "refueling" process for the human infant can be accomplished, at least to a limited extent, by mechanisms other than contact comfort. The mother seems to contribute to an enhanced emotional security even from a distance through her facial expressions and her words. However, the most effective method of comforting the child still seems to be through direct physical contact.

Each individual child seems to respond to internal motivational signals that define his preferred pace for reducing his dependence on the mother. If the mother tends to be aloof, the child will make great efforts to gain more of her attention; if the mother is too attentive, or tends to "smother," the child will try to compensate by ignoring the mother and thus increasing the emotional distance. Although the innate motivations of the human infant are undoubtedly somewhat different from those of other primate infants, it seems clear that the motivation involves innate appetitive emotions that operate on principles similar to those described for the infant primate.

The Mahler group has identified a number of rather distinct phases in the development of the human infant. Some of the phases may be associated with simple maturation of the brain as certain embryonic parts become functional. Some of the phases seem to correspond to characteristic stages in the development of the child's mental model of the world. Still others seem to represent biologically programmed changes in the innate value structure (or motivation system) that are designed to motivate an appropriately cautious exploratory behavior for each phase of development. Although Mahler and co-workers allude to such factors in their discussion, the prevailing theories in psychoanalysis do not encourage such distinctions between innate values and the mental model, so these distinctions are not included as a formal part of their theory. It would be interesting to review their results to determine to what extent the experimental data might justify such a classification of the developmental phases.

CHILDHOOD EXPERIENCE AND PERSONALITY

It is now widely recognized that both personality and intellectual potential seem to be largely determined before an individual reaches normal school age. The relative importance of heredity versus experience is still very unclear but (to the extent that experience has an influence on personality) the greatest effect must occur in infancy and early childhood. The deprivation experiments with other primates rather strongly suggest that early childhood experiences should also have an important effect on human personality. Early childhood experience centers on two key emotional developments, the relationship of the infant to the mother and the relationship of the child to siblings and other playmates. It seems probable that these two experiences may be crucial in the early formation of personality.

Evidence both with humans and other primates indicates that the experience with the mother can have an important effect on the child's sense of personal security. A child whose relation with the mother has been warm and loving will tend to be secure and confident. A child who has experienced frustration in that relationship may be insecure and unduly cautious.

The second pivotal experience common to most primates concerns the struggle for dominance or leadership with siblings and playmates. The success experienced in these early sibling competitions may have an important effect on the balance between the submissive and dominant personality traits that will emerge in the adult personality.

The formation of personality is a complex process and many independent factors, both genetic and environmental, are involved. Obviously there are many different personality characteristics that can be used to analyze (or represent) differences in personality types (8, 9, 10, 11). Nevertheless, in considering the effect of early childhood experience these two dimensions (personal security and dominance) seem to be of particular relevance.

On this basis, M. M. Benyamin of the University of Maryland has recently proposed a "branch" representation of personality types (see Table 8.1) which highlights these two dimensions of childhood experience (3). These two personality dimensions are closely correlated with some of the major factors identified in formal personality studies, but as always there is no perfect correspondence between the factors in different personality classifications.*

The characteristics indicated in the table of course represent extreme or relatively "pure" personality types. Obviously most real personalities fall at more intermediate points. The chart suggests that a surprising number of personality traits may be correlated with these two key factors—"personal security" and "degree of dominance."

---

* Unfortunately the nonlinear and qualitatively different *manifestations* of Benyamin's "dominance" factor depending on the "security-insecurity" factor precludes an easy comparison of the "branch" hypothesis with results of formal factor analysis that otherwise might be obtained by a simple rotation of coordinates (for example, see Kasselbaum [15]). The nonlinear manifestation of one of the major dimensions of personality may help to explain some of the ambiguity of the factor analysis results. The best present evidence supporting the branch hypothesis comes from experience in psychiatric counseling where family environment is correlated with personality disorders. For example, see Storr (21).

TABLE 8.1 *

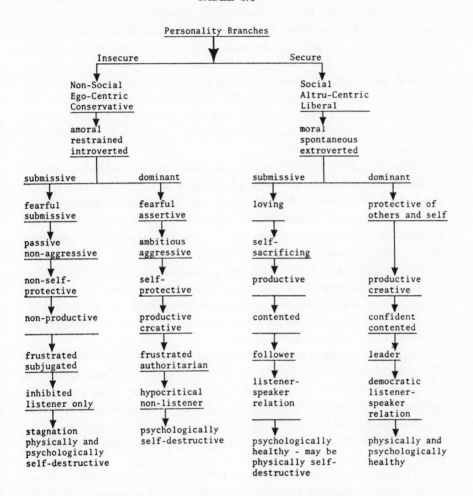

* Reproduced by permission of M. M. Benyamin.

This branch representation was originally proposed by Benyamin for the analysis of political systems and political personalities. Naturally most political leaders fall closest to the dominant-secure branch. A few, however, correspond to the dominant-insecure or introverted branch. De Gaulle, Nixon, and Stalin were notable examples of this type. In such leaders, fear of criticism, insecurity, and excessive ambition combine to produce a restrictive authoritarian approach to leadership. Obviously very few leaders correspond to either of the two submissive branches.

The personality branches in Table 8.1 correlate both with sex and with age position in the family. The dominant types are more likely to be male and an oldest or only child. The submissive types are more likely to be female or a younger child. The insecure types are somewhat more likely to be male.

Extreme cases of the insecure-submissive branch account for much of the pathological depression found in mental hospitals. Extreme cases of the insecure-dominant branch produce another pathology called the schizoid personality. Such individuals may be highly successful and creative, but they are very unlikely to be content or socially well-adjusted. An excellent description of some of these extreme personality types can be found in the book *Human Aggression* by Anthony Storr (21).

The effects of early experience on personality implied by the branch representation is particularly interesting because of its superficial "rationality." For example, frustration in the relationship with the mother discourages spontaneous behavior and teaches caution and suspicion. Frustration in the early competition for dominance teaches that efforts at dominance are unrewarding and encourages a submissive and loving behavioral strategy. Of course the personality differences seem to be far too profound to be attributable to learning in the rational mind. The effects are more nearly as if the innate human "value system" itself undergoes a learning process similar to *operant conditioning*. If such "learning" occurs within the "value system" it is probably most important in the early formative years.

In Part I of the book we treated the innate value system as if it were completely determined by the genetic "system design." But in principle there is no reason why the value system itself should not be capable of learning and adaptation. The operant conditioning concept may provide a useful model to describe this type of "learning." As we proceed, we will find other evidence which suggests the existence of "learning" within the human value system itself.

The question may arise whether such responsiveness of the primary value system to environmental experience may destroy the concept of an innate built-in value structure. But the value system does not respond passively like soft clay that can be arbitrarily shaped by the environment. To the extent that such responsiveness exists, it takes place within limits and in accordance with rules that are determined by the basic genetic design. For example, the response of a system to operant conditioning depends on which experiences are defined as "aversive" and which are defined as "reinforcing." These fundamental distinctions between aversive and reinforcing experiences are a vital part of the original system design. In formal mathematical terms the introduction of such a responsiveness does not make any real difference in the basic design concept; it simply increases the complexity of the "value function" required to calculate the values from the experience of the organism.

IMPLICATIONS FOR INFANT CARE

In recent years there has been a tendency to assume that the baby's "real" needs are limited to his physical needs. If he is not wet, not hungry, and not stuck with a pin, then it is assumed that his "real" needs have been met. The experiments with other primates show conclusively that this is not so. The need to be held, to be rocked, and perhaps (for the human infant) to be talked to may be just as real as the need to be fed. Like the infant monkey, the human infant needs the comfort of a "mother" to quiet his fears, and to give him the courage to explore so that further normal development can occur.

The uneducated peasant mother and most mothers in primitive societies respond naturally to the psychological needs of the baby. In many primitive societies the mother provides some sort of back carrier so that the infant can ride with the mother as she goes about her work. The peasant woman provides a cradle to rock her baby when he is restless and not in her arms. She can be patient and watchful as the baby begins to explore. Often the modern woman provides a stationary crib and a confining playpen which restricts exploration. She may even provide a mechanically supported bottle so that the baby is emotionally deprived even in suckling. These mistakes illustrate the potential danger of partial or incomplete scientific knowledge. Many of the mistakes may be the result of "objective analysis" by male doctors and "child experts," who lack a woman's maternal instinct.

The emotional attachment of the infant to the mother as well as the mother to the infant is facilitated by repeated and frequent body contact. This two-way bond provides the emotional foundation on which the child's personality can mature. If the child is offered the security of such a bond, the child's dependence on the mother will gradually wane during maturation so that a confident and secure personality will emerge. But if the bond is denied and normal attachment fails to occur, the child will be fearful and insecure and may remain overly dependent on his mother.

With the growing tendency of women to engage in full-time employment away from home, there has been pressure to provide baby-sitting help and day-care centers so that the mother can return more quickly to her career. Our growing understanding of the attachment process suggests that, over the long run, both mother and infant may pay a high emotional price for such services. Extensive psychiatric research on the effects of mother-child separations seem to confirm this concern (1, 2, 4, 5, 6, 7, 13). In the human infant the child-to-mother bond is extremely intense until the child reaches about three years of age. Prior to this time, the normal infant will refuse to accept any separation from the mother. Forceful separation will result in extreme distress. If the separation is prolonged it will result in anger against the mother and the consequent weakening of the attachment.

Human emotions include no tie stronger than the mother-child bond. This bond and the potential consequences of disturbances in the bond are of great importance in understanding both normal and abnormal human behavior. Probably the best and most detailed treatment of this subject is provided by John Bowlby in his two-volume work *Attachment and Loss*. The first volume, *Attachment* (4), was published in 1969. The second, *Separation, Anxiety, and Anger* (5), was published in 1973. In the course of his analysis of the processes of attachment and loss, Bowlby has developed a rather complete theory of behavior. Although his approach to the problem is completely different from the one taken here, the structure of his theory seems remarkably similar. The Bowlby books are strongly recommended for readers wishing to pursue this aspect of instinctive human values.

The human infant is motivated by instinctive values that are a legacy of the primitive past. The values may not be entirely appropriate to our modern society, but they are genetically fixed in our human inheritance. They cannot be changed without genetically redesigning man. While these innate drives may seem inappropriate or irrational in our modern environment, we can ignore them only at our peril.

REFERENCES

1. Ainsworth, M. D. "The Development of Infant-Mother Interaction Among the Ganda." In *Determinants of Infant Behavior*, vol. 2. London: Methuen; New York: Wiley, 1963.

2. Ainsworth, M. D., *Infancy in Uganda: Infant Care and the Growth of Attachment*. Baltimore, Md.: The Johns Hopkins Press, 1967.

3. Benyamin, M. M. "The Branch Theory and Mental Health." An Address to Maryland Eastern Shore Pharmaceutical Association, Jan. 20, 1974.

4. Bowlby, John. *Attachment and Loss*, vol. 1: *Attachment*. New York: Basic Books, 1969.

5. Bowlby, John. *Attachment and Loss*, vol. 2: *Separation, Anxiety, and Anger*. New York: Basic Books, 1973.

6. Burlingham, Dorothy T., and Anna Freud. *Young Children in Wartime*. London: Allen and Unwin, 1942.

7. Burlingham, Dorothy T., and Anna Freud. *Infants Without Families*. London: Allen and Unwin, 1944.

8. Cattel, Raymond Bernard. *The Scientific Analysis of Personality*. Baltimore: Penguin, 1965.

9. Eysenck, Hans J. *The Structure of Human Personality*. London: Methuen, 1953.

10. Gough, Harrison G., and Alfred B. Heilbrun. *The Adjective Check-List*. Palo Alto, Calif.: Consulting Psychologist's Press, 1957.

11. Guilford, Joy P., and Wayne S. Zimmerman. *The Guilford-Zimmerman Temperament Survey.* Beverly Hills, Calif.: Sheridan Press, 1955.

12. Harlow, H. F., and R. R. Zimmerman. "Affectional Responses in the Infant Monkey." *Science* 130 (1959): 421–32.

13. Heinicke, C. M., and I. J. Westheimer. *Brief Separations.* New York: International University Press, 1966.

14. Hinde, R. A. *Development of Social Behavior.* New York: Academic Press, 1971.

15. Kasselbaum, C. G., A. S. Couch, and P. Slater. "The Factorial Dimensions of the MMPI." *Journal of Consulting Psychologists* 23 (1959).

16. Mahler, Margaret S., Fred Pine, and Anni Bergman. *The Psychological Birth of the Human Infant.* New York: Basic Books, 1975.

17. Mason, William A. "The Effects of Environmental Restriction on the Social Development of Rhesus Monkeys." *Primate Social Behavior,* pp. 161–173. New York: Van Nostrand Reinhold Co., 1963.

18. Mason, William A. "Early Social Deprivation in the Non-Human Primates: Implications for Human Behavior." *Environmental Influences,* pp. 20–100. New York: Rockefeller University and Russell Sage Foundation, 1968.

19. Mitchell, G. "Abnormal Behavior in Primates." *Primate Behavior: Developments in Field and Laboratory Research,* edited by L. Rosenblum. Vol. 1, pp. 195–249. New York: Academic Press, 1970.

20. Morris, Desmond, ed. *Primate Ethology.* Chicago: Aldine, 1967.

21. Storr, Anthony. *Human Aggression.* New York: Atheneum, 1968.

Chapter 9

Structure of "Selfish Values"

> The human mind has certain innate or inherited tendencies which are the essential springs or motive powers of all thought and action.
> WILLIAM McDOUGALL (2) *

T HIS CHAPTER is concerned with innate "Selfish Values." These values deal with obvious and uncomplicated aspects of our lives and they are functionally related to the survival and well-being of the individual. Because of their simplicity they provide good examples for developing our theoretical understanding of the innate human values.

ILLUSTRATIVE VALUES IN THE HUMAN ADULT

The previous chapter introduced the subject of innate biological values by discussing the infant primate. One difficulty with that discussion was that it did not make contact with our introspective experience as human adults. How can we convince ourselves that the same motivation principles are applicable to adults?

* Chapter 2.

In the adult, the innate values take the form of *emotions, feelings,* and *physical sensations;* but because of our routine familiarity with these valuative sensations, we tend to lose sight of the systematic way they are programmed to provide the necessary motivation. To remind ourselves that adult motivation also operates in accordance with value principles, let us consider the complex sequence of feelings involved in even the simple act of bodily elimination. This provides a particularily useful example because the experience is universal and frequent.

As the physical need for excretion increases, we experience a degree of discomfort which becomes progressively more intense as the need persists. Thus we experience an increasingly negative value which persists and can be relieved only by the appropriate act of elimination. When we engage in the act of elimination (either defecating or urinating) we experience relief as the discomfort or negative value is removed. One might logically expect that this part of the motivation system by itself would be adequate, but nature did not stop there in the "system design." During the process of excretion we usually experience a pleasurable sensation which seems to be appropriately located in either the anus or the urethra. Thus there is an additional positive value associated with the act itself. Finally, as an aftermath we may experience a generalized feeling of well-being or accomplishment (i.e., a positive value) which fades very gradually thereafter. Older people who are unable to completely void their bladder will often complain of the loss of this sense of completion of the act. Thus, the motivation system for excretion includes three separate components which are orchestrated in time to provide the most effective motivation. Obviously, from an evolutionary perspective, this is one of the oldest and therefore one of the most specialized and well organized of the human motivation systems.

One other point needs to be made with regard to the conscious control over elimination. In this case, the conscious control serves to suppress an almost automatic act of elimination which in more primitive species may not be subject to rational control. As the physical need increases it requires progressively stronger signals from the conscious mind (i.e., more conscious effort) to override the instinctive or natural process. If effort is made to delay the process too long, the conscious control is likely to fail. That is, it will be overridden by an automatic instinctive process. For this reason, sudden fear, surprise, or other emotional disturbances that distract the conscious mind can lead to an embarrassing loss of control.

This example, where conscious control serves only to inhibit an automatic instinctive process, is not unique. In the case of body functions that are essential to health, good design principles dictate that erroneous "rational" decisions must not be allowed to endanger the physical system. This example illustrates one way that nature has found to provide a gradual evolutionary transfer of control, from built-in "reflex" type responses to a more "rational" control by the conscious decision system. Very probably a similar mechanism is used in the infant, to convert from automatic reflex responses to a value-driven behavior.

The system of values involved in bodily elimination is unusually straightforward. The motivating values are designed to respond to simple sensory data from well-defined parts of the body. The resulting motivation system is simple, reliable, and easy to understand.

The "value systems" used to motivate many other types of human behavior are much more complex. Nature has had to resort to much more ad hoc and heuristic design concepts to provide the motivation needed for some of the more complex aspects of human behavior. The "objectives" of these motivation systems are less obvious, and presumably it was more difficult for evolution to find a really satisfactory "value structure." Observation of the behavior of any species (primate or otherwise) will reveal many ways in which even the present complex system of values is not working in a really satisfactory way. The design of a system of values that will produce completely satisfactory behavior is extraordinarily difficult. Even the evolutionary process has not yet found fully satisfactory solutions.

EVOLUTION'S DESIGN PHILOSOPHY

Complex engineering systems usually are based on one of two basic design approaches. On the one hand there are systems designed from start to finish by a solitary genius. Such designs are usually characterized by simplicity, elegance, and the absence of irrelevant or incongruous components. The early Model T Ford and the early computers designed by A. M. Turing or John Von Neumann are examples of such designs. On the other hand, there are designs that have been developed bit by bit over the years, as a result of a long series of committee decisions. These designs usually seem disorganized and unsystematic. They include many irrelevant and nonfunctional complexities. Most modern automobiles and computers reflect this type of design. A detailed examination of these systems will usually reveal a complex and apparently disorganized "rat's nest" of linkages and interconnections.

The instinctive value systems developed by evolution seem to be of the latter type. A succession of interacting value components have been added one after another (as needed to improve behavior) without any unifying design concept. Thus the values are not easily classified or organized. Indeed the *only* organization in the human value structure seems to be what is imposed by *unavoidable* laws of logic and system efficiency.

In the past, many efforts have been made to understand and organize the values, drives, or emotions that underlie human behavior. Most such efforts have failed to produce convincing descriptions of the value structure, because they have tried (unsuccessfully) to fit all of the value components within the framework of a single logical plan. Our approach here will be quite different. We will begin by recognizing that there is no reason to expect the instinctive values to fit any unifying concept. Any and all design concepts that will produce a useful time dependence (or interrelation among the values) are likely to be found within the system. Thus, our approach will be to try to understand the drives as we find them. If any unifying concepts emerge, it will probably be because the concepts were imposed on the "system design" by fundamental principles of logic and efficiency.

THE ANALYSIS APPROACH

The previous chapter discussed the role of instinctive values in the behavior of the infant primate. The discussion was somewhat awkward because we lacked a really appropriate language. It is evident that there is a need for a more systematic and orderly way of thinking about the problem. The present chapter develops an improved terminology for organizing information about instinctive values.

As we mentioned earlier there seem to be three natural classifications that are imposed by logical principles. Values can be classified by function, by time phasing, and by sign. Functionally, we can ask which of three categories the values are associated with:

    I. *"Selfish Values"* for individual survival and welfare.
    II. *"Social Values"* for group survival and welfare.
    III. *"Intellectual Values"* for the motivation of intellectual activity.

Second, we can classify values by their *time* relationship *relative* to the activity they are supposed to motivate. Again there are three alternatives:

    A. *"Preceding values"*—If a value sensation precedes the activity it is intended to motivate we call it a "preceding value."

B. *"Concurrent values"*—If a value sensation is experienced during the activity we call it a "concurrent value."

C. *"Trailing values"*—If a value sensation follows the activity we call it a "trailing value."

Finally we can classify the values by their sign. Values can be *positive* or *negative*, or they may be *complex*, as in the case of taste or smell that can carry either positive or negative components.

To illustrate the classification scheme, consider the values that comprise the motivation system for body elimination. These are all Category I, or "Selfish Values," because they are associated with individual welfare. But this simple motivation system includes examples of "preceding," "concurrent," and "trailing" values. The discomfort that precedes the act is a "preceding" value. The pleasurable sensation that accompanies the act is a "concurrent" value, and the sensation of well-being and accomplishment that follows the act is a "trailing" value.

Our goal here in discussing the human value structure is certainly not to develop definitive conclusions with regard to the actual structure. The present goal is much more modest. It is simply to suggest concepts and principles that should be useful during the initial phase of analyzing and classifying human values.

There appears to be no single traditional name which encompasses all of the innate values. One group of the values has traditionally been identified as the system of "biological drives." The word "drive" traditionally has referred to values that fluctuate in response to obvious *physical needs* of the organism. Another group of values has been identified as "emotions" or "affects." The word "emotion" traditionally has referred to values that fluctuate in response to less obvious *social* or *environmental stimuli.* Finally there are a group of evaluative *physical sensations* such as pain, taste, and smell which are neither emotions nor drives. In our present treatment emotions, drives, and valuative sensations will all be treated under a common functional classification.

CLASSIFICATION OF INNATE "SELFISH VALUES"

In this chapter we will consider only those values whose evolutionary function is the welfare of the individual. Table 9.1 shows a tentative organization of some illustrative "Selfish Values." Each of the values shown is followed by a sign which indicates whether the value is positive (+) or negative (−), or both (+ −). Where no word for the value exists we have avoided coining a new word and have simply substituted with an asterisk (*) the name of the activity that actualizes the value.

TABLE 9.1
Illustrative Table of Selected "Selfish Values"

CLASS A (PRECEDING)		CLASS B (CONCURRENT)		CLASS C (TRAILING)	
Hunger	−	Swallowing *	+		
Thirst	−	Taste	+ −		
		Smell	+ −		
Fear	−	Clinging *	+		
				Well-being or sense	
Rocking urge	−	Rocking *	+	of Accomplishment	+
Sucking urge	−	Sucking *	+		
Anger	−				
Itch	−	Pain	−		
Excretion urge	−	Excretion *	+		

* Where no word for the value exists we have substituted with an asterisk (*) the word for the action the value is intended to motivate. Thus the words "swallowing, sucking, clinging, and rocking" in the table really refer to the pleasure of these activities.

The first thing that should be apparent from the table is that the value elements in the three columns are quite different. If we are to provide a standard format for describing the values, the format will have to be quite different depending on whether we are dealing with "preceding," "concurrent," or "trailing" values. It is also apparent that our traditional terminology for discussing motivation—i.e., the concept of drives and the satisfaction of drives—is really applicable only to the "preceding" values. One of the worst features of this traditional language (in the context of either the "concurrent" or the "trailing" values) is that it encourages a confusion between the process of achieving value from a drive and the process of satiating or diminishing the drive. When we speak casually about "satisfying a drive" we inevitably link these two different concepts.

In an effort to minimize the confusion we will hereafter adopt a different terminology. We will speak of the achievement of value in terms of a drive as "actualizing" the value. This word has been used in the past by a number of psychologists and psychiatrists.* It is particularly appropriate for our purpose since it suggests the "act" or "activity" that *actualizes* the value, and at the same time it implies the conversion of a *potential* value into an *actual* value.

* See particularly Frankl (1).

The total value "actualized" by a decision is given by the difference between the *actual* time-integrated value achieved and the value that would have been achieved in the *absence* of the appropriate decision.

DEFINING THE "VALUE-FUNCTION" FOR SPECIFIC VALUES

To provide an adequate description of a specific value-function we need to be able to predict the "value" as a specific *function* of other system parameters. We need to know how the value is turned on and how it is turned off. Ideally we would like to be able to specify how the value is calculated for any combination of circumstances.

According to Table 9.1, the "preceding" values are always negative. They are turned on as a consequence of circumstances over which the decision system has little control; they are turned off as a consequence of action taken by the decision system. A description of the value function for a "preceding" value therefore requires a definition of the types of environmental circumstances that will cause the value to be turned on. It also requires a definition of the types of action that will result in decreasing or "turning off" the value. "Preceding" values are negative or unpleasant. They motivate the system to do whatever is necessary to get them "turned off." The more quickly and completely they can be "turned off," the more *value* the system will succeed in "actualizing."

Class B, or "concurrent," values can be either negative or positive. The "concurrent" values are actualized immediately as a consequence of activity which is directly enjoyable or unpleasant. The amount of value achieved per unit of time by any specific activity may depend on other system parameters. For example, the same food will taste better or worse depending on the level of hunger. The description of a "concurrent" value will usually require a specification of environmental circumstances that will influence the *potential* values available. The description will also require a specification of the types of activity necessary to actualize the value.

Class C, or "trailing," values are normally "turned on" by an action of the decision system. They will fade thereafter following a more or less routine timetable. The description of "trailing" values requires a definition of the types of action that will "turn on" the value and a specification of how the value will usually fade. The "trailing" values are actualized during the period after the act but before the value fades.

The preceding simplified description implies somewhat more regularity than actually exists in the structure of the values. The actual drives are interrelated in what can only be described as an arbitrary ad hoc, or heuristic, design scheme.

Although the design scheme may seem arbitrary and ad hoc it is also a very familiar part of our commonsense experience. With the foregoing theoretical framework in mind, most readers would be able to construct from their own experience a fairly good description of the way these values operate. Since scientific knowledge has not yet progressed very far beyond such a basic intuitive level, our discussion of the value structure will be very brief. The purpose is not really to provide new information, but rather to utilize some familiar examples to confirm our understanding of the principles. The review deals qualitatively with the structure of the values shown in Table 9.1, so the reader may wish to refer occasionally to that table.

CLASS A, OR "PRECEDING," VALUES

Hunger and thirst: Among the "preceding" values, hunger and thirst are two of the most simple. These values seem to depend almost entirely on the nutritional or chemical state of the body but are modified by the fullness of the stomach.

Fear: Fear seems to be triggered by a multiplicity of complex external stimuli. If the stimulus is withdrawn the fear will gradually wane. At least in the small child, it appears that fear can be "turned off" much more quickly as a result of the specific activities of clinging, rocking, or sucking. Fear depends on how far the individual's "model building" has progressed. Things outside the scope of the current world model are strange and therefore fearful. As things are brought within the model they may either be confirmed as fearful, or they may become familiar and not feared.

In the design of artificial decision systems we have generally adhered to a rule of simplicity which requires that the "primary values" be *completely* independent of rational thought or analysis, but it is apparent that evolution has not respected this simplifying restriction. Thus, the cause of fear depends on the status of a "model building" process, which determines which things are familiar.

Rocking and sucking urges: The existence of these Class A value compo-
nents in addition to the Class B enjoyment of sucking and rocking is obviously
speculative. Although it seems certain that innate motivation for these activi-
ties exists, the balance between Class A and Class B components is very
uncertain.

Anger: Anger is perhaps the most complex of the five listed "preceding"
values. It is "turned on" as a result of frustration, when normally successful
methods fail to actualize other values. It can be "turned off" by a vigorous at-
tack on the presumed source of frustration. We often speak of "giving vent to
anger" as if, in this way, we achieve relief from a pent-up *negative* value. It is
possible that there is also a "concurrent" value which is actualized during the
course of an attack. However, if such a concurrent value exists it does not
seem to have a name. In casual conversation we often distinguish between
"anger" and "frustration" as if they were two different emotions. However, it
is unclear whether there is really any difference in the subjective emotional
sensation. Possibly the distinction is concerned simply with two different
kinds of causes for the *same* emotional sensation.

As in the case of fear, anger also depends on the state of the "world
model." For example, if the work required to earn a reward (such as food) is
routinely very great, the work will be performed, and the reward will be
earned without incident. But exactly the same amount of work will generate
frustration and anger if the work usually required is much less. Thus, the de-
velopment of anger depends on a relationship between what is expected and
what is actually encountered. Similarly, the target of anger depends on an in-
terpretation of what is frustrating the desired satisfaction, and this also de-
pends on the status of "model building."

Of course, this does not mean that anger is under rational control. The
automatic and uncontrollable nature of anger is illustrated in the familiar plea,
"Stop it! You're making me angry!" The victim is saying that he does not want
to be angry (he does not like to be angry, because it is a negative value), but if
the tormentor continues he will not be able to avoid it. Although "reason"
does not control anger, the *results* of reason can have a direct effect on both
the development of anger and the choice of a target for the anger.

Itch: The discomfort of the itch motivates the individual to give attention
to specific parts of the body. It motivates a slap, a scratch, or an inspection
which protects against insects.

Excretion urge: The discomfort associated with this urge is a negative
value, very similar in structure to hunger or thirst.

CLASS B, OR "CONCURRENT," VALUES

The Class B, or "concurrent," values show a diversity comparable to that found for the "preceding" values. The first three in Table 9.1, taste, smell, and the pleasure of swallowing are part of the same motivation system as hunger and thirst, but they operate in quite different ways.

The pleasure of swallowing: If our previous hypothesis is correct there is a "concurrent" value associated with the act of swallowing, but the value seems to arise as a consequence of a temporary reduction in the hunger sensation. If this is correct, then the motivation for swallowing is *automatically* correlated with the degree of hunger. The act of swallowing also seems to occur almost as a "reflex" response to liquid in the back of the mouth. The automatic response can be suppressed or delayed by conscious effort, but in the absence of conscious effort it will automatically occur.

Taste and smell: Whereas hunger and thirst provide the "preceding" values which motivate nutritional intake, taste and smell provide steering values which largely determine *what* is eaten. Given a choice, we will usually select food which tastes "good." The "taste" of food is determined by a combination of sensory data from the tongue ("taste") and the olfactory nerves ("smell"). It is commonly believed that the "goodness" or "badness" of a taste is inherent in the food, i.e., that it is an invariant property of a specific "taste." There is compelling evidence, however, that the matter is much more complex.

Animals that are low on salt content will go to great lengths to lick salt. Experiments on free choice food selection by small children have shown that given a selection of *natural* foods to choose from, their choice will instinctively converge toward an approximately balanced diet. (This may not occur if the cafeteria selection includes artificial food with unnaturally high sugar content or processed foods that may have had nutrients processed out.) It seems clear that the "goodness" of any specific taste is dynamically adjusted by the "value system" to reflect the *nutritional* needs of the body. There is also evidence that an individual can learn to like nourishing foods which are initially unappealing. Thus the "goodness" of taste may be subject to a learning process much like operant conditioning, so that preferences can be adjusted to reflect actual nutritional experience with specific foods.

Clinging, rocking, and sucking: The motivations for clinging, rocking, and sucking all seem to involve some mix of preceding and concurrent values. At present we can only speculate about the actual mix. Indeed it is possible that there is no Class A, or "preceding," value involved at all. If so, however, then it seems clear that the *potential* value available for the "concurrent" value must exhibit "appetitive" characteristics. If the individual abstains, the potential value grows very large, but as value is actualized the potential value gradually decreases. Each of the three values seems to follow its own law of satisfaction.

The pleasure of sucking, rocking, and clinging are interesting values in another respect. As previously noted, the same activities that serve to actualize these values seem also to have the effect of diminishing fear.

Pain: Pain, of course, is a very specific value whose purpose is to motivate the system to avoid physical damage to itself. As would be expected the pain "threshold" is set so that it is normally felt *before* actual physical damage occurs. Pain is described as a concurrent value because it usually occurs simultaneously with the activity that it is intended to deter.

The pleasure of excretion: This pleasurable sensation which has been previously discussed adds some additional motivation favoring excretion.

CLASS C, OR "TRAILING," VALUES

Little needs to be said about the single "trailing," or Class C, value in Table 9.1. The sense of accomplishment seems to have been very widely used in evolution's design to "reward" many types of activity. Perhaps the most interesting observation is the small number of Class C values. From a "system design" point of view this is not surprising. If a value is not to occur until *after* the act, it is functionally less important that it be unique or easily identified with a specific act. Conversely values which precede an act *need* to be distinguishable, so that the *appropriate* act will be motivated.

Logically, the sensation of aching muscles might be added as a "trailing" value whose purpose is to discourage excessive or unusual exertion. However, it might also be classified, like pain, as a "concurrent" value intended to discourage *use* of tired muscles.

A more detailed study would undoubtedly reveal many more Category I, or "Selfish," values, but no purpose would be served by treating them here in more detail. Our primary purpose in analyzing the "Selfish" values has been to develop basic principles that will be needed to understand and interpret the more complex "Social" and "Intellectual" values.

EVOLUTION AND THE DISTINGUISHABILITY
OF VALUES

The foregoing analysis of the Category I, or Selfish Values illustrates another important trend that will become more pronounced as we move into the more sophisticated Social and Intellectual Values. The simple primitive values (hunger, thirst, pain, itch, taste, smell, excretion urge, etc.) can be generated by quite *simple* value systems. The factors that influence these values are few, and they are comparatively easy to understand. Presumably these values can be generated satisfactorily with a minimum of cybernetic resources. Because these values can be produced so economically, it is practical to have a very large number of such distinguishable values. For example, there are numerous qualitatively different forms of "good taste" or "good smell." Similarly there are hundreds of forms of pain or itch that are separately distinguishable by physical location. This distinguishability allows us to respond almost instinctively to different sources of stimulation.

Conversely, the specific stimuli which generate the valuative sensations of "fear" and "anger" are very complex. Presumably the generation of these emotions requires a much more complex cybernetic system. Economy of cybernetic resources dictates that we cannot have hundreds of separately distinguishable "fear" or "anger" sensations. The subjective sensation produced by these emotions seems quite similar *regardless* of the cause of the emotion, even though the appropriate response to the emotion may vary widely depending on the circumstances that produced it.

Thus, the simple primitive values tend to be very numerous. Usually the action that is motivated by these simple valuative sensations is equally simple and obvious. A very simple form of "rational mind" could respond quite appropriately to these *primitive* values because their message is so direct.

On the other hand, the more sophisticated values such as anger, fear, and the sense of well-being or accomplishment are quite nonspecific. The ability to respond appropriately to these more sophisticated emotions requires a much more capable "rational mind." Thus, as we move from the primitive to the less primitive values, the extent to which the value structure provides detailed information tends to decrease. More and more reliance is placed on the *analysis ability* of the "rational mind." One obvious reason for this trend is that the more sophisticated values evolved later, at a time when the rational mind was more highly developed.

From the perspective of our *subjective* mental experience, it makes a great deal of difference whether the values are separately distinguishable. But in a certain abstract theoretical sense, it does not make any difference at all. For example, suppose the two emotions of fear and anger were combined (or added algebraically) outside the mind and then delivered to the "conscious mind" as a *single* valuative sensation. Theoretically it should make *no difference* in the choices we should make between different action alternatives.

An external observer monitoring the resulting choices therefore might have no way of knowing whether we were consciously responding to one valuative sensation or two. He might be able to deduce the combined numerical value function from the choices, but he would be very hard pressed to draw any conclusions at all about which value components are separately distinguishable when they are delivered to the conscious mind.

Thus the distinction in our value structure between "fear" and "anger" is primarily an issue of internal design. It has an important effect on our subjective experience, but is of relatively minor importance in determining the action decisions. The only real reason for the distinction is to reduce the analysis burden on the rational mind. If they were not distinguishable, we would be forced to *deduce* from the context of a situation whether the proper response is to attack or flee.

RATIONAL "CONTROL" OVER VALUE SENSATIONS

On several occasions we have emphasized that the primary value system is not controlled by the rational mind. But the rational mind does have an ability to shift its attention, so that within certain limits we can reduce our discomfort by deliberately choosing to ignore pain, hunger, or sorrow. Conversely, when we are ready to eat we may deliberately focus attention on our hunger so that we can more fully enjoy a meal. But the ability to shift attention away from a value sensation does not really change the value. In a certain objective sense the value signal is still there even though we may try to ignore it. The ability to ignore or emphasize certain value signals increases the ability of the system to concentrate on problems that are rationally perceived to be more important. But on the other hand the ability to shift attention is actually quite limited. As cancer victims can testify, if the pain is sufficiently severe, there is no way it can be ignored.

DESIGN LIMITS OF A BIOLOGICAL SYSTEM

Here and in the chapters that follow we will be primarily concerned with what we usually think of as "normal" behavior, but before proceeding it is appropriate to discuss briefly the problem of abnormal behavior. Biological decision systems are the product of an evolutionary process which has gradually refined the design to make it efficient and durable within a normal (or anticipated) range of environmental circumstances. In discussing the concept of abnormal behavior, it is helpful to use an analogy with mechanical systems that are developed on the basis of a good engineering design. Such systems can fail to function properly for three basic reasons:

1. Faulty construction
2. Normal wear and tear
3. Stresses which exceed the design limits or circumstances which were not anticipated in the engineering design

For example, a bridge can fail because critical rivets were omitted in the construction. It can fail in old age as a result of wear and tear which weakened the structural components. Or it can fail because of an earthquake, a flood, or exceptionally heavy loads which exceed the design limits.

In an exactly parallel way a biological system can fail to function properly because of faulty construction (i.e., genetically inherited defects), because of old age, or from exposure to circumstances and stresses outside the range contemplated in the "system design." A good engineering design will be robust in the face of a broad range of normal use and abuse. It will recover from stresses and continue to function essentially as intended. However, if the system is exposed to circumstances outside the range anticipated in the design, the results may be unpredictable. It may survive without damage, or it may experience temporary or permanent failure.

Normal maturation of the human mind depends on a complex sequence of events which were anticipated in the "system design." Just as normal development of vision requires the presence of light and normal development of speech requires an environment in which language is used, so also the normal development of motivational and emotional response requires an environment in which emotional and motivational experiences fall within the "design limits" of the system. So long as these experiences remain within the broad normal range anticipated in the evolutionary design, the system will recover from stress and continue to develop along essentially normal lines. Although experience always has some effect on future behavior (and may influence future motivation) the motivation will remain within normal limits and behavior will continue to be purposeful and adaptive.

If circumstances that are encountered early in life exceed design limits, the system may be thrown off the course of normal behavioral development, and subsequent responses to the environment may be nonadaptive. The abnormality may feed on itself and become magnified. Once this has happened, the evolutionary design for the development of the system has been destroyed. The development that will follow thereafter is unpredictable, but it is likely to be nonadaptive.

This dependence of the development of biological systems on environmental influences makes it more difficult to distinguish experimentally between genetic and environmental influences. The modern technological environment is certainly very different from the environment in which the human decision system evolved. It includes many kinds of experiences which could not have been anticipated in the evolutionary design. On balance, it is surprising how little these seem to have interfered with the normal development of the human personality.

After reflecting a while on the principles of a value-driven decision system, some readers may begin to wonder how such a concept can be compatible with certain types of abnormal behavior such as masochism, which at least superficially appears to be in direct conflict with the principle of pain avoidance. It seems likely that in such individuals a sexual pleasure response has become somehow linked with the sensation of physical pain, so that under certain circumstances the sexual urge can dominate and obscure the pain. Although it seems clear that such behavior is abnormal, it is unclear to what extent the problem is usually of genetic or environmental origin.

In the following chapters as we discuss human Social and Intellectual values we will be confining our attention to the structure of human values which falls within the mainstream of normality. Although various forms of abnormal behavior are a very important part of our human experience, they are beyond the scope of the present discussion.

REFERENCES

1. Frankl, Victor Emil. *The Doctor and the Soul.* New York: Knopf, 1955.
2. McDougall, William. *An Introduction to Social Psychology.* London: Methuen, 1908.

Chapter 10

Social Values
and Our Primate Heritage

> The basic virtues of love, honesty, sacrificial be-
> havior, etc. are generally not doubted, even
> among sophisticates today. What is lacking is an
> effective aura of credibility for motivating them.
> RALPH WENDELL BURHOE (1)

A NUMBER of controversial issues must be faced in our discussion of
the Category II, or "Social Values," which are concerned with survival and
well-being of the group. First, the mere existence of an altruistic (or social)
component of the innate human values may be controversial, particularly
among readers outside the behavioral science field. Second, the existence of
innate human motivations that are more complex than the simple emotions of
fear or anger is still a controversial issue even among psychologists and other
behavioral scientists. The present chapter discusses some of the evidence for
the existence of a rather detailed set of innate Social Values. The following
chapter will build on this evidence to develop a more unified theory concern-
ing the structure of the Social Values.

THE GENETIC ORIGINS OF ALTRUISM

The real implications of the doctrine of "survival of the fittest" have been
widely misinterpreted in the popular press. The genetic selection process is
not concerned with just individual survival, but rather with the effective re-

production and survival of the species. The genetic pool for any species always includes a wide mixture of genetic characteristics. As the species evolves, those characteristics that are most effective in reproducing themselves will be enriched in the genetic mixture, while those which are least efficient in reproducing themselves will be depleted. Thus the evolutionary forces inevitably favor those behavioral characteristics that are most effective in contributing to the survival and multiplication of an individual's descendants. If this multiplication of descendants requires self-sacrifice for the welfare of the next generation, then behavioral characteristics which motivate such self-sacrifice will be favored. Conversely, if the evolutionary objective requires a selfish attention to personal survival in order to be able to reproduce, then just such selfish behavior will be favored. The actual mix of selfish and altruistic behavior that will be most favored is the one that is most effective in maximizing an individual's own descendants.

The mother's compulsive concern for the welfare of her children is an obvious example of an altruistic characteristic that has been strongly favored by this principle. E. O. Wilson has observed that the operation of this principle leads almost inevitably to a certain amount of conflict in the innate objectives of male and female. The evolutionary forces tend to favor males that are motivated to impregnate a maximum number of females. At the same time, the same forces will tend to favor females that are each motivated to secure the exclusive support and protection of a male for herself and her offspring.

Obviously the genetic selection process favors only those forms of altruism that are effective in increasing an individual's own descendants. General altruistic behavior with regard to unrelated individuals (or with regard to other species) is not specifically favored. Although this conclusion about the scope of genetically favored altruism is certainly valid with regard to the statistical process of genetic selection, it does not follow that behavioral altruism is necessarily limited to immediate descendants. There are a number of reasons why this simple statistical argument does not really define the limits of behavioral altruism.

1. *By-product Altruism.* Specific altruistic behavioral traits that are genetically favored because of their contribution to the survival of direct descendants will very frequently motivate a similar altruistic behavior toward others whose genetic relationship is more distant. Such by-product altruism is a common occurence because of the difficulty of designing real-world behavioral mechanisms that cannot be triggered by more distantly related individuals.

The cowbird exploits just such by-product altruism in the behavior of other birds. It leaves its eggs in the nests of other birds which altruistically hatch and feed the cowbird chicks. The willingness of the host birds to act as foster parents is an accidental by-product of their altruistic parental behavior. This behavior by the host birds is sometimes cited as evidence that they cannot distinguish between the cowbird chicks and their own. But this is not necessarily so. For example, a small girl has no difficulty in distinguishing between young puppies and a human baby. Nevertheless, the "cute" characteristics of the puppies excite her maternal instincts, and her resulting behavior toward the puppies is maternal and altruistic. Such by-product altruism is, of course, even more prevalent in the relationships between individuals of the same species, and it can lead to many forms of rather unstructured cooperative behavior within a species.

2. *Genetic Adaptation to a Social Environment.* Once such cooperative behavior has begun, the genetic fitness of the individual can be enhanced by exploiting the cooperative social environment provided by others within the species. Thus there is a tendency to evolve interlocking behavioral traits which are interrelated much like a deliberate system design. In effect, the society itself becomes an essential part of the environment to which individuals of a species must adapt. If the socio-cultural environment is so structured that anti-social behavior can lead to exclusion from the society (or so that socially altruistic behavior can lead to reciprocal altruism, which contributes to the genetic survival of the individual), then subsequent genetic evolution will have a positive tendency to favor socially altruistic behavioral traits. Burhoe has provided a rather detailed description of the way in which this type of "symbionic" relationship between man and his society can lead to the evolution of social altruism (2).

3. *Group Selection Mechanisms.* Where there is such cooperation within social groups, there may nevertheless be hostility and effective genetic competition between groups. Those groups that are most effective in cooperating for mutual protection and support will therefore tend to be favored in this type of genetic competition. The fact that modern man seems to have been successful in killing off all competing genetic neighbors, suggests that this type of group selection mechanism has been an important factor in the recent evolution of the human species.

Thus, as a practical matter, innate altruistic behavior mechanisms are not necessarily limited to the individual's direct descendants. One cannot use theoretical statistical arguments to determine the actual limits of behavioral altruism. To develop a real understanding of the mechanisms of social behavior there is no substitute for careful observation of the actual patterns of behavior. The actual structure of social motivation is much more complex than any simple dichotomy between selfishness and altruism, and it can only be understood in the context of the evolutionary experience of the species.

The remainder of this chapter is divided into five sections. The first section uses the human sexual response as an example of innate "Social Values" to illustrate some general principles concerning such values. The second section develops a "systems design" perspective with regard to the motivation of generally cooperative behavior in a social species. The third reviews the structure of nonhuman primate societies, with emphasis on those aspects of behavior that offer a model for human "social motivation." The fourth section examines some of the diversity of sexual behavior within the primate order, as a background for interpreting the evolutionary basis of human sexual behavior. The complex structure of the innate social motivations raises some new methodological problems which are discussed in the final section of the chapter. This examination of our primate heritage sets the stage for a more theoretical analysis of the structure of human "Social Values" in the following chapter.

SEX—A PROTOTYPE FOR "SOCIAL VALUES"

It is a common popular view that all "innate" human motivations are directed toward the survival and "selfish" welfare of the individual. A large fraction of human striving is directed toward goals that have nothing to do with the survival of the individual. Nevertheless, some skeptics will surely refuse to believe in the existence of the "Social Values." For this reason, we will begin with a brief discussion of the human sexual drive, which demonstrates the reality of "Social Values" and also serves as a prototype for the other Category II values. Somehow, in the past, sex has been popularly treated as if it were a "Selfish," or Category I, value. When sex is properly classified as an unselfish, or "Social Value," it provides a compelling demonstration of some important design principles concerning the innate Social Values.

THE "ALTRUISTIC" BASIS OF SEX

Sexual copulation has been absolutely necessary to the survival of the species, but it is totally irrelevant to individual survival. In the broad evolutionary scheme of things, the innate desire for copulation works for the benefit and survival of society; it is not selfishly oriented for the benefit of the individual. Thus, the instinctive values that lead to copulation are clearly Category II, or "Social," values.

Usually when individuals engage in sexual activities they are motivated by personal desires and personal enjoyment. They engage in sex because they like it; they do not need to take a consciously altruistic view to be sexually motivated. The innate value system is so designed that individuals instinctively want to engage in sex. Uusally they do it for the pleasure of the experience, not for the purpose of conception.

IRRATIONALITY IN "SOCIAL VALUES"

The sexual drive also illustrates the arbitrariness (or lack of "rationality") that can exist in the structure of our innate values. Almost everyone has essentially the same innate Category I, or "Selfish," values, and these have a superficial "rationality" about them because they appear to be related to the obvious needs of the body. Consequently, when we consider only the Selfish Values, it is easy to get the mistaken impression that all innate values are "logical" or "rational." It is only when we see different individuals motivated by dramatically different innate values that we are likely to notice how really arbitrary the values can be. The sexual drive is a particularly good example because the associated male and female values are so obviously different.

The typical male finds it hard to believe that women can find men attractive, for he subjectively perceives no inherent attraction in other men. Similarly, the typical female perceives little attractiveness in other women. Young lovers are often amazed that the object of their love can find them attractive. Thus, sexual attraction illustrates the arbitrariness and lack of "rationality" that is possible in the innate values of a human adult. Almost every adult retains a vivid recollection of the magic of sexual attraction as it first appears in adolescence. The perception of the attractiveness of the other sex at this age is exceedingly intense. Girls may see boys as brighter, bigger than life, almost as if they were surrounded by a magic halo. The boys will see the girls as brighter, more perfect than life. When the girl smiles, there is an automatic and intense emotional response in the boy. This instinctive attractiveness of the opposite sex is without any foundation in logic, but it seems just as real as if it were literally a part of the visual image. The "attractiveness" seems to be added subconsiously to the visual image before the image is delivered to the conscious mind.

This attractiveness of the opposite sex can vary widely depending on the circumstances. A mature man who has been sexually isolated will find that women again take on an aura of extreme attractiveness, and he may find it almost impossible to see women in any way except as sex objects. The same man, when sexually satisfied, will find that it is easy and natural to talk to women simply as other human beings.

A man who is sexually unsatisfied will find that women are unusually fascinating. Initially he may enjoy simply talking to them, but after a brief period

of conversation, he will be motivated to closer contact.* This chain of succes-
sive desires is frequently misinterpreted by women, who assume that the suc-
cessive advances are part of a deliberate plot, designed from the beginning to
get the woman into bed. Although in some cases this may be true, it is equally
common that the desires unfold naturally and automatically. The man may be
perfectly honest when he says at the outset that he just wants to talk. But after
engaging in conversation a while, he may develop new motivations which he
did not anticipate.

Thus, the sexual motivation consists of a chain of desires or values. The
action motivated by one step in the chain leads naturally to a situation in which
the next value or drive in the sequence is activated. Because of the shifting in-
stinctive value scale, the individual may make decisions in the heat of passion
that he would ordinarily consider irrational.

THE BEHAVIORAL FUNCTION OF SEXUAL EMOTIONS

In most primates, the sexual embrace is briefer and less involved with
emotions than in man. The human embrace apparently invokes many of the
same emotional responses that the infant experiences when he is able to cling
to his mother's body. There is a sense of security, safety, and tranquillity that
can abolish all worldly fears and concerns. The response transcends the ra-
tional and is one of the most rewarding aspects of the human sexual experi-
ence. The sense of security and tranquillity in the experience is compounded
by the sexual orgasm and its subsequent release of emotional and physical ten-
sions.

The human is unique among primates because the female is receptive
most of the month and because the female as well as the male may experience
an orgasm. The experience of orgasm, particularly in the female, works to
produce a strong emotional bond to the sexual partner. The prolonged physi-
cal contact seems to enhance this bond. Most of the ways that the human sex-
ual response differs from other primates seem contrived to make the experi-
ence more intense, more prolonged, and more enjoyable.

It seems likely that these changes in the human sexual response were
devised by evolution to generate a strong emotional bond between man and
woman. Such a bond may have been necessary in a primitive society to pro-
vide the woman with the protection and support needed for prolonged child-
rearing responsibility. Of course, the emotional link operates both ways, but
the emotional tie of the woman to the man seems in most cases to be the most
intense.

* It is possible that some of the responses in this sequence are specifically built-in (as in the
case of bodily elimination) so that the rational mind may have to actively intervene to inhibit the
natural response.

In some ways, this solution seems clumsy. It might seem preferable to provide the stronger emotional link bonding the man to the woman. But nature does not always find ideal solutions. The existing emotional ties between man and woman, combined with the emotion of anger which inhibits taking another person's spouse, seem to have been adequate to stabilize primitive male-female relationships.

RELATION OF INNATE VALUES TO SOCIAL TABOOS

The daily behavior of individuals is guided both by the innate or primary values and by secondary or derived values such as taboos or moral and ethical principles. Often the traditional "moral values" seem to be *direct* reflections of related "innate values." Two examples will serve to illustrate this point.

Almost all human societies, no matter how primitive, have taboos that limit sexual relations between close relatives. Sometimes the taboos deal with social or geographic relationships that are only indirectly correlated with blood kinship, but they nevertheless serve essentially the same function. Such taboos have an obvious social benefit, since they reduce the frequency of genetic deformities, but the fact that the taboos are so universal suggests that they may be based on innate values as well as reason.

The behavior of children raised in the intimate communal life of an Israeli kibbutz provides a striking confirmation of this assumption. Children of both sexes in the age range from seven to ten are raised in a communal children's house. They live together in intimate daily association like brothers and sisters in a large family. Since the children are not blood relatives, there is no cultural taboo against intermarriage when they mature. Indeed, we might expect such marriages to be very common because marriages are usually more likely between individuals who in childhood have lived in the same neighborhood and have shared a wide range of common experiences. But an analysis of the marriage patterns of children raised in the kibbutz discloses an astonishing result. Individuals raised together in the kibbutz almost never intermarry (14). They do not even court. Apparently the intimate proximity during those formative years activates an innate inhibition, which in evolution's design is intended to inhibit sexual relations between siblings. The normal sexual attraction between maturing males and females in these cases seems to be automatically inhibited as a consequence of early childhood intimacy. This conclusion is consistent with the observation that most brothers and sisters fail to experience any mutual sexual attraction, and may even feel a sense of revulsion at the idea. Nevertheless, although the basic incest prohibition seems to be innate, the inhibition is socially *reinforced* with a formal "taboo." Of course, the validity of the taboo is undoubtedly confirmed by the "anger of the gods," in actual genetic experience, when the taboo is violated.

STRUCTURE OF HUMAN VALUES

In those nonhuman primate societies where the female is very attractive to the male, the existence of some kind of behavioral mechanism which tends to inhibit taking another male's mate seems to be quite common. These mechanisms tend to stabilize relationships and reduce the frequency of deadly combat. It seems possible that a similar mechanism (which goes beyond a rational fear of the female's angry mate) may also operate, although quite imperfectly, in man. When a woman is emotionally committed to a single man she tends to be less interested in other men. Consequently she emits fewer sexual signals and thus other men perceive her to be less attractive. Usually when it becomes known that a girl is firmly attached to a particular boy, other boys will cease to compete. The girl's general attractiveness to other boys seems gradually to be restored if the couple separates. The innate reduction in the "attractiveness" of a woman who is already "committed" may underlie the traditional cultural view that a woman is "spoiled" when she becomes sexually attached to another man.

SEX DEPENDENCE OF OTHER VALUES

Almost everyone will acknowledge the obvious differences in the structure of male and female sexual values. But there is a surprising reluctance to recognize other differences between the typical male and female value systems. We see the differences so regularly that it is easy to overlook their significance and to convince ourselves that they are only the result of social influences that operate differently for male and female members of society. Obviously social forces play a role, but basic differences in innate male and female values are probably far more important. One of the clearest demonstrations of the pervasive effect of innate values is provided by the transsexuals.

The transsexual has the physical characteristics of one sex, but the emotional and motivational characteristics of the opposite sex. Experience in the treatment of these individuals has shown that neither psychiatric treatment nor psychological counseling can change the motivational structure (5). Although the individual's innate values are abnormal, they seem firmly and irretrievably built into his system.

The transsexual personality generally shows up at a very early age, perhaps as early as three or four years, but almost certainly by age eight or nine. A transsexual girl will be interested in boys' sports and uninterested in dolls and will tend to be aggressive and competitive. A transsexual boy may be timid, interested in dolls and housework, and uninterested in (or afraid of) rough play with other boys. The effects of the unconventional motivation in these cases are so striking that the family almost always feels that something is seriously wrong.

The traditions of society, however, will normally force these individuals to live a role corresponding to their physical sex. The reactions of the individuals to the unnatural role are most revealing . One male transsexual found that he was intensely jealous of his wife. This was particularly true when she became pregnant—because *he* wanted to be pregnant! *He* wanted to nurse the baby and look after the house.

The typical transsexual is obsessed by the desire to make his physical body conform to his emotional and motivational sex. He feels more comfortable in clothes of the opposite sex. Both male and female transsexuals are often willing to undergo long and painful sequences of operations to being their bodies into conformity with their mind. In most cases after the operations the individuals are able to adjust to a satisfactory life in their new sex, whereas in their native sex they had been very discontented.

The experience of the transsexual carries an important message. The pattern of innate human values is very complex indeed. The typical female value profile is very substantially different from the typical male profile. The magnitude of the differences are dramatically illustrated when we see an entertainer imitate a person of the opposite sex. While both male and female seem to operate with much the same rational intelligence, their behavior is very different because their innate values are different. From an evolutionary design perspective, high intelligence has universal survival value, so genetic improvements in intelligence tend to be incorporated in both sexes. But sexual specialization in motivational patterns is clearly advantageous for species survival. Male values are never likely to be really understood by females, just as female values are not likely to be fully understood by males.

Of course, there are wide variations in the value profiles of individual men and women. An individual woman may have many characteristics that are more masculine than the average male, and a male may have many characteristics that are more feminine than the average female. Ideally we might hope that society would allow each individual to be true to his own innate values, but unfortunately because of social pressures individuals often feel compelled to match a stereotype which is not their own.

At this point, it should be apparent not only that "Social Values" exist, but also that the ensemble of human Social Values is complex. Before considering the Social Values in more detail, it may be helpful to look at the problem of social behavior from the broad perspective of an evolutionary system design.

INSTINCTIVE SIGNALING: A FOUNDATION FOR COOPERATIVE MOTIVATION

Groups of animals such as a troop or herd can be more successful if the individuals within the group are motivated to work cooperatively for their mutual benefit. To accomplish this with nonverbal animals, the animals must be designed so that they "like" to cooperate. Somehow interactions between the animals must be devised so that cooperative activities will occur.

Nature seems to have found a quite successful method for generating the necessary interactions between members of a species. The technique involves a signaling method in which each animal instinctively communicates (for example by facial expression) his own emotional or motivational state. This signaling system can then be used in the evolutionary design to trigger a diverse set of instinctive social motivations. The expression of one animal can automatically generate emotions (i.e., innate drives or values) in another animal. A threat or anger face can engender fear. An expression of appeasement or submission can cause anger to fade. An expression of discomfort or agony can engender sympathy and a desire to help. The animals can be designed so that they "like" certain types of expressions and "dislike" others. For example, we may enjoy seeing people smile and thus may be willing to go to considerable effort to evoke a smile or a laugh. To use the terminology of the ethologists, the social signals can serve as "releasers" for specific forms of *emotional* response.

Almost all social animals have a repertoire of expressions or signals. It is unclear what fraction of the signals are incorporated as specific "releasers" in the innate motivation system, but it seems clear that many of them operate in this way. From a design perspective, the existence of a rich repertoire of expressions provides a foundation of information on which more sophisticated motivation systems can be built.

As every dog lover knows, the dog uses his tail to signal his moods. His signaling is instinctive and automatic. From an evolutionary point of view, it probably serves no function except to communicate his mood to other dogs in the pack. Ethologists have recently given considerable attention to these "social signals."

Figure 10.1 shows some examples of social signaling in the wolf. The wolf also conveys a considerable amount of information through facial expressions. Figure 10.2 illustrates some of the expressions involved. Although the wolf shows a rich repertoire of expressions, the wolf's expressions are not nearly as

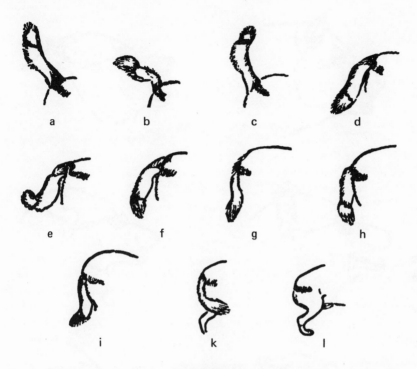

FIGURE 10.1
Use of the Tail for Expression in the Wolf

A Self-confidence in social group B Confident threat C With wagging; imposing carriage
D Normal carriage (in a situation without special tension) E Somewhat uncertain threat
F Similar to D, but specially common in feeding and guarding G Depressed mood H In-
termediate between threat and defiance I Active submission (with wagging) K and L
Complete submission.

Reproduced from *Science Man & Morals* (1965) by W. H. Thorpe. Reprinted by permission of Methuen & Co., London (after
Schenkel [12]).

rich as those of a chimpanzee, or of man. In the next chapter we will deal in
more detail with the role of facial expression in primate societies. It seems ap-
parent that in almost all cases the transmission of the facial signal is an in-
nate, instinctive, or automatic response. The individual does not choose to
take on an expression. The expression happens automatically as the mood
occurs.

The expressions serve two important functions. First, other animals can
gradually *learn* the meaning of the expressions. In this way the expressions can
provide a rudimentary form of communication for a nonverbal society. But

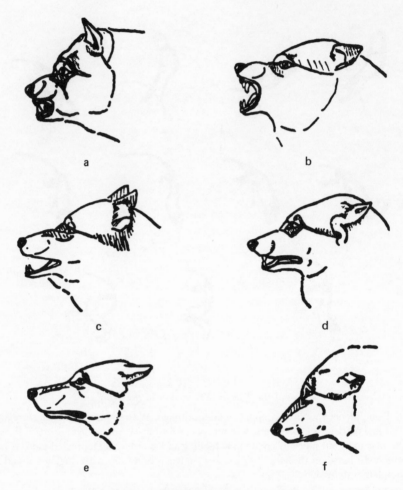

FIGURE 10.2
Facial Expression of Emotion in the Wolf

A Fully confident threat B High intensity threat with slight uncertainty C Low intensity threat with uncertainty D Weak threat with much uncertainty E Anxiety F Uncertainty with suspicion in the face of an enemy.

Reproduced from *Science Man & Morals* (1965) by W. H. Thorpe. Reprinted by permission of Methuen & Co., London (after Schenkel [12]).

the expressions appear to play an even more significant role by *releasing* or stimulating emotional or *motivational responses* in other members of the society. Thus the expressions provide a basic "mechanism of interaction" on which cooperative patterns of motivation can be built.

THE STRUCTURE OF PRIMATE SOCIETIES *

To understand the evolutionary origin of human social values it is necessary to see how such motivations operate in a typical nonhuman primate society. It seems fair to assume that the social structure we observe in these societies is a rather direct reflection of the instinctive social motivations of the individual animals. Of course the young animals are molded by their experience within the troops, so theoretically cultural traditions can exist. But troops of the same species at widely separated locations seem to exhibit essentially the *same* social structure. We must therefore conclude that the basic social structure is an equilibrium state of the society, which results directly from the innate motivations of the species. Because the animals are nonverbal, we can be assured that they are not motivated by complex myths and similar inventions of human civilization.

Examination of primate societies for *different* species reveals a large amount of diversity in the social structure. Nevertheless, the building blocks of the society seem to be quite standard. The diversity of social structures among the species seems to be achieved through differences in the emphasis placed on the various building blocks.

PRIMATE SOCIAL GROUPS

The smallest cohesive group in primate societies is the maternal group. This group consists of the mother, her infant, and possibly some older siblings. This group, which is built on maternal attraction between the mother and infant, is apparently universal in primate societies.

The next larger group we will call the family group. It consists of a single male and one or more females and their offspring. Thus the family group may include one or more maternal groups. The family group is very important in some primate societies but nonexistent in others. The family group is based on heterosexual attraction between mature males and females. The attraction may be only one-sided or it may be mutual. It may be permanent or it may exist only when the female is sexually receptive. Nature seems to have used variability in the form of sexual attraction to adapt the social behavior patterns to a wide variety of environmental situations.

The next level of social organization we will call the troop. Most species of

---

* Our present knowledge of these societies is a result of careful field work by numerous observers. The treatment of the subject here is necessarily brief and oversimplified. Readers who wish to pursue the topic further are referred to the following sources which were particularly helpful to me. Southwick (15), Kummer (8), Devore (3), Jay (7), Gavin (4), Van Lawick-Goodall (17, 18, 19), Schrier and Stollnitz (13), and Rosenblum (10).

primates that have been investigated live in more or less closed social groups. These troops range in size up to as many as 800 animals, but the typical troop size seems to range from 20 to 100 animals. The members of a troop recognize each other as individuals and are generally intolerant of outsiders. Where several troops range over overlapping territory, the troops tend to avoid each other. Occasional meetings between troops are usually accompanied by hostility. In some species, the troops seem to recognize territorial boundaries which they defend from encroachment by other troops. The cohesiveness of the troop involves many innate social responses, but the interactions between the mature males of the troop seem to be the most important factor in troop solidarity.

All primate societies include the maternal group, but in any given society the family level of organization and/or the troop level can be missing. When the troop bonds are particularly weak, the prevalent form of organization is usually the family or one-male group. Where the family bonds are weak or transient, the troop organization seems to dominate and a promiscuous sexual relation exists between males and females of the troop.

SOCIAL STRUCTURE OF THE TROOP

Anthropological evidence suggests that organizations such as the clan, the band, and the tribe were important throughout the early history of man. Thus the innate human social values probably correspond most closely to those of other primates that live in a troop society. Troop life has been carefully studied for a number of species. The macaque and baboon appear to have been studied in the greatest detail and they seem typical of this mode of organization.

The social structure of a troop becomes most apparent when the troop is on the move in relatively open country. The center of the troop is occupied by the dominant or leader males together with their females, infants, and juveniles. A little way from the center of the troop there is a region that is occupied by less dominant males and associated females and infants. Toward the edges of the troop there is an area that is occupied by mature young but nondominant males. Some of these young males are only loosely attached to the troop. They may wander away for periods of many days.

If the troop is threatened by a predator, the young adult males will place themselves between the center of the troop and the predator. They will attack as a mob and frighten away the predator. Thus the troop structure clearly has survival value for the society. It provides for protection of females and juveniles and tends to preserve the most experienced males as leaders.

TROOP LEADERSHIP AND RESPONSIBILITY

The leader males of a troop play an important role in settling disputes between other members of the troop. They are also particularly influential in determining the group activities of a troop. Of course, without a language, the nonhuman primate cannot actually discuss alternatives.

Hans Kummer, in his book *Primate Societies* (8), provides a detailed description of the decision process within a troop of baboons as they select a direction for the day's foraging. Individual males near the edge of the troop initiate particular directions of motion. The effectiveness of these initiatives depends strongly on the status of the initiator. An initiative by a particular old male may be followed or supported by many neighboring males, while a similar initiative by a young male may receive no support. As a result of multiple initiatives, the troop shifts its shape rather like an amoeba. Here and there the periphery of the troop protrudes in a kind of pseudopod which may persist or withdraw again. After about half an hour of such indecisive activity, several dominant males near the center of the troop will rise and move decisively toward one of the pseudopods. The troop then moves purposefully in the direction initiated by that pseudopod. The interpretation seems clear that the pseudopods were proposals for a direction of march, and that some influential male near the center finally made a decision. The troop, however, is not led by a single leader. On successive days the decisive move may be made by different dominant males. It seems apparent that the older males acquire a position of influence or leadership within the troop and that they respond to the position with something akin to a sense of responsibility.

DEVELOPMENT OF THE INDIVIDUAL

To really understand the primate social structure, we need to know how the innate motivations of the *individuals* operate to produce the integrated social structure of the *troop*. Perhaps the best way to develop these ideas is to follow the development of an individual primate from infancy to maturity.

The infant is extremely attractive to female primates. The newborn infant is a subject of great interest, not only to the mother, but also to other females. There may be competition between other females to hold or fondle the infant. The mother will usually limit the degree to which others may handle the baby, but she may allow her older daughter or a female grooming partner more than usual privileges. Thus, in the event of death of the mother there is a substitute mother ready to adopt the infant.

The intense interest of the females in the infant is in marked contrast to the response of the males. The males of the troop may exhibit some interest, but it seems more like an idle curiosity to find out what is of such interest to the females. The female's interest in holding and caring for the infant seem to reflect truely innate social values. The adult female engages in such activity because she enjoys it.

As the infant begins to mature, the behavioral differences between the sexes begin to be apparent. The female ventures less from her mother. She engages in less play activity, and when she plays it is usually with a smaller group, perhaps only one other playmate. The young male engages in rougher or more aggressive play with more individuals. When play gets rough the mother may intercede to protect her young one from other youngsters.

As male juveniles approach adulthood, the rough and tumble play becomes more seriously aggressive. The outcome of fighting at this stage is important in settling the dominance relations between young males of the same age group. Dominance relations (particularly among the males) are very important to the structure of primate societies. Once the issue of dominance has been settled between two individuals it tends to remain as an almost permanent part of the relationship. The submissive individual will allow the dominant one first access to food or a resting place. On the other hand, the dominant animal is inhibited from taking undue advantage of the submissive one. Once the dominance is established, the dominant animal seems willing to be friendly and often assumes a protective relation to the submissive animal. The submissive animal will express the submission by presenting his rump in a gesture of appeasement to the dominant animal. The dominant animal may confirm his dominance by mounting the submissive animal much as he would a sexual partner.

While the dominant animal is inhibited from unduly exploiting his advantage, it is clearly advantageous to be dominant. The dominant animals have the advantage both in feeding and in access to sexual partners. The striving for dominance or social status seems to reflect an innate social value which is particularly important for males.

While the infant primate is attractive primarily to females, the more mature juveniles also seem to be attractive to adult males. Young males will sometimes assume a protective fatherly role relative to juveniles. The relationship seems reciprocal, and the juveniles will actively seek out an adult that takes this type of interest in them. In the case of juvenile females this may later develop into a consort relationship, whereas in the case of young males it seems to be strictly a fatherly relation, which serves to provide an adult model for the juvenile.

As the young male matures, he enters a phase where his interests are relatively nonsocial. He engages in less social grooming than the female juveniles or the adult males. His limited social interests seem to be focused more on other young males of a similar age. At this stage he may join an all-male group separate from the troop, or he may live for a period as a solitary male. This period of "independence" in the male apparently has no counterpart in the female, but the male "independence" phase may have important survival advantages for the species. It facilitates crossbreeding between troops, since the male may not return to the same troop he left. It enables the male to develop a broader experience with the terrain and greater self-confidence that will serve him well if he survives to become a leader. The males at this stage experience a higher mortality than females, and this contributes to the rather high ratio of females to males of breeding age within the troop.

It is unclear to what extent the social isolation of the young males is a matter of choice. Probably the desire for independence combines with the lack of acceptance for low-ranking males within the troop to produce the overall pattern of social isolation.

Usually after the period of independence, the male will intermittently and then permanently attach himself to a troop. The male then begins a struggle for acceptance within the troop. He can achieve acceptance by physical domination of some of the males in the troop. There also is evidence that he may be able to increase his acceptance by acting as a protector for a juvenile. On a priori grounds one would expect that he could also improve his acceptance by helping to protect the troop against predators, but at present there seems to be no concrete evidence for such an effect.

As the dominant or leader males become older and less effective, they may gradually be replaced by mature subleaders. Thus the young male can move gradually toward a position of dominance. Fighting almost never occurs between the dominant males of a troop because the issue of dominance between them has long since been settled.

SURVIVAL BENEFITS OF PRIMATE SOCIAL STRUCTURE

The overall pattern of troop behavior has obvious survival benefits. The troop itself provides safety in numbers, which tends to protect the group from predators. The hostility toward alien troops tends to spread the population of troops more evenly over the terrain to use available food resources more efficiently. The pattern of established dominance between individuals provides a routine mechanism for settlement of disputes with a minimum of actual fighting. The inhibition of the dominant animal with regard to taking undue advantage of the submissive ones minimizes the disadvantages of the dominance system for the younger nondominant members of the troop. The tendency of the troop to cluster around (and follow the lead of) the older dominant males gives the troop the benefit of experienced leaders and preserves the experienced leaders by keeping them within the safety of the troop center. The location of the females and juveniles toward the center of the troop provides protection for the child-rearing mothers. The tendency of the young males to gravitate to the edges of the troop provides protection for the rest of the troop. The higher mortality of weaker males exposed at the edges of the troop selects the strongest and smartest males for breeding and leadership in the troop.

It is obvious that this general structure of primate society is not an accident, but the result of selective evolutionary forces. The structure of the society reflects the instinctive motivations of the individual animals. These motivations depend on the sex of the animal, and the motivations change as the animal matures. Of course the detailed structure of the underlying innate values is far from clear. For the present, however, our objective has been only to call attention to behavior patterns that illustrate the existence of a complex structure of innate social values. The impressive parallels between human and nonhuman primate societies is consistent with our assumption that man's innate values have much in common with those of other primates.

PARALLELS WITH HUMAN BEHAVIOR

Before proceeding it may be worth summarizing some of the innate motivations that man seems to share with other primates:

1. The fascination of adult and juvenile females with the infants.
2. The relative disinterest of adult males in infants but increasing interest in juveniles.
3. The sense of comradeship between young adult males.
4. The desire for adventure and independence exhibited by young adult males.
5. The tendency of females to relate most strongly to the immediate family group. First, their relation to their mother, and then their relation to their own offspring and possibly a male consort.
6. The tendency of the male to interact more widely with other males in the troop.
7. The intense sense of competition between males for status, acceptance, or dominance.
8. The inhibition with regard to taking advantage of a submissive individual. "Don't kick a man when he is down."
9. The sense of responsibility that goes with status or leadership. For example, the interest in settling disputes.

Undoubtedly a much longer list could be developed, but it already seems evident that human behavior is governed by a complex structure of innate values that we are only beginning to understand. The study of other primates is useful because it gives us a better understanding both of the degree to which behavior is guided by instinctive values and of the level of detail included within the innate value structure.

VARIABILITY IN PRIMATE SEXUAL MOTIVATIONS

The previous section described the social structure of a primate troop to provide an evolutionary perspective concerning the origin of human social motivations. We deliberately omitted any description of the sexual motivations, because the nonhuman primates do not seem to include any direct parallel with human sexuality.

A review of other primates suggests that the patterns of sexual attraction are rather easily modified from species to species. Apparently evolution has used the variation in sexual attraction both to limit crossbreeding between species and to adapt each species to its own specific environment. The way this adaptation has occurred in other species may provide clues to the origins of human sexuality.

In the typical primate society there is no permanent bond between the male and female. The female becomes sexually receptive for a period of a few days during each month. During this time her sexual capacity exceeds that of any individual male. The receptive female may interact sexually with most of the mature males in the troop. Copulation is brief and does not appear to have much emotional significance for either the male or female. This typical sexual pattern applies to most species of primates. It is applicable to most monkeys and all of the great apes, which are man's nearest relatives. Deviations from the typical pattern occur in widely scattered species and the particular form of the deviation does not seem to follow any standard pattern. Hans Kummer has carefully studied a number of such deviations in primate mating patterns. Some of his conclusions are well worth repeating.

Most deviations from the typical primate sexual pattern result in a more or less permanent attachment between males and females. The development of such bonds seems to be an evolutionary adaptation to specific environmental pressures. When the food supply is scattered so that small groups of two or three adults are needed for efficient foraging, a permanent bond between the male and female can be functional in several ways. It places the foraging females under the protection of one specific male. Since the females in most species are considerably smaller than the males, this provides better protection of the females against predators. The bond also improves the female's position in the competition for food, since she is no longer in direct competition with the males. Such relatively permanent male-female bonds seem most likely to evolve in situations where predators threaten the very small bands that are needed for efficient foraging.

ILLUSTRATIVE DEVIATIONS FROM THE PRIMATE NORM

The devices that nature has used to accomplish such male-female bonding are surprisingly diverse. Kummer (8) reports on two species of baboon, the anubis and the hamadryas. These two species are almost identical except for the sexual bonding pattern. The anubis baboons follow the traditional promiscuous primate sexual pattern. They normally live in a forested area and spend their night sleeping in trees. The hamadryas baboons live in an adjacent cliff and desert area with meager vegetation, and they typically sleep in the cliff areas. The hamadryas baboons have developed a sexual relationship in which the adult male maintains possession of one or more females. The difference in behavior seems to reflect a genetic change in the male baboon.

Evidently the male hamadryas baboon has been genetically redesigned so that he finds the female baboon very attractive regardless of whether she is in the sexually receptive part of her period. Because the male baboon is larger and stronger than the female (he weighs about twice as much) he is able to enforce his possessiveness on the females. The male continuously watches to make sure his females are following him. If a female fails to follow, the male bites her on the back of the neck and forcefully herds her back into line. The female soon learns that she can avoid such conflicts only by obediently following the male.

The female hamadryas baboon may occasionally try to escape. During her sexually receptive period, she may try to mate with other males, but she only dares to do so if she can escape the view of her possessive male. The arrangement is clearly less than satisfactory for the female, but it serves the evolutionary function of providing the female with a male protector. It is as if evolution had found a simple "quick fix" for the baboon behavior pattern to accomplish this goal. Whether the arrangement was satisfactory to the females was immaterial, so long as the basic survival objective was served. The present bonding mechanism requires the male to waste considerable effort herding his females. Obviously, it would be more efficient if the bonding were mutual, so that the female would voluntarily follow her possessive male. Presumably the hamadryas baboons should now be ripe for such a genetic change in the female, but the random mutations of evolution have not yet provided this solution for them.

The more efficient solution has already emerged in a very close relative of the baboons, the gelada (which is sometimes called the "gelada baboon"). The geladas occupy a relatively treeless area in the highland meadows of Ethiopia. The gelada male does not have to herd his female. One-male "family groups" of geladas develop voluntarily by action of both sexes.

Kummer reports on an experiment in which a group of geladas, previously mutual strangers, were introduced into a common enclosure. Immediately after the introduction a male would pair off with a dominant female. The female would present her rear, he would mount her, and she would then groom his cape. From this point on the female was intensely possessive. She would viciously attack any other female that approached her mate. But the sexually mature females typically outnumber the mature males by two or three to one, so the one-to-one pairing is not stable. Ultimately other females are allowed to enter the one-male family group, but only after they have begun by approaching the dominant female. The *number two* female must first go through a "pair forming" ritual with the *number one* female in the role of male. The second female presents her rear and is mounted by the number one female who she then proceeds to groom. After a secure dominance relation is established between the two females, the second female may be allowed to interact with the male. Thus the gelada one-male "family groups" actually consist of a chain of animals in which each in turn dominates its immediate subordinate. The gelada solution to the bonding problem seems to be based on a combination of female possessiveness and a change in the male which makes the females seem continuously attractive.

The patas monkeys have evolved a totally different solution. These monkeys form no troops, and the one-male groups live far apart. The groups seem to be maintained simply by the fact that they stay so far apart. Tests with the patas monkey in captivity show that the males are extremely intolerant of each other. Males will viciously attack each other and the victorious male will relentlessly pursue the loser. Thus the patas monkey seems to have achieved one-male groups simply as a result of increased mutual hostility in the male.

To casual observation, it seems that among all primates, the gibbon has a "marriage" pattern that is most similar to the human. The standard gibbon family includes just two adults, a male and a female, together with their offspring. The female gibbon is almost the same size as the male, and there seems to be no strong pattern of male dominance. The gibbon sexual attraction seems to be voluntary and mutual. The gibbon family appears quite durable and may in fact be a lifelong monogamous commitment. There is some evidence that when the female is sexually receptive she will occasionally interact with other males, but without diagnostic experiments we can only guess at the underlying structure of the gibbon sexual bond.

HUMAN SEXUAL ADAPTATIONS

Each of the preceding departures from the "standard" primate mating pattern appears to be an independent evolutionary experiment. The human sexual attraction also seems to be a unique evolutionary experiment. It seems clear that at some time in the past strong evolutionary pressures worked to provide human society with a durable sexual bond. It is probable that in modern human societies the original evolutionary forces no longer exist. Indeed the real need for the bond may not even be present in existing primitive societies. Nevertheless, as a result of primitive evolutionary forces, the human sexual pattern is now strikingly different from our nearest relatives, the great apes.

It seems possible to accomplish sexual bonding by any strategy that will make the attraction between the sexes permanent rather than transitory. In the hamadryas baboon the bonding was accomplished simply by changing the subjective preferences of the male. In humans a wide variety of changes have occurred. Many have been physical changes in the female designed to encourage a more permanent male-female relationship. It is instructive to list some of the obvious changes:

1. Extension of the period of sexual receptivity to cover almost the entire sexual period.
2. The development of a female orgasm, which makes it easier for a female to be satisfied by one male, and which also operates psychologically to produce a stronger emotional bond in the female.
3. The development of the female buttocks and breast as important sexual symbols. Figure 10.3 illustrates the primitive role of buttocks as a sexual signal. The analogy to the modern breast is obvious. The prehuman male apparently developed a psychological attachment to the female hindquarters. As the frontal sexual approach became more natural, the breasts offered a way of duplicating the attraction of the buttocks for the frontal approach.
4. The development of the hymen, which tends to discourage sexual experimentation by the female until there is a strong emotional involvement.
5. An increase in the typical time duration of the human sexual encounter, which increases the opportunity for physical contact and resultant emotional attachment.
6. An intensification of the emotional element in the human sexual embrace. The intense emotional content of the human sexual embrace may be the result of retaining the infantile emotional response to clinging.*

The sexual behavior of the modern male and female also suggests that there have been some important adaptations in the innate human "value structure." Both the human male and the female have developed a considerable sexual possessiveness or jealousy, so that a sexual competitor is likely to be met with hostility. This jealous response (which seems quite similar to the possessive behavior of the gelada female) is most intense in females but it occurs in both sexes.

Why did nature go to such lengths to introduce stability into human sex-

---

* Many human adaptations seem to have been accomplished by retaining infantile primate characteristics into adulthood. Some examples include hairlessness, curiosity, adult playfulness, and the shape of the human foot and face.

a

b

c

FIGURE 10.3
Sociosexual Signals

    (a)   Stone-age figurine from Savignano, Italy.
    (b)   Hottentot female from Bersaba.
    (c)   Bustle dress (1882).

From *Primate Ethology* (1967), ed. Desmond Morris. Reproduced by permission of Weidenfeld & Nicolson, Ltd.

ual relations? Actually a large number of separate pressures can be identified. The evolution of man as a hunter increased the dependence of the females on the males. The movement of human society from the forest to open country for hunting placed early man in considerable danger from predators. This also increased the importance of male protection for the females. The exceptionally long period of dependence of the human infant certainly taxed the endurance of the primitive female and increased the benefits to be obtained by providing her with male assistance.

But there remains an important puzzle. Why would evolution have chosen to protect the female by the apparently awkward method of bonding her emotionally to a male rather than vice versa? The genetic statistical concepts mentioned at the beginning of the chapter suggest one answer. The review of other primate societies suggests another factor that may have been important. In a primitive environment, the male mortality is often much higher than the female mortality. Where there is a shortage of males, the society must be able to adapt to a life-style in which each male is a protector for several females. This adaptation to a polygynous family structure would be more difficult if the males were linked by an *exclusive* emotional bond to a *single* female. Evolution may therefore have found it more satisfactory to concentrate the intense and exclusive emotional bond in the female. To make this approach successful in providing protection for the female, the male motivation system would have to be designed to respond appropriately to the emotional commitment of one or more females. Our everyday experience with the way couples normally fall in "love" seems generally consistent with this concept. Usually the real emotional commitment of the man follows the commitment of the woman, as if it is a response to the woman's love.

A "DESIGN CONCEPT" FOR THE HUMAN SEXUAL BOND

We were able to "explain" some of the sexual adaptations of other primates in terms of rather simple "system design" concepts with obvious survival value for the species. It seems likely that at some time during human evolution the human adaptation reflected a similar kind of "system design" concept. It is interesting to speculate about how the system may have functioned in such a prehuman (probably preverbal) society to provide efficient male-female bonding appropriate to the survival of the species. By combining information from other primates with observations about modern human behavior it is possible to develop a rather plausible hypothesis.

The bond between the prehuman male and female probably developed in a two-phase process. The first phase served to bring the pairs together, while the second phase cemented a relatively permanent bond. In the first phase the initiative belonged primarily to the male, but in the second phase the initiative belonged to the female. In both cases the initiative was motivated by an innate, genetically determined value structure.

In the first phase the male's attention is directed toward the female in response to a rather superficial physical attraction. Desmond Morris has suggested that the modern male has an innate response to certain key features of the female anatomy—such as the hips, the curve of the waist above the hips, and the breasts. The initial male response may be nothing more than a positive "valuative" signal which triggers "interest," * so that the male "enjoys" watching the "attractive" female. The key features in the female anatomy that trigger this response appear to have been chosen by evolution to favor the slim young female figure in preference to the matronly figure of a mature female. This tended to focus male attention where it was likely to be most "productive" from an evolutionary view.

The intensity of the response to such superficial female features seems to be strongest in males that are sexually unsatisfied. This ensured that the least committed males would be the ones most active in the pursuit of unattached females, and this in turn tended to *equalize* the number of females that might become attached to each individual male.

In this initial phase of the process, the female response to the male's interest probably depended both on her own emotional state and the "suitability" of the male. A female who was already attached or bonded to another male would have little or no interest in the male because of the *exclusivity* of her existing bond. This "disinterest" of the already bonded female was necessary for the stability of the existing bond, and it also served to reduce the likelihood of violence between competing males. But an *unattached* female would be "pleased" by the male's interest and would respond with instinctive "social signals" such as a smile or a lowering of eyes which generate an enhanced interest in the male and encourage his approach.

The response of the female was probably relatively insensitive to the physical appearance of the male. After all, from an evolutionary perspective it does not matter much whether he is young or middle-aged. But it is important that he be both willing and able to protect the female and to help care for a family. The innate response of the female therefore tended to favor males who showed evidence of strength, courage, intelligence, and consideration for the females. This "selectivity" of the female response of course tended to favor the more dominant but least attached males.

The initial emotional response of the female was probably rather similar to "admiration," and the male was designed to respond favorably to such admiration. When the male interest continued, the female response would gradu-

---

* The main discussion of valuative signals dealing with "interest" or "curiosity" will be postponed until the discussion of intellectual values in Chapters 12 and 13.

ally deepen into a more *exclusive* and enduring "love." This transition of the female response from "admiration" to "love" was probably accelerated by physical and sexual contact. With this transition to the intense and *exclusive* form of female attachment the relationship entered its second phase. The emotionally committed female inevitably transmitted numerous social signals which reflected her commitment, and the male was designed so that he responded emotionally to these signals and returned the commitment. But the male's commitment remained nonexclusive, whereas the female commitment tended to be exclusive.

In this mature phase of the relationship the male was less influenced by the superficial physical characteristics of the female, because his attachment was now based on *her* emotional commitment. Although the male at this stage was firmly committed to the female, he maintained his superficial physical interest in other "attractive" females and remained emotionally susceptible to any other females that exhibited a sufficiently intense emotional interest in him. This insured that (even when all available males were bonded to one or more females) "attractive" maturing females would still be able to find a male protector.

Obviously, the preceding description parallels closely the popular stereotyped conception of the modern male-female bond, but it also illustrates the adaptability and efficiency of this familiar mechanism in very primitive societies where male mortality could be high. It seems probable that the human male-female bond in its typical modern form still reflects many of the features illustrated in the foregoing description.

On the other hand, the existence of substantial homosexual and transsexual populations makes it clear that there are particularly wide variations among individuals in the intensity of different aspects of these innate sexual motivations. The wide variations in the ease and frequency with which women achieve orgasm is also well documented. Since this is a very important psychological factor in the human sexual bond, it may be indicative of a similar diversity in other elements of the bond. Even primitive man operated in a much wider range of environments than any other primate. A diversity of instinctive sexual values in the genetic pool and the resultant genetic adaptability of the species could itself be an important survival asset for a species which must operate in a diversity of environmental circumstances.

STABLE MOTIVES—A METHODOLOGICAL DILEMMA

Many of the behavioral characteristics discussed in the previous sections reflect stable motivational patterns which provide rather standard, species-dependent responses to common situations. These behavior patterns do not seem to be motivated by obvious transient emotional states such as fear or anger.

The existence of rather stable instinctive motives (or values) poses a serious methodological problem. Where the instinctive motives are subject to rather large natural changes (as is the case with fear and anger) it is possible to observe changes in behavior. The observed behavioral changes can then be correlated with indicators of motivational state such as facial expression, pulse rate, skin conductivity, muscle tension, and measured brain waves. This provides a fairly systematic method for investigating these variable "value" components. Unfortunately, this simple approach cannot be used to study the more stable components of motivation. We cannot quantitatively measure the effects of these stable motives on behavior because we have no way of "turning off" the values, to provide a controlled comparison.

We may be able to gain some insight with regard to the structure of these value systems by making comparisons between individuals who may have inherited very different value systems. For example, we can obtain some information about the effects of stable motives by making behavioral comparisons between male and female, in situations where there are reasons to believe that their instinctive values are different. We can also gain insight by observing behavioral differences between normal individuals and individuals with motivational abnormalities. For example, the transsexuals provide one excellent source of comparative behavioral data.

There is reason to believe that some of the classical personality disorders may in fact be caused by distortions in the innate value system. Some psychopathic personalities are almost completely devoid of feeling for others, as if they lack a large segment of the innate "social values" that are part of the normal human inheritance. The schizoid personality may have abnormally strong innate values associated with dominance and at the same time may be abnormally weak in other social values. Mental patients suffering from "depression" experience a general loss of motivation, almost as if the whole instinctive "value system" were failing to function.

Of course, many mental disorders involve much more than just a distortion of the innate value system. For example, in cases of severe depression there are usually a series of related problems, such as obsessions which involve much more than a simple loss of motivation. In schizophrenia there is a loss of contact with reality and an inability to distinguish hallucinations from actual experience. Such disorders probably would not be very useful in analyzing the "value structure" because they involve at least a partial or temporary loss in the capacity for rational thought.

Although many mental disorders involve much more than just a distortion of "innate values," there is an extremely wide variety of human personality and behavioral disorders. As scientists became more aware of the motivation system as a possible source of behavioral disorders, we can expect to learn much more about the role of this innate "value system" in both normal and abnormal behavior.

MEASURING SOCIAL PREFERENCES—THE "CHOICE BOX"

How can we systematically study the values that motivate a biological decision system? Gene Sackett of the University of Wisconsin has devised one simple and obvious method for analyzing stable motivations. You place the decision system in a "choice box." By presenting the animal with a well-controlled series of choices and measuring the frequency of decisions, it is possible to learn more about the instinctive values that motivate behavior. In 1970, Sackett (11) reported on a series of such experiments which provide additional evidence for the existence of innate social motivations.

The "choice box" used by Sackett was shaped like a hexagon. When a test animal was placed in the center of the choice box, he was surrounded by six separate transparent Plexiglas panels. Through each of these panels, the test animal could view a different cage which might contain one or more stimulus animals. After an introductory period of about five minutes, all of the transparent panels were raised so that the animal could enter any one of the six choice compartments beyond the Plexiglas panels and approach the cage of one of the stimulus animals. The ensuing choice period lasted about ten minutes. During this time, the experimenters recorded the amount of time the test animal spent in each of the six choice compartments. In some cases, the test animals were too timid to actually enter the choice compartments. In these cases, the experimenters recorded the length of time the animals spent facing toward specific stimulus animals.

One of Sackett's objectives in the experiments was to demonstrate the existence of innate patterns of social preference which depend on age and are not the result of learning experiences. For this reason the experiments were carried out using animals which had been deprived of normal social experience. The experiments were very successful in demonstrating the existence of innate preference patterns. Unfortunately, the test series was sufficiently limited that a number of ambiguities remain in the interpretation of the results.

Moreover, because the test animals were reared in abnormal isolation, it would not be justified to assume that the observed preference patterns necessarily correspond to the normal maturation of social preference. Nevertheless, the patterns of preference that he observed, as well as the age dependence of the preferences, correspond very closely with the behavioral patterns of animals in the wild. Consequently, it seems to be a very reasonable interpretation that social behavior observed in the wild is in fact motivated by innate patterns of social preference very similar to the ones observed in Sackett's experiments.

Sackett found that socially isolated infants up to the age of about one year showed a strong preference for adult females of their own species (as opposed to adult females of other species). This preference must be instinctive since the animals had been separated from adults of their own species since birth, and the adults of any monkey species are so different from the infants that the result could hardly reflect similarities to themselves or to age-mates in nearby cages.

Unfortunately the experiments were not designed to determine whether the infant response was triggered by species specific physical characteristics or by social signals transmitted by the adult female. The strong preference for adults of the same species was less pronounced in older, more socially experienced juveniles. The apparently reduced preferences shown by the experienced juveniles might reflect an increased curiosity by the experienced juveniles concerning unfamiliar animals, or it might reflect reduced social signaling by the adult female, when the infant is older and less "cute."

To confirm the existence of purely instinctive responses, a number of monkeys were reared in isolation except for projected color slides showing monkeys in a variety of activities. The monkey pictures were interspersed with a number of nonmonkey control pictures. It was found that the monkeys generally responded with greater interest and activity to the monkey pictures. Exploratory and play type of responses seemed to be initiated particularly by pictures of other infants. When the subjects reached the age of about 80 to 120 days, they began to respond "fearfully" to threat pictures, even though these pictures had not previously produced a fear response!

Sackett also conducted tests in the "choice box" to determine the preference of the animals for adults of different sex. One might logically expect that very young animals, dependent on female care, would show a preference for adult females over adult males and that this preference should wane as the animals mature. The results are particularly compelling because they were obtained with test animals that had had little or no opportunity to interact with adults of either sex, so the observed preferences are almost surely instinctive. Figure 10.4 shows the observed shift of preferences with the age of the test animals. If the preferences had been purely random one would expect that 50 percent of the time the animals would select their own sex. The large departures from the 50 percent line in Figure 10.4 therefore appear to be indicative of real "innate" preferences.

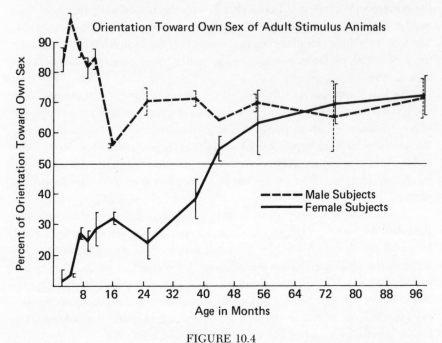

FIGURE 10.4

From *Primate Behavior: Developments in Field and Laboratory Research*, vol. 1 (1970), ed. L. Rosenblum. By permission of Academic Press.

During infancy both sexes show an overwhelming preference for female adults, which continues for about the first six months in male infants and for the first year in females. Apparently at that time curiosity becomes important and a more balanced distribution is shown. Nevertheless, both sexes continue to prefer female adults until they reach maturity. With the onset of adulthood at about forty months of age, the male interest shifts more to adult males, while the female interest in female adults continues to predominate. These instinctive changes in interest correspond closely to the role each animal plays in the troop society at the ages in question. While the experiment seems to confirm the role of the "innate values" in governing age appropriate behavior, it remains unclear whether the "preference" of the infants was simply an innate response to social signals transmitted by socially experienced adults. To remove such ambiguities control experiments would be needed in which the stimulus animal is also an "isolate," or in which stuffed or dummy animals are used to provide the stimulus. Nevertheless, the conclusion that the behavior of the infant reflects an "innate" response to *something* seems unambiguous.

Using the same group of socially deprived subjects, Sackett also tested their "innate" preferences for age-mates of different sexes. Figure 10.5 shows the results of these experiments. The figure shows that female adolescents shift their interest toward the opposite sex at an age between sixteen and twenty-four months. Male adolescents continue a predominant interest in their own sex until they are young adults, around forty months. But the adult male eventually shows an even more intense interest in the opposite sex than the female. The slower maturing of the male sex interest may help to explain the independence phase of the male adolescent, at an age where the female adolescent is likely to become attached to an adult male consort. A somewhat similar age differential in the maturation of sexual interest seems to occur in a large portion of human adolescents.

One important characteristic of monkey behavior in the wild is the tendency to distinguish between "them" and "us." Strange monkeys and members of other troops are usually treated with hostility, whereas members of the same troop are treated with tolerance. It is natural to ask whether any basis for this behavior would appear in "choice box" experiments.

Subject monkeys were placed in the choice box and were given the opportunity to choose between stimulus monkeys that had been raised in dif-

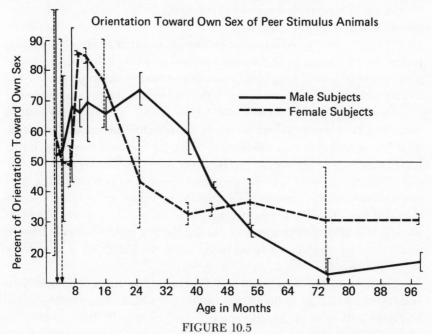

FIGURE 10.5

From *Primate Behavior: Developments in Field and Laboratory Research*, vol. 1 (1970), ed. L. Rosenblum. By permission of Academic Press.

ferent types of environment. Even when all the stimulus monkeys were total strangers to the test animal, the test animals showed a marked preference for animals that had been reared in a similar environment! Animals that had been reared in isolation preferred other isolates. Those reared in wire cages preferred other wire-cage-reared animals. Those reared in isolation except for one age-mate preferred animals with a similar rearing experience. It appears that they were able to recognize animals whose social responses were similar to their own, and they showed a marked preference for animals whose response they "understood." This innate preference for individuals with familiar responses seems consistent with the suspicious human response to strangers, outsiders, and members of other ethnic groups.

The "choice box" was also used to measure interactions between infants and mothers. Infants were given a choice between their own mother and other female adults. Except for infants raised by several different "mothers," all the infants showed a marked preference for their own mother. This preference was strengthened in infants that had been previously separated from their mother. But the strongest attachment by far was shown by infants raised by inadequate rejecting mothers (mothers who themselves had been raised without a mother). This strong attachment of the infant to an inadequate mother probably reflects the unfulfilled hunger of the infant for adequate mothering. This interpretation is supported in other experiments in which infants of about ten months of age were given a choice between an adult female or an age-mate. Infants that had received adequate or normal mothering showed a general preference for the age-mate, but infants raised by multiple mothers or inadequate mothers showed a preference for adult females, and this preference was by far the strongest for infants raised by the inadequate (motherless) mothers.

The "choice box" seems well suited for evaluating the relative strength of instinctive motivations. Where no strong innate motivations exist, the curiosity drive spreads attention rather evenly over the choice alternatives, but as specific motivations become strong they overcome the diffusing effect of curiosity. Thus the time spent in specific choice compartments is probably a good rough measure of the intensity of the underlying "values" or motives. It seems clear that the "choice box" type of experiment can provide a powerful tool for investigating animal motivations, but to use the tool most effectively, experiments must be carefully designed to minimize the likelihood of alternative interpretations.

Having established the existence of instinctive (or unlearned) social preferences which change as the animal matures, it would be interesting to see how these instinctive preferences evolve in normal animals. Thus, a repetition of many of the experiments using subject animals raised in more normal conditions would seem to be warranted. If such experiments were combined with others using dummy or isolate stimulus animals they should help to clarify the structure of the innate motivations that underlie normal behavior in the life of the troop.

MEASURING INSTINCTIVE HUMAN INTERESTS

The choice box is not likely to be very useful in measuring innate human values. It is too likely that human subjects will want to influence the results of such an experiment in one way or another. The measurement of human motivations will probably require instrumentation for measuring subconscious responses to stimuli. If our hypothesis about the structure of the brain is correct, the instrumentation should be aimed at detecting activity in the frontal lobes, the midbrain, and the hypothalamus. One set of experiments, however, has already shown that at least in principle it is possible to learn quite a bit about the innate human motivations.

It has been observed that the pupil of the eye dilates measurably when looking at "interesting" or "pleasant" materials as opposed to neutral ones. Because the response is completely unconscious, it can be used as a rather objective indicator of an individual's subjective response. Figure 10.6 illustrates some typical results showing how men and women respond to various pictures. The difference in the responses of men and women to pictures of a baby, or mother and baby, is particularly revealing. It matches the observa-

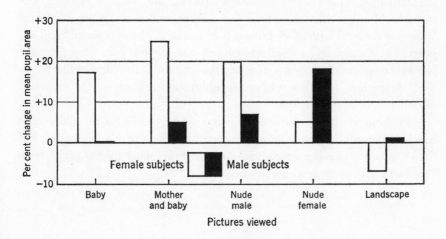

FIGURE 10.6
Pupil Size Variations

Changes in mean pupil size, in terms of percentage of decrease or increase in area, from size during viewing of control patterns, in response to various pictures.

From "Pupil Size As Related to Interest Value of Visual Stimuli" (1960) by E. H. Hess and J. Polt. Copyright © 1960 by the American Association for the Advancement of Science.

tions on other primates and confirms the common stereotype about reaction of men to newborn infants.

There is clearly a need for a wider spectrum of measurement techniques so that we can obtain a more complete understanding of the structure of innate human values.

SUMMARY

From an evolutionary design perspective it is obvious that the survivability of a species can be greatly enhanced through the introduction of "Social Values." Human sexual behavior provides unambiguous evidence that complex and detailed human social motivations do in fact exist. The chapter shows how the social signaling mechanism can provide a communication system to trigger other social motivations and thus coordinate cooperative behavior. Observations of other primate societies suggest that the innate Social Values associated with this signaling system are quite complex and detailed. The innate motivations seem to depend in very important ways on both the age and sex of the individual. The extensive parallels between the behavior of other primates and human behavior suggest that a large fraction of our innate "Social Values" are shared with other primates. Although the human sexual bond appears to be a uniquely human development, comparisons with other primates provide considerable insight concerning the origin and structure of this bond. The "choice box" experiments provide additional evidence for the existence of "innate" social motivations.

In the next chapter we will focus our attention first on the differences which distinguish human social motivations from those of other primates, and then proceed to develop a more specific hypothesis about the valuative structure that underlies human social motivations.

REFERENCES

1. Burhoe, Ralph Wendell. Editorial in Zygon: Journal of Science and Religion 9, no. 1 (March 1974).
2. Burhoe, Ralph Wendell. "The Source of Civilization in the Natural Selection of Co-

adapted Information in Genes and Culture." *Zygon: Journal of Science and Religion* 11, no. 3 (September 1976).

3. Devore, I. *Primate Behavior: Field Studies of Monkeys and Apes.* New York: Holt, Rinehart and Winston, 1965.

4. Gavin, J. A. *The Non-Human Primates and Human Evolution.* Detroit: Wayne State University Press, 1955.

5. Green, Richard, and John Money. *Transsexualism and Sex Reassignment.* Baltimore: Johns Hopkins Press, 1969.

6. Hess, Eckard H., and James Polt. "Pupil Size as Related to Interest Value of Visual Stimuli." *Science* 132 (1960): 349–50.

7. Jay, Phyllis C. *Primates: Studies in Adaptation and Variability.* New York: Holt, Rinehart and Winston, 1968.

8. Kummer, Hans. *Primate Societies.* Chicago: Aldine-Atherton, 1971.

9. Morris, Desmond, ed. *Primate Ethology.* Chicago: Aldine, 1967.

10. Rosenblum, Leonard A. *Primate Behavior: Developments in Field and Laboratory Research,* vol. 1 (1970); vol. 2 (1971). New York: Academic Press, 1970.

11. Sackett, Gene P. "Unlearned Responses, Differential Rearing Experiences, and the Development of Social Attachments by Rhesus Monkeys." *Primate Behavior: Developments in Field and Laboratory Research.* L. Rosenblum, Editor. New York: Academic Press, 1970.

12. Schenkel, Rudolf. *Ausdrucks-Studien an Wölfen* Leiden: E. J. Brill, 1948.

13. Schrier, Allan M., and Fred Stollnitz. *Behavior of Non-Human Primates,* vols. 3 and 4. New York: Academic Press, 1971.

14. Shepher, J. "Voluntary Imposition of Incest and Exogamic Restrictions in Second Generation Kibbutz Adults." Ph.D Thesis, Department of Anthropology, Rutgers University.

15. Southwick, Charles H. *Primate Social Behavior.* New York: Van Nostrand Reinhold, 1963.

16. Thorpe, W. H. *Science Man and Morals.* London: Methuen, 1965.

17. Van Lawick-Goodall, Jane. "A Preliminary Report on Expressive Movements and Communication in the Gambe Stream Chimpanzees." *Primates: Studies in Adaptation and Variability.* New York: Holt, Rinehart and Winston, 1968.

18. Van Lawick-Goodall, Jane. "The Behavior of Free-living Chimpanzees in the Gambe Stream Reserve." *Animal Behavior Monographs,* 1968.

19. Van Lawick-Goodall, Jane. *In the Shadow of Man.* Boston: Houghton Mifflin, 1971.

Chapter 11

Human "Social Values"

All that makes life seem worthwhile
Dwells in your eyes, and
The spell of your smile.

HERZER AND LÖHNER (6)

THE PRECEDING CHAPTER focused primarily on patterns of social motivation common to many primates. This chapter is concerned specifically with the structure of *human* "Social Values." The first three sections develop evidence concerning the evolution and general structure of human social values. These sections include a discussion of the structure of primitive human society, an analysis of human facial expression as a "social signaling" system, and a review of the *evolution* of facial expression from other primates to man. Following the presentation of evidence, the next four sections develop a more detailed theory or hypothesis about the structure of the Social Value system. The final sections are concerned with some of the practical and theoretical implications of the theory.

How much difference should we expect between human social motivation and the social motivations of other primates? The line of evolution leading to man separated from the other primates about 10 to 20 million years ago. This span of time is sufficient to permit substantial modifications in the innate motivations of the species.

The large frontal lobes of the human brain are one of the newest developments in human evolution. Substantial regions of the frontal lobes seem to be even newer than the speech areas. Like the speech areas, these areas distinguish the human brain from the brains of the great apes. The frontal lobes play an important role in subtle aspects of social and intellectual behavior. Individuals who have suffered damage to these areas are lacking in intellectual motivations, social sensitivity, and appreciation of the feeling of others. The frontal lobes appear to operate, together with the hypothalamus, to provide modern man with a more sophisticated and delicately tailored set of social and intellectual values. How should we expect human motivations to differ functionally from those of the other primates?

Probably the most important distinctions developed during the evolution of primitive man as a hunter. When our prehuman ancestors shifted to hunting as a way of life, they began with the physical and behavioral heritage of a primate species. They lacked natural weapons such as fangs and claws, so their success in the hunt depended on mutual cooperation and the use of tools and weapons. For example, when a primitive tribe hunts, one group will frighten the prey and drive it toward another group that waits in ambush. The kill is accomplished with weapons, and tools are used to prepare the carcass for eating.

This basic strategy for success in the hunt required intelligence, communication, and cooperation. Over the ages, evolution responded with the necessary genetic modifications. The brain increased in size. Improvements in communication were achieved at first in the traditional primate way, with a richer inventory of facial expressions, to reflect moods and motivation, and through the enrichment of vocal calls and cries. As intelligence improved (and the vocal inventory of sounds increased) a dramatic improvement in communication occurred. Language began to be used to communicate specific ideas and concepts rather than just a general emotional state. With this breakthrough in language, facial expression no longer had to carry the full burden of communication. Specialized primate expressions that had served detailed communication functions were no longer necessary, so a reserve of capacity in the facial signaling system became available that could be adapted to other purposes.

Although language made cooperation technologically possible, it would not have led to cooperative *behavior*, without the parallel development of the *motivation* to cooperate. Thus the hunting strategy also generated evolutionary pressures which strengthened cooperative motivations. The existing innate values motivating social cooperation were strengthened, and new, uniquely human social values began to emerge.

Although language (as a tool of the rational mind) was able to take over much of the responsibility for specific communication, it could not take over the *signaling* role required for social *motivation*. Because languages are a product of the rational mind, they evolve too rapidly and are too variable to provide a reliable mechanism for triggering the innate motivations. As the role of facial expression in communication decreased, the role in *motivating* social behavior increased. The facial signaling system was modified to meet the human need for better motivation of cooperative behavior.

The need for cooperation in a primitive hunting society extends beyond just cooperation in the hunt. The society must develop methods for sharing food so that the individual is not completely dependent on his own success in the hunt. When a large animal is killed the food must not be monopolized by just the dominant animals or the ones that actually accomplished the kill. So the success of a hunting society requires the development of sharing. The development of mechanisms of sharing seems to be common among hunting animals. For example, a wolf that has participated in a large kill may return to his mate and instinctively regurgitate undigested food. No such innate sharing mechanism exists in any of the primates.

It has been suggested that one of the key reasons that the great apes were unable to evolve into a hunting society is that they failed to develop a capacity for sharing. Within a vegetarian and insect-eating society, each adult can be responsible for his own feeding. When resources are scarce, the rule of dominance prevails.

To provide the necessary motivation for sharing, evolution faced a challenging design task. The problems were much like those faced in the operation of a socialist commune. How can each individual maintain a motivation to work if others will willingly share the products of their labor? Nevertheless, the problem had to be solved if the human animal was to be successful as a hunter. The next few sections are concerned with the methods evolution used to solve this basic problem.

To provide some clues concerning the nature of the solution we will discuss briefly the social structure of primitive societies that have survived into modern times. Although many of these societies include a rudimentary agriculture (and in this sense are not typical of a truly primitive hunting and gathering society), they still depend heavily on hunting, and they are the closest modern approximation to a primitive human society.

PRIMITIVE HUMAN SOCIETIES

Human cultural evolution has moved progressively toward larger and more formal social structures. Elman R. Service (15) of the University of California has defined a number of stages in this evolution. Like all such classifications, the boundary between the stages is somewhat ambiguous, but it nevertheless provides a useful organization for analysis of cultural evolution. His cultural sequence includes the following stages: band, tribe, chiefdom, and primitive state. The band is the most primitive and loosely organized form of society. It operates as a cooperative, loosely defined group of families without a formal leader. Tribal organization differs from the band organization in that there is a well-defined leader and unambiguous membership in the tribe. There is also some specialization of function such as a medicine man or shaman. The chiefdom involves more specialization of functions in government, and usually includes at least a loose definition of territorial boundaries. The primitive state or kingdom is the direct predecessor of the modern nation-state.

For our present purpose, the very primitive band society is of greatest interest. It is the most useful for an evolutionary comparison, since it most closely resembles the societies of the nonhuman primates. Given the generally accelerating pace of cultural evolution, it is probable that this most primitive form of society was predominant during most of human evolution. It is from this type of society that we should expect to learn the most about the original function of man's innate "Social Values."

The primitive human societies that have survived into modern times are located in remote and isolated areas that are widely scattered over the globe. They appear to be remnants of almost unrelated primitive cultures that were adapted to different ecological environments. Nevertheless, there are remarkable similarities in the behavioral characteristics of these primitive cultures. The societies are built around the family. The basic social unit is the man, his wife, and their offspring. The dominant family pattern is the monogamous marriage, but polygyny occurs in situations when there is a shortage of eligible males. Polygyny also occurs in individual cases when a widow marries her husband's brother as a way of providing suitable support for the children.

The universality of this basic marriage pattern in primitive societies testifies to the success of evolution in instinctively binding the human male and female. Of course, the innate basis of this bonding may be less effective within the framework of modern society; but until modern times, the marriage bond has been one of the remarkable constants of human society. Even where the custom has not required sexual fidelity, the institution of marriage has existed as a durable human partnership.

The individual families in such primitive societies see themselves as a part of a larger group that lives and hunts in the same general area. This larger group is linked by complex ties of intermarriage. It seems to correspond closely to the troop in nonhuman primate societies. Depending on the environmental pressures, the overall group may be subdivided into smaller bands of families that maintain intimate social and economic ties. In the most primitive societies there is no formal or legal structure to organize the bands and maintain unity in the larger group. The unity is maintained informally, and intermarriage between neighboring bands contributes to a sense of community with others of a similar cultural heritage.

This primitive human society shows many similarities with other primates. The leadership is informal and it is exercised primarily by mature males that have achieved a position of respect in the society. In almost all primitive societies the older generation is respected for their experience and wisdom, and the respect is formalized in the language and traditions of the society.

The striving for dominance that is so typical of other primate males continues in the human society. Individuals are very concerned with their social status and their general level of acceptance within the society. Social status is based on much more than sheer physical force. It involves a judgment about the individual's generosity, his wisdom, his contributions to society, his bravery, and his acceptance of the social norms and customs of the society. A large fraction of the striving in primitive human societies is motivated by concern for social status and acceptance. This concern itself appears to be an "innate" human trait. It is possible that some of the basic criteria used to judge the quality of an individual—wisdom, generosity, bravery, and conformity with the norms—may also reflect innate human values.

Primitive human societies differ greatly from other primates in the *amount* of cooperation that is achieved. Within primitive human societies "sharing" is a way of life. The exchange of gifts is so routine that in many societies it is considered bad manners to express appreciation for a gift, since this might imply that the gift was not expected or that the generosity of the giver was in doubt. The sharing is not limited to food, but extends to all types of resources. The practical result is that scarce resources are shared within the societies approximately in proportion to need. This behavior may reflect some innate and uniquely human values that evolved during the transition to a hunting economy.

SOCIAL MOTIVATION AND FACIAL EXPRESSION

Basically, facial expressions are a system for signaling the mood or motivational state of an individual. What types of information are of highest priority for transmission by such a communication system? Obviously there is no reason to transmit static or constant moods or motivations. Because such motivations do not change, there is no information to send. Among those types of motivation subject to change, the highest priority should go to those that are important in "social" as opposed to "individual" behavior.

Traditionally, the innate values which motivate human behavior have been classified into two categories: biological drives and emotions. From our present perspective, biological drives are simply instinctive motivations that vary in response to obvious *physical* needs. Emotions are motivations that vary in response to the less obvious *social* and environmental stimuli. For this reason, it is natural that most facial expressions are connected with emotions rather than biological drives. Obviously there are some exceptions. For example, intense pain or agony is reflected in a characteristic facial expression. Apparently facial expressions are not limited to "emotions" but can be used to convey information about any motivational factors that are subject to variation. In general we should expect that facial expressions would be reserved for those motivational factors whose transmission is most important in "social" behavior.

In our previous discussion of emotions such as fear or anger, we considered the motivational consequences only for the individual who experiences the emotion. But when an "emotion" is displayed in a facial expression, it can serve as a motivational trigger for other members of the society. To understand the survival value of the expression (and the emotion that causes it), we must consider the impact of the expression on the behavior of others, as well as the direct effect of the emotion on the individual who experiences it. Indeed, it is even possible that the primary purpose of an emotion could be simply to *generate* a facial expression, so that other individuals can respond!

The close link between facial expressions and emotions has led to a recent revival of interest in facial expressions as a way of studying emotions (1, 2, 4, 5, 7, 8, 13, 14, 16, 19). As a result of this work a number of conclusions seem rather firmly established. As we would expect, the facial expressions associated with specific emotions appear to be "innate" or instinctive. The *same* basic expressions are identified with the *same* emotions (and can be recognized by all human races) in all societies. Although there is some variability in the reliability of recognition in different cultural and racial environments, the similarities seem far more important than the differences.

Although facial expressions are normally "innate" or instinctive responses to emotions, it is possible to achieve a high degree of conscious control over the facial expression. Natural facial responses can be suppressed, and an individual can learn to counterfeit artificial expressions that are very similar (if not identical) to the natural expressions.

There appears to be a reinforcing feedback process between the facial expression and the emotions. A person who deliberately forces a smile is soon likely to feel more happy. A person who feigns an angry face is soon likely to feel angry. Conversely, if the expression for an emotion is suppressed the emotion itself may fade more quickly. But there is some evidence that such conscious interference with the expression of emotions is likely to produce in-

ternal tensions or anxiety. Facial expressions also tend to be socially con-
tagious. A smile generates other smiles, a sad or dejected face tends to make
others sad.

The conscious identification of expressions with specific emotions is nor-
mally learned during childhood. The success of children in recognizing emo-
tions in pictures of facial expressions improves gradually with maturity. This
obviously is not inconsistent with "innate" or instinctive *emotional* responses
to facial expressions. It only means that the association of facial expressions
with specific emotional states and the *names* of the emotions is a learned capa-
bility.

Because of the obvious relationships some investigators have assumed a
one-to-one correspondence between facial expression and emotion (7, 19).
The present theoretical framework does not support this assumption. Because
the number of easily distinguished facial expressions is quite limited, emotions
whose communication is not critical to social behavior may not appear in the
facial expressions. Moreover, where the distinction *between* two emotions is
not socially important (or where other signals, such as a laugh or cry, can be
used to distinguish) then *two* emotions may produce essentially the *same*
expression. Finally, facial expressions may serve to communicate other infor-
mation about the motivational state, such as "pain" or "anguish," which is not
usually considered to be of "emotional" origin.

Because facial expressions serve as the main information channel for
transmitting information concerning social motivation factors, it provides some
of the best clues available for deciphering the structure of human emotions
and social motivation. For our present analysis, we will assume that the facial
expressions will include those emotions whose communication is most impor-
tant for motivating social behavior. Although different investigators differ on
many details there is remarkable agreement that at least the eight facial ex-
pressions shown in Table 11.1 can be distinguished with quite high reliability.

TABLE 11.1
Some Easily Distinguishable Facial Expressions

Fear, terror, apprehension
Anger, rage, determination
Surprise, startle, amazement
Interest, attention, expectancy
Joy, mirth, pleasure, happiness
Sorrow, distress, anguish, pensiveness, sadness
Contempt, disgust, scorn, revulsion
Shame, blush,* humiliation

---

* Sometimes not considered an expression.

The range of emotional states encompassed within each type of facial expression is illustrated by the list of words on the line. In some cases these may reflect different intensities of the same basic emotion. In other cases they may reflect separate emotions that produce a very similar expression. To begin our investigation of the way these expressions operate in evolution's design concept for human social behavior, we will review briefly the motivational purpose of these expressions and their associated emotions.

FEAR AND ANGER

These emotions were discussed earlier in connection with the "Selfish Values." In that context, fear motivates a withdrawal from a fearful object and a retreat to a location of security. Anger motivates an attack on an object which is frustrating the achievement of an objective. The fact that these emotions also show up as facial expressions suggests that they also play an important role in social behavior, and thus serve both as "Selfish" and "Social" values.

There are at least two obvious social functions that could be served by a facial expression that portrays fear. (1) When there is a threat external to the social group, an expression of fear can serve as a warning to others. For such external threats we might expect that fear should be accompanied by vocal cries or screams. (2) Fear can also arise in situations internal to a troop. When a weaker animal is attacked by a stronger animal it is important to terminate the conflict before too much damage occurs. It is possible that an expression of extreme fear or terror on the face of the victim may serve as a triggering signal to reduce anger and thus inhibit the stronger animal. Thus there are some obvious social benefits in the expression of fear, but what about anger?

It may seem paradoxical to consider anger as a Category II or "socially beneficial" value. What survival benefit can be achieved by a motivation which leads to fighting between members of the species? Perhaps the best way to address the question is to ask what type of behavior within a species might be expected if anger had been omitted from the "Social Values." From a purely "rational" point of view (in the absence of anger) the weaker animals would be wiser to let a stronger one take their food, rather than risk physical injury by resisting. But such behavior could ultimately lead to starvation for the weaker animals, who would become progressively weaker. Anger motivates the weaker individual to fight, even when it may seem "irrational" or even suicidal to do so. It serves as a deterrent to the stronger animal who knows he cannot achieve his goal without a fight. Thus, the presence of anger helps to achieve a more "equitable" distribution of resources.

Recent work on two-person "nonzero sum" games has helped to clarify the importance of anger in social bargaining.* Originally mathematicians were disturbed that they could not produce "rational" or logical solutions to "nonzero sum" games similar to those developed for simpler "zero sum" games. But it is now widely accepted that no simple "rational" solutions exist. To illustrate the essential features of a "nonzero sum" game, consider the problem of dividing a large inheritance between the wife and brother of a deceased. The will specifies only that they must agree on a division, otherwise the entire inheritance will go to charity. The brother stubbornly insists that he should receive 90 percent of the money, and the wife must decide whether to accept 10 percent or let the entire inheritance go to charity. The situation is clearly not "zero sum," because the total amount of money to be divided depends on whether they can agree. From a strictly rational point of view, the wife should accept 10 percent since she will be much better off with 10 percent than none. But in all probability the wife will get angry. She will insist on a "fair" share, and in her anger she will let the money go to charity rather than allow it to go to the greedy brother. In practice, the threat of an "irrational" angry response serves as a deterrent to greedy behavior by the stronger individuals, and provides a social basis for resolving nonzero sum situations.

It seems probable that the primitive origins of our concepts of "fairness" and "equity" and "natural law" can be traced to the emotion of anger. A "fair" solution to a problem is one that will not make either party angry or that will leave the two parties about equally angry. From this point of view, we should expect that the situations that stimulate anger would have been adjusted by evolution so that the interplay of anger between members of a species will resolve disputes in ways that are *functional* for the survival of the species.

This social role of anger makes it obvious why it is so important to have the emotion *show* in a facial expression. When an expression which portends anger appears, even a dominant animal knows he is risking a fight. He can alter his behavior before an actual fight develops.

The emotion of anger also serves a similar role in the relations *between* species. The fact that animals of any species are likely to get angry when their "rights" are violated serves as a deterrent to other species. For example, even though an individual bee that stings in protecting the hive may die, the bee's anger is functional for the survival of the species. Clearly the choice of the circumstances that will give rise to anger must be one of the most important considerations in evolution's design for a species.

---

* A "nonzero sum" game is characterized by the fact that the total payoff to the competing players is not fixed, but can depend on how well they cooperate.

SURPRISE AND INTEREST

The surprise or startle reaction occurs in response to a sudden and unexpected (but potentially significant) event, such as a loud noise, or the sudden appearance of a large animal. The reaction diverts attention from prior activity and automatically makes all intellectual and sensory resources available to deal with the unexpected event. A facial expression denoting surprise can serve a "social" function by alerting others. The specific facial expression for surprise, eyes wide open and mouth open, also serves a direct functional role by making vision available and preparing to attack if necessary.

Interest is not usually considered an emotion, although it does correspond to a subjective mood. Logically, interest and surprise are related. Interest is interrupted by surprise, and surprise results in at least a temporary transference of interest. A facial expression which shows interest is particularly useful in a nonverbal society where it can call attention to phenomena of "interest." Probably for this reason, the facial expression showing interest is considerably more obvious in nonhuman primates, although the expression continues in humans in a somewhat attenuated form.

JOY AND SORROW

Joy and sorrow play a vital role in the motivation of human "social" behavior. Joy is a positive value, sorrow is a negative value, so the individual is motivated to avoid sorrow and to seek joy. Success in those social activities which serve a beneficial evolutionary function is rewarded with joy; failure to engage in such activities or to succeed in them is penalized with sorrow. Unlike the emotions of fear and anger which are as ancient as the crocodile, "joy" and "sorrow" are relatively recent evolutionary developments that are almost exclusively "social" in their function. They do not seem important in herd animals—neither sheep nor cows seem to exhibit any evidence of joy or happiness. Even among the primates, joy and sorrow seem to be quite muted, except in the most socially cooperative species.

The ability to experience these two emotions seems to be linked both to intelligence and to an evolutionary strategy of cooperative social behavior. Probably the most striking example, other than man himself, is the common dog. The vigor with which the dog expresses joy is one of his most endearing characteristics as a pet. The widespread emotional attachment between man and dog seems to be based more than anything on the fact that both species experience intense emotions of joy and sorrow. There is reason to believe that in both species this emotional structure grew out of an evolutionary strategy of cooperative hunting. But why is the expression of joy and sorrow so important for the motivation of cooperative behavior?

Suppose we wanted to design a general-purpose system to motivate cooperative behavior. How would we go about it? We probably would try to design a system such that benefits achieved for one individual would be valued as a benefit to all. The expression of joy and sadness serves this basic function. The expressions are contagious, so that one individual's experience of joy or sadness tends to be shared by all. This automatically produces a motivation for cooperation.

Joy or sorrow usually occurs as a consequence of success or failure in dealing with other objectives or values. The dog wags his tail with the arrival of food. The return of a sex or grooming partner produces joy. Loss of such a partner produces sorrow. Thus activities that might appear to be adequately motivated by other values are *nevertheless* accompanied by changes in the degree of joy or sorrow. From the narrow perspective of individual motivation, the occurrence of joy or sorrow in such situations tends to *increase* the motivation to restore a loss. For example, the intense sorrow experienced with the loss of a loved one produces a large and immediate negative value that can be turned off only by finding the lost one. For an animal with a limited ability to think ahead, this immediate value which effectively anticipates future unhappiness can intensify immediate efforts to recover the loss.

But the most important advantage of this linkage of joy and sorrow with other motivations comes from the *social* consequences. When joy and sorrow are linked in this way to other value elements, it automatically links the facial expression to the *other* objectives. In this way, the cooperative motivation produced by the facial expression is extended to include *all* activities of the species. Because the experience of joy or sorrow is contagious, each individual can contribute to his own happiness by assisting another who is in distress. Like all human motivations, the importance and effectiveness of this basic "altruistic" motivation varies widely between individuals. Nevertheless, the basic evolutionary purpose of these emotions and expressions seems clear.

CONTEMPT AND SHAME

The presence of these two emotions on a list of only eight key facial expressions is particularly striking. The functional purpose of these emotions in social behavior is intuitively obvious. Their purpose is to motivate conformity with prevailing norms of group behavior.

An expression of contempt serves to notify an individual that he has violated accepted social standards. His experience of "shame" as a consequence serves to punish his transgression. The fact that this mechanism of social control is included in the very limited inventory of facial expressions tells a great deal about the structure of human motivation. An expression of contempt could not be an effective social control unless concern for the opinions of others were also an innate human motivation.

THE EVOLUTION OF FACIAL EXPRESSION

In the preceding section we used the evidence from human facial expressions to provide some clues about innate Social Values. In this section we will check to see if the evidence concerning the evolution of facial expression is consistent with our functional interpretation.

As we mentioned earlier two broad trends should be expected in the transition from nonhuman primates to man. First, with the introduction of language, facial expression should be relieved of much of the burden for detailed social communication dealing with the nuances of "social" motivation. Second, the need for better cooperation and sharing in a hunting society should produce changes designed to motivate more cooperative social behavior. How do these expectations correspond to the evolutionary evidence?

Figure 11.1 illustrates some typical facial expressions for one of man's nearest relatives, the chimpanzee. Figure 11.2 shows a chart of the probable evolutionary development of major facial expressions in primates. The chart is based on one prepared by Suzanne Chevalier-Skolnikoff (3) and published in 1973. Obviously, one cannot ask the nonhuman primates what emotions go with their expressions. But careful observations of the circumstances in which the expressions occur have produced considerable confidence that the interpretations are essentially correct.

The chart in Figure 11.2 shows a continuous elaboration of facial expression to convey more accurate nuances of emotion as we progress from a relatively primitive primate (the lemur) to the macaque monkeys and apes. The available signaling capacity in the monkeys and apes is devoted largely to elaborations of three basic emotions—anger, fear, and affection. The elaborations seem to be concerned with the details of the social dominance hierarchy. With the advent of human language some of the elaboration of these emotions seems to disappear.

The expressions for submission, reassurance, and affection types 1 and 2 converge to a single expression of happiness and joy. In monkeys these expressions tend to occur in response to a dominant animal, but they also occur between animals that are almost equals in the dominance hierarchy. An animal that is normally greeted with this expression has achieved a high level of domi-

a. "Glare"; anger, type 1.

b. "Waa bark"; anger, type 2.

c. "Scream calls"; fear-anger.

d. "Silent bared-teeth"; type 1, horizontal bared-teeth, submission.

e. "Silent bared-teeth"; type 2, vertical bared-teeth; fear-affection (?).

f. "Silent bared-teeth"; type 3, open-mouth bared-teeth; affection.

g. "Pout face"; desiring-frustration (?).

h. "Whimper face"; frustration-sadness (?), type 1, or type 1-2 transition (infant).

i. "Cry face"; frustration-sadness, type 2 (infant).

j. "Hoot face"; excitement-affection (?).

k. "Play face"; playfulness.

FIGURE 11.1
Some Facial Expressions of Chimpanzees

Note—These drawings are presented for illustrative purposes only. They are diagrammatic and do not claim to precisely depict actual expressions of emotion. They are drawn after photographs and descriptions from van Hooff, 1971; and van Lawick-Goodall, 1968a, b. All expressions were drawn from the same angle in order to facilitate comparisons. Reproduced from Chevalier-Skolnikoff (1973), permission of Academic Press.

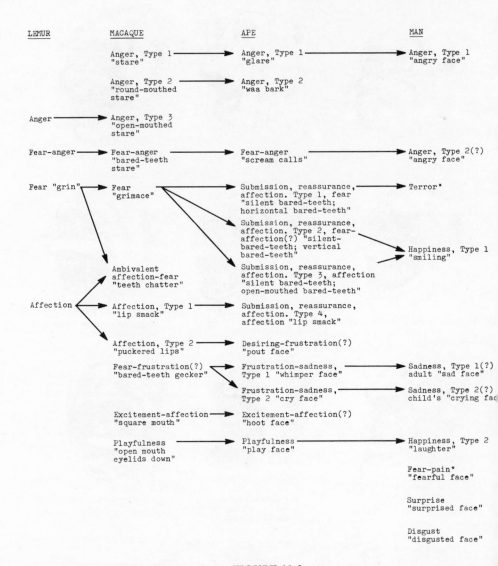

LEMUR	MACAQUE	APE	MAN
	Anger, Type 1 "stare"	Anger, Type 1 "glare"	Anger, Type 1 "angry face"
	Anger, Type 2 "round-mouthed stare"	Anger, Type 2 "waa bark"	
Anger	Anger, Type 3 "open-mouthed stare"		
Fear-anger	Fear-anger "bared-teeth stare"	Fear-anger "scream calls"	Anger, Type 2(?) "angry face"
Fear "grin"	Fear "grimace"	Submission, reassurance, affection. Type 1, fear "silent bared-teeth; horizontal bared-teeth"	Terror*
		Submission, reassurance, affection. Type 2, fear-affection(?) "silent-bared-teeth; vertical bared-teeth"	Happiness, Type 1 "smiling"
	Ambivalent affection-fear "teeth chatter"	Submission, reassurance, affection. Type 3, affection "silent bared-teeth; open-mouthed bared-teeth"	
Affection	Affection, Type 1 "lip smack"	Submission, reassurance, affection. Type 4, affection "lip smack"	
	Affection, Type 2 "puckered lips"	Desiring-frustration(?) "pout face"	
	Fear-frustration(?) "bared-teeth gecker"	Frustration-sadness, Type 1 "whimper face"	Sadness, Type 1(?) adult "sad face"
		Frustration-sadness, Type 2 "cry face"	Sadness, Type 2(?) child's "crying fac[e]"
	Excitement-affection "square mouth"	Excitement-affection(?) "hoot face"	
	Playfulness "open mouth eyelids down"	Playfulness "play face"	Happiness, Type 2 "laughter"
			Fear-pain* "fearful face"
			Surprise "surprised face"
			Disgust "disgusted face"

FIGURE 11.2

Probable Evolutionary Development of Some Facial Expressions in Primates

Reproduced from Chevalier-Skolnikoff (1973), by permission of Academic Press.
* Elaboration of original chart based on private communication from Chevalier-Skolnikoff.

nance and acceptance in the primate society. In all probability the dominant animal finds it quite rewarding to be greeted in this way. In humans the smile (a descendant of this rewarding expression) is used as a greeting between friends, to express approval or admiration, and to express joy or gratitude for a favor. From a behavorial point of view the most important aspect of the "smile" is its reward value to the recipient.

The chart also shows three human expressions which are not obvious descendants of any older primate expressions. First there is a new type of fear expression which is also used to express pain or anguish. This expression may serve to appease an attacker or to motivate assistance for an individual in pain. Second, there is the surprise face. It is unclear why this expression should have evolved in man, assuming that it did not exist in other primates. Finally, there is the expression for disgust or contempt which serves a purpose in "social" control as previously described. It is noteworthy that three of the expressions that seem new in man—joy or smiling, pain or fear, and disgust or contempt—all seem to be associated with the motivation of cooperative behavior. On the basis of the evolutionary evidence, it seems probable that the smile of joy or gratitude and the expression of contempt are both essential parts of a motivation system which produces cooperation and sharing in primitive human societies.

THE SEARCH FOR A VALUE STRUCTURE

In many ways it should be easier to develop a theory concerning human social values than it would be if we were dealing with nonhuman primates. Although the simplicity of the nonverbal societies makes it easier to be objective about their innate values, we cannot ask the animals why they behave as they do, and we cannot share their subjective experiences. As we begin to deal directly with human values we should be able to use many new sources of information. We can use judgment based on personal (subjective) experience, we can use evidence of literature and mythology, and we can use insight gained in psychological and psychiatric counseling.

Unfortunately, until very recently, the concept of innate or instinctive motivations has been generally in disfavor among psychologists. Consequently, there is little recent theoretical work by psychologists that is relevant to our approach. To provide a starting point for our discussion, we are forced to look to earlier theoretical efforts. In 1908 a remarkable little book, *An Introduction to Social Psychology,* was published by an eminent British psychologist, William McDougall. McDougall was a careful experimenter and an astute observer of human nature. He was influential in placing British psychology on an experimental and physiological basis. In his book he developed a theory of human social behavior based on the concept of inherited instinctive tendencies (motivations). Although his theory was nonquantitative and in embryonic form, it had much in common with the ideas developed here. His theory encompassed the role of emotions and sentiments in social behavior, but with the rise of behaviorism and the emphasis on quantitative rather than qualitative concepts, McDougall's ideas lapsed into disuse.

A revival of emphasis on innate motivations now seems to be in progress in behavioral science. Wilson's work, *Sociobiology* (1975), provides a unified framework for understanding the genetic origins of social behavior. In 1973, the Nobel Prize in medicine was shared by three ethologists, Konrad Lorenz, Nikolaas Tinbergen, and Karl von Frisch, whose work demonstrates the evolutionary basis of behavior. The implications of ethology in understanding human behavior have recently been emphasized by a number of respected authors: Desmond Morris, *The Naked Ape* (11) and *Intimate Behavior* (12); Konrad Lorenz, *On Aggression* (9); Anthony Storr, *Human Aggression* (17); and Lionel Tiger and Robin Fox, *The Imperial Animal* (18). But none of the recent authors has attempted a theoretical synthesis comparable to McDougall's effort in 1908. To illustrate how the concept of instinctive values can be applied to human social behavior we will build on McDougall's original framework. McDougall's language and ideas have been modernized to fit them into the value-theory concept, and a number of specific changes have been made to provide a better match with recent experiments.

McDougall's original theory was vulnerable to a number of criticisms which seemed to make it incompatible with the scientific views of the time. His work was criticized as nonoperational because he did not propose any plausible mechanism that could convert his "instinctive motives" into behavior. In the present framework the value-driven decision system provides this missing mechanism. McDougall's work was also considered ad hoc because he failed to relate his innate motivations to any identifiable motivational force. Consequently, there appeared to be no way to check the theory or to decide what might constitute a complete or compehensive list of innate motivations. The present approach corrects this serious problem by identifying the innate motivations with specific emotions and other valuative sensations. The relevant valuative sensations are subjectively recognizable and specific cybernetic systems have been identified that appear to be responsible for generating the sensations. At least in principle, therefore, the resultant innate motivations are objectively measurable. Moreover, if a model were developed that could account for the time-dependent variation of all the valuative sensations, then presumably the list of innate motivations would be complete.

Because the structure of the innate social values is comparatively complex, the discussion of these values at present is necessarily more speculative than was the case for the selfish values. As always, the specific value structure has been chosen only for illustrative reasons, to demonstrate the principles of value motivation.

COMPONENTS OF THE SOCIAL VALUE SYSTEM

We postulate that the innate value system that motivates social behavior includes three kinds of components: "emotions," "sentiments," and "instinctive motives." The "emotions" provide the actual valuative sensation which is the ultimate criterion of value for the conscious mind. When we view the brain as a decision system, the emotions can be interpreted simply as "valuative signals" that carry information from the motivation system to the conscious mind. Decisions of the conscious mind are ultimately evaluated in terms of an innate "value scale" that is defined by the emotions.

The other two components, the "sentiments" and the "instinctive motives," are internal parts of the motivation system. They are part of a computational mechanism that determines the intensity of the final "valuative signals." Their influence is perceived in the conscious mind only because of the *effect* they have on the valuative signals (the emotions) that are ultimately delivered to the conscious mind. "Sentiments" and the "motives" influence the emotions in quite different ways.

The "sentiments" become associated with specific persons, places, and objects within an individual's environment. As "sentiments" develop, the individual acquires a tendency (or predisposition) to experience certain types of emotions in the presence of the person or object with which the "sentiment" is associated. The word "sentiment" was originally used by McDougall in almost this same way to refer to an emotional predisposition, so we have simply adopted McDougall's name for these emotional components.

We will use the phrase "instinctive motive" to refer to the rather stable underlying factors which contribute to the calculation of the values. Where a single emotion (or value sensation) serves to motivate many different types of activity, it is useful to think in terms of a separate underlying "motive" associated with each such activity. We can think of each "motive" as an independent contributor to the intensity of the final emotion, or value sensation.

With the foregoing definition of a "motive" one might logically include "sentiments" as a form of motive, but there is an important distinction between these two components of the motivation system. The "motives" corre-

spond to *generalized* but genetically inherited valuative factors which influ-
ence the emotions in fairly stable and predictable ways. The sentiments are
associated with specific people and places, and they develop as a consequence
of experience. Of course, the general underlying rules that govern the forma-
tion of sentiments might still be identified as a special kind of "motive," but we
will not do so. The word "motive" will be reserved for those rather stable fac-
tors which seem to influence emotions in predictable ways. The rules which
govern the growth of sentiments will be treated as part of the "sentiment" sys-
tem.

The phrase "instinctive motive" was chosen both because the "motives"
move the emotions, and because the resulting changes in the emotions serve
to *motivate* specific types of activity. Although McDougall recognized the exis-
tence of such underlying motives (which he described loosely as "secondary
instincts") he did not suggest any plausible way of fitting these "instincts" into
a motivation system. In our present context his phrase "secondary instinct"
seems too vague and conveys meanings that are quite inappropriate to the
present theoretical structure. For this reason, we have chosen to change this
part of his terminology.

The social "motivation system" depends on detailed interrelationships
between these three kinds of components, as they operate in *different* individ-
uals in the society. As with most systems, the individual components of the
system make sense only in the context of the other components. For example,
within a clock or watch an individual gear or wheel cannot be understood in
isolation. Its real function is to mesh with the other gears, so it makes sense
only in the context of the rest of the system. The components of the human
social "motivation system" fit together in an analogous way.

Because of the intimate interrelationships there is no satisfactory place to
begin a discussion of the system. For somewhat arbitrary reasons we have
chosen to proceed in the following order:

1. Emotions
2. Sentiments
3. Instinctive motives

After all the components have been considered in a preliminary way, we will
return to consider their interaction as components of an integrated system.

THE EMOTIONS

Table 11.2 provides an illustrative list of human emotions. Although the
specific choices of emotions to include in the list have been influenced by the
experimental information on facial expression, no attempt has been made to
limit the list to those emotions that are easily recognized in the face. On the

TABLE 11.2
Illustrative List of Human Emotions

NEGATIVE	——	POSITIVE
*Sorrow	——	*Joy
*Shame [1]	——	Pride [3]
Scorn	——	Admiration [4]
Revulsion	——	Acceptance
*Anger	——	Tender Emotions [5]
	——	Pity, Sympathy [5]
*Fear	——	
*Surprise	——	*Interest, Curiosity
	——	*Amusement [2]

(Negative rows "Scorn" and "Revulsion" are braced together and marked 6)

[1] The blush is functionally a facial expression although it is sometimes not so classified.

[2] The facial expression is similar to joy but the difference in emotion is apparent subjectively and in the associated laughter.

[3] Presence of this emotion (called self-display by McDougall) is apparent in body carriage which tends to reverse the symptoms of shame.

[4] Facial expression may exist but is apparently not distinguishable from smile of joy.

[5] Apparently no obvious facial expression.

[6] Revulsion or disgust is an avoidance emotion applied to food or inanimate objects; scorn or contempt is reserved for individuals of the same or similar species. The facial expressions seem to be very similar.

contrary, an effort has been made to complete the apparent pattern by filling in generally recognized emotions that may not be represented in facial expressions. Those emotions that are clearly included in a well-defined facial expression are noted with an asterisk. No serious effort has been made to be exhaustive in the list, so in all probability there are a number of other emotions that have been omitted.

It is rare to have an experience in which only one of the basic emotions is involved. Almost all actual emotional experiences involve some combination of the basic emotions. But some emotions seem to be mutually exclusive, as if they are opposite poles of the same emotion. Where this seems to be the case, the emotions are placed on a single line, separated by a dash. Some authors have treated "fear" and "anger" in this way because superficially they seem to produce opposite behavior (attack versus retreat). But the data on facial expression, particularly in animals, where fear and anger seem to *mix* in expressions, suggests that these two emotions are really independent. The chart suggests an alternative pair of opposites which may be more nearly mutually exclusive: anger—which generates a desire to hurt or destroy; and the "tender emotion"—which produces a desire to help, to hold, or to protect.

Each of the nine lines in Table 11.2 is intended to represent a separate and independent emotion, so that each of the nine emotions can occur in any combination with any of the other emotions. Assuming that this is correct, the space of *possible* emotions is at least a *nine*-dimensional vector space. (If Table 11.2 has omitted some important basic emotions the dimensionality is even larger.) Just as any position in a room (which is three-dimensional space) can be defined by specifying three coordinates (distance from the floor, distance from the south wall, and distance from the east wall), so also any real emotion could be defined by specifying a numerical value for each of the nine basic emotions. Some authors have attempted to display the range of the emotional space in two- and three-dimensional charts (14), but according to the present view, such efforts are doomed to failure. It is not possible to depict a nine-dimensional system within a two- or three-dimensional space.

SENTIMENTS

As the infant primate gains in experience, his generalized clinging desire is conditioned or reinforced so that it applies with exceptional strength to the infant's mother, in preference to all other soft furry objects. Thus the infant becomes "attached" to the mother. This illustrates in very simple form the process that seems to be involved in the development of sentiments.

The existence of sentiments motivates the individual to respond differently to different individuals depending on the nature of the sentiment that has developed. "Love" and "hate" are examples of sentiments. Basically a sentiment is a predisposition (or tendency) to develop certain combinations of emotions in the presence of a particular individual. If we hate an individual, we are more likely to become angry or afraid in his presence. If we love an individual we are more likely to experience joy, security, and the "tender emotions" in his presence. If the love or hate is sufficiently strong, then just the mention of the name (or the thought of the individual) may be sufficient to trigger the corresponding emotions.

The exact mechanism by which sentiments develop is far from clear. It certainly involves an associative learning process and it is possible that the mechanism is much like a simple "operant conditioning" of the emotional response. For example, if a certain emotional response (in the presence of a specific individual) repeatedly produces rewarding results, the emotional response may be reinforced. It will respond to smaller and smaller stimuli, until finally the simple presence of the individual or even the mention of the name may be sufficient to trigger the emotion. The emotion, of course, is not a rational response. It occurs automatically in response to certain stimuli. This type of "operant conditioning" differs from the traditional operant conditioning of behaviorist psychology in that it is the *emotions* (not the resulting actions) that are being "conditioned." This need for a fairly elaborate associative learning mechanism in the human motivation system may have been one of the reasons why the associative cortex of the frontal lobes was needed by the human motivation system.

Of course, it is possible that the development of each sentiment may follow a different set of "prewired" rules. But as an initial hypothesis, the "operant conditioning" concept seems particularly general and simple, so that it is probably a good hypothesis to consider in initial experiments. Even within the "operant conditioning" hypothesis there are many unanswered questions. What types of outcomes are rewarding or reinforcing for each emotion? How easily does reinforcement occur? How does the development of a "sentiment" affect either the magnitude of stimulus required to trigger the emotion or the intensity and duration of the resulting emotion?

Ideally we would like to list or classify the important human sentiments much as we previously listed some of the important emotions, but the problem is not so simple. The sentiment one feels toward any individual involves *many* emotions. The extent of involvement of each emotion is different, and probably almost unique for each individual who holds a place in our lives. As a consequence, our language does not contain ready-made names for most sentiments. It does provide a few concepts that are suggestive—respect, esteem, love, affection, like, dislike, and hate. But the list is short and apparently inadequate for our purpose.

What should we require of an adequate list of sentiments? There are probably three important requirements:

1. The list should be *consistent* with the list of emotions. That is, for each sentiment, we should be able to identify the specific emotions involved.
2. The list should be *rich* enough to encompass most of the standard emotional attachments that develop between individuals in various human societies.
3. The list should be a *primary* list of sentiments in the sense that all other sentiments can be represented as "mixtures" of the primary sentiments.

It is unclear whether such a general list is possible, but we may get some additional clues when we consider the third component of our motivation system, the "instinctive motives."

INSTINCTIVE MOTIVES

Table 11.3 displays an illustrative list of "instinctive motives." The structure of this list is far more uncertain than either the emotions or the biological drives. There are several reasons for the uncertainty. First, there has been very little research dealing with "instinctive motives." Second, the motives are difficult to study because they are essentially constant, so one cannot observe the behavioral consequences of *changes* in the intensity of these motives. Finally, because the *many* "motives" all contribute to only a small number of emotions (joy, sorrow, pride, and shame) the "motives" are not

TABLE 11.3
Illustrative List of Instinctive Human Motives

---

1. Desire for dominance (rivalry)
2. Desire for approval
3. Desire for social acceptance
4. Gregariousness
5. Enjoyment of conversation
   a. Talking
   b. Listening
6. Activity motive (desire to exercise body and exploit one's physical skills)
7. Enjoyment of humor in conversation and play
8. Social preferences
9. Team motive (desire to work with others for common goals)
10. Constructive motive (desire to make or build something)
11. Contribution motive (desire to contribute or do something meaningful for society)

---

perceived subjectively as separate or distinct, so we cannot identify them reliably from personal introspective experience. For these reasons, the evidence which suggests the structure of the "instinctive motives" comes more from behavioral observations than from introspective experience.

The list in Table 11.3 was constructed by beginning with McDougall's list of "secondary instincts." Additions and deletions were made to reflect recent information on primate behavior and to provide a better *theoretical* fit with the structure of the "emotions" and "sentiments." Nevertheless, the list must be considered as purely illustrative. It is certainly incomplete, and it probably includes some "motives" which may later be shown to be *learned* rather than *innate*.

Observation of the "social" behavior of different animal species makes it clear that innate motives similar to those in the table must in fact exist. Nevertheless, until recently there has been great skepticism about the existence of such motives in humans. If the motives exist, why do we not sense their existence? We subjectively experience other components of value in very obvious ways: We feel pain and sensual pleasure; we experience fear and anger; we sense hunger and thirst. Why then are the instinctive motives not experienced in similar ways?

The answer is that these motives are in fact experienced in very similar ways, but whereas hunger and thirst both have separate and distinct value sensations, many of the instinctive motives are linked to the same value sensations. For example, being liked will make us happy, participation in play or conversation will make us happy, working with a team can make us happy. As success is achieved with regard to any of these "instinctive motives" the emotional value sensation is moved toward happiness or joy. As failure is experienced the value sensation is moved toward sadness or sorrow. In the period following such a success or failure, the value sensation gradually drifts back to the normal neutral (or slightly negative) position. Almost all of the "social motives" seem to be concentrated in just *two* of the nine emotional dimensions, specifically the emotions of *joy versus sorrow* and *pride versus shame.*

Since the proposed list of instinctive motives is sure to be controversial we will discuss briefly the rationale that leads to the inclusion of each of the motives in the list:

1. *Desire for dominance:* The dominance structure is so widespread in the animal world that it almost certainly reflects a truly innate motive. The wide variations in the intensity of the dominance drive between primate species also suggest an inherited trait. In humans, the dominance drive seems to be reflected in a general competitive rivalry involving a desire for power, for athletic success, for business success, or for social leadership. Whenever men gather in informal groups there seems to emerge a clear leadership hierarchy. As would be expected from our primate heritage, the leadership hierarchy is much less evident in female social groups.

What specific sensations of value are involved in the dominance motive? It seems likely that the key one is "pride" and secondarily "joy." An improvement in the dominance position probably generates pride and joy, whereas a loss in dominance generates shame and sorrow.

2. *Desire for approval:* The clearest evidence for this as an "innate motive" is found in the behavior of a small child and his desire to please. The child will respond to a smile of approval as he would to a reward of candy. Most adults can subjectively sense their own desire to be "admired," and their intense dislike for being "scorned." The facial expression of scorn or contempt would not be such an effective social control unless the desire for approval was an almost universal motive. Therefore, the existence of scorn and shame in a very limited list of human facial expressions provides strong evidence of an instinctive desire for approval.

The individual seeks approval because approval contributes to pride and joy, whereas disapproval contributes to shame and sorrow. Thus the value sensations involved in the approval motive are much the same as those involved in the dominance motive.

3. *Desire for social acceptance:* While this desire to be liked rather than disliked is very similar to the desire for approval, there seems to be a real difference between being "liked" and being "admired," or between being "disliked" and being "scorned." Because of this difference, the desire for social acceptance is also included as a separate "innate motive." Almost all individuals, both children and adults, show a strong desire to be liked by their peers.

The valuative sensations in this motive are similar to those in the dominance and approval motives, but the emphasis seems to be more on joy versus sorrow than on pride versus shame.

4. *Gregariousness:* This is a very primitive instinct. It is the motive that keeps the herd or the flock together. It amounts to a fear of being alone. This motive may be partly responsible for keeping primate troops together. Although it may not be a very strong motive in humans, it probably exists.

5. *Enjoyment of conversation:* Even a casual look at the facial expressions of a group engaged in conversation makes it clear that they "enjoy" the conversation. In most gatherings, the fact that they are talking is far more important than what they are talking about. An innate enjoyment of conversation is undoubtedly esential to the development of good communication skills, and these skills have obvious survival value. Observation of such groups also suggests that the enjoyment of talking may be separate from the enjoyment of listening. Some people clearly like to talk; others seem to prefer to listen.

The innate enjoyment of conversation however is almost surely a "unique" human characteristic. As the phrase "enjoyment of conversation" implies, the main valuative sensation associated with this motive seems to be joy or happiness.

6. *Activity motive:* The desire to exercise the body and exploit one's physical skills seems obvious in the play behavior of all the higher animals. This motivation does not seem to occur in lower animals where physical movements are more nearly prewired and do not need to be learned. But in the higher animals, which operate on the decision-system principle, practice is essential to good coordination. The enjoyment of practice (i.e., play) is therefore an essential innate motive for such higher animals.

The desire to exercise the body and exploit physical skills seems to be connected primarily with the value sensation of joy. We are happier when we are doing something. We are happiest if it is difficult and we are doing it well. In the latter case, and especially if others are watching, we are rewarded as well with the value sensation of pride.

7. *Enjoyment of humor:* The instinctive nature of the enjoyment of humor is particularly obvious. We do not decide to laugh; we laugh automatically when something funny occurs. Humor produces a short-lived joy. The joy contributes to the enjoyment of social activities, and in this sense it has "social" survival value. But humor is fundamentally an "Intellectual Value" whose real purpose concerns the motivation of mental processes. Therefore, the main discussion of humor (and its primary survival function) must be postponed to later chapters that deal with the "Intellectual Values."

8. *Social preferences:* The choice box experiments suggest that individuals have certain innate social preferences. They prefer individuals with a sim-

ilar rearing background. Infants prefer females to male adults. Young adults prefer others of the opposite sex, etc. Such "social preferences" are probably fundamental in motivating the individual behavior which determines the *structure* of society for any primate species.

These innate preferences may respond to certain prewired identification keys, so that the individual automatically feels happier or less fearful in the presence of certain types of individuals. Of course, many social preferences develop through experience as sentiments are formed, but such *acquired* or *learned* social preferences do not belong on a list of *innate* motives. We are concerned here only with that component of social preference which is instinctive.

Because of reinforcement in subsequent experiences and because of the social pressures to conform to a behavior "norm," the innate preferences (even when initially not very strong) are likely to be magnified in the actual species behavior.

9. *Team motive:* The desire to work with others for common goals probably evolved as a cooperative hunting instinct. As would be expected this instinctive motive seems stronger in men than in women. The motive is apparent in small businesses, and in the project organization of larger businesses. It is also apparent in the formation of juvenile gangs. The value sensation involved in this motive again appears to be primarily joy versus sorrow. We "enjoy" working with a team.

10. *Constructive motive:* The desire to make or build something may have emerged as an "instinctive motive" with the development of a tool-using culture. Certainly such motives would facilitate the use and development of tools, once tool using had begun to appear. The desire to paint, crochet, knit, sculpture, or putter in a hobby shop seems to be evidence of such a motive. We enjoy working on something or building something, so the value sensation is apparently joy or happiness. When the task is complete we may experience another value sensation which is a generalized sense of well-being or accomplishment.

11. *Contribution motive:* A desire to do or contribute something meaningful for society would certainly have survival value. The evidence for the actual existence of such a motive seems to be most clear in the "frustration" expressed by those in modern society who feel they are denied the opportunity to contribute. The desire may also be reflected in various volunteer, political, and charitable activities. The value sensations associated with this desire to contribute meaningfully to society seem to be almost exactly the same as those involved in the constructive motive, although there may be a greater mixture of pride in the final sense of accomplishment.

STRUCTURE OF THE SOCIAL MOTIVATION "SYSTEM"

Having introduced the three components of the social motivation system (emotions, sentiments, and motives), we are now ready to consider their interactions as a system.

INTERACTION OF "MOTIVES" AND "SENTIMENTS"

The first three motives (dominance, desire for approval, and desire for social acceptance) provide a particularly good illustration of the kinds of interaction involved. Each of these "motives" is defined for the individual in terms of the *sentiments* of *other* members of the society. The facial expressions of others in the society provide the individual and his innate value system with information about the "sentiments" by displaying their emotions. In this way each individual can recognize the sentiments of others and can assess his level of achievement relative to his "innate motives." The assessment takes place at two levels. It takes place *subconsciously* within the "motivation system" itself where it generates a *direct* change in the individual's level of "pride" and "happiness." It also takes place *consciously* as the individual attempts to understand in his rational mind the reasons for his success and failures.

To provide an adequate description of a "motive" we need to know how "success" with regard to the motive is defined within the motivation system. For example, what does it mean to have a desire for "dominance"? The position of each animal in a dominance hierarchy is measured by the "sentiments" of the other animals. With regard to a more dominant animal, the individual is likely to feel fear, respect, and possibly admiration. With regard to a less dominant animal that has not accepted the submissive position, the individual is likely to feel anger and perhaps scorn. With regard to a submissive animal he may feel a sense of responsibility, tolerance, and possibly the "tender emotion." The objective of dominance therefore is defined in terms of the "sentiments" of other animals. The male undoubtedly finds it extremely rewarding or pleasurable to be treated as dominant. This provides the basic motivation which generates the desire for dominance.

The desire for approval also provides a good example of such social interactions. It is possible that in nonhuman societies "social approval" and "dominance" may be synonymous, but in human society there appears to be a real distinction. Approval or disapproval can be felt with regard to a more dominant as well as with regard to a less dominant person. Success with regard to the "approval motive" is measured in terms of the "sentiments of approval" that are felt by other individuals, and these sentiments are observable in their facial expression and behavior. The intensity of the desire to be "approved" by another person seems to be proportional to that person's importance in our lives. The more interactions we have with the person, and the more dominant he is, the more important his approval becomes.

The desire for social acceptance is also measured in terms of the sentiments of others. In this case the key sentiments are the sentiments of "liking" and "disliking" in a social sense. If we like a person we enjoy being with him. People like to be liked (i.e., to be socially accepted). When we are socially attractive we are happier than when we are socially unattractive. Social acceptance contributes to joy; lack of social acceptance produces sorrow.

The fact that the social "motives" are defined in terms of human "sentiments" (which are also simply components of the human motivation system itself) helps to explain the extreme adaptability of the human "social system." Human society has been able to adapt to an exceptionally wide range of physical and technological environments because (to a first approximation) the motivation system does not depend on the physical environment. It is dependent only on internal interactions within the framework of human society. (At present the impersonal nature of modern urban society seems to be adversely affecting the operation of this human motivation system, but that is a problem which will be discussed in more detail in Part III.)

Of course, it does *not* follow that *all* "innate motives" are based on sentiments. The herd instinct, or the desire to be with others which we called "gregariousness," does not involve the sentiments of others; it is concerned simply with their presence. There may be an optimum amount of interaction with others that produces the greatest happiness, so that gregariousness beyond this level may contribute to sorrow (and perhaps anger) rather than joy. In a primitive hunting society such an effect would operate to break up troops when they become too large to be ecologically viable.

TIME-PHASE CLASSIFICATION OF SOCIAL VALUES

During the analysis of the "selfish values" in Chapter 9 we found it useful to classify values by their time-phase relationship to the activities being motivated. Values were classified as "preceding," "concurrent," or "trailing" depending on the time-phase relationship. How do these concepts fit into the structure of the social values? In the case of the "innate motives" just *two* dimensions of value (pride and joy) must serve for a *large number* of separate underlying motives. These same two basic "value sensations" also serve in all three of the time-phase relationships.

A "successful" social experience may be accompanied by intense joy. This is the "concurrent" (or Class B) component of the reward. It may be followed by a feeling of pride and happiness which is the "trailing" (or Class C) component. If an individual fails to engage in such activities, the value sensation gradually drifts toward sadness and depression. This is the "leading" (or Class A) value component which is intended to motivate appropriate social activity. If the individual is particularly unsuccessful in some social activity, the experience will be followed with a sense of sadness or shame which constitutes a trailing (or Class C) penalty. The built-in drift of the value sensation toward the neutral (or slightly negative) position is a necessary design feature, to ensure that it will stay on the useful part of the scale and will not be pushed into saturation as a result of a long series of successes or failures.

THE INDISTINGUISHABILITY OF SOCIAL VALUES

It is appropriate to ask why each of the instinctive motives is not provided with a separate and distinct "value sensation." In an earlier section, we mentioned two basic reasons. The social "motives" developed quite late in the evolutionary process after the design of the hypothalamus had been rather firmly established. The introduction of new valuative sensations would have required an important "redesign" of this ancient system. It was easier to add "motives" in the newer more malleable part of the brain, and communicate the information via preexisting "valuative channels" in the hypothalamus.

Second, these new "motives" developed sufficiently late in the evolutionary process so that the "rational" mind had already acquired impressive powers of association and analysis. With this capacity the mind could learn to respond quite appropriately to a complex value system, even though the individual "motives" were not separately identified with a different emotion or valuative sensation. We can now add a third important contributing factor. The association of joy with the rewarding facial expression (the smile) makes it a preferred emotion on which to build a cooperative motivation system. Thus, in this later evolutionary phase when new motivations were needed, the easiest "quick fix" was to connect the new motivation to an *existing* value sensation. The sensation of joy and sorrow became a sort of catchall or miscellaneous "value sensation," and a large fraction of the more recent human motives became associated with this single emotional dimension. Thus, the importance of joy and sorrow relative to the other emotions has been greatly enhanced during the evolution from the higher primates to man.

In an earlier chapter we noted that the use of separately distinguishable sensations of value simplifies the problem of associating the value with the specific activity that is being motivated. We might therefore expect that this use of a single value sensation for a large number of motives might cause confusion, so that the individual might have difficulty recognizing what type of activity is being motivated. This indeed proves to be the case. In our modern society, individuals often feel sad, depressed, despondent, and frustrated—without purpose or motive in their lives, and they may have no real idea of the source of the problem. In many cases, they are simply failing to involve themselves in those types of activity which produce joy, so they suffer the penalty of sadness which also is intended to motivate these activities.

If the motivation is so unclear, it may seem surprising that this type of ambiguous motivation system evolved, but we must remember that modern society is very different from the primitive societies in which the system evolved. The informal structure of a primitive (and perhaps preverbal) society should have made it easier for the individual to learn what type of activity he liked. In such a society the maturing individual quite naturally gained experience with social activities involving all elements of his instinctive motivation system. As a consequence, he learned what kind of activities he enjoyed, and he was largely free to participate as he liked. In the impersonal and verbally inhibited urban environment, it is easy to fall out of the habit of social participation. The elementary schools often do not encourage it. A child raised in an urban apartment can grow to maturity without ever participating in normal childhood social activity and without even contributing meaningfully to the family, much less to the wider society. Thus they may fail to gain the experience required to respond naturally to their own innate motivation system. They can grow depressed, despondent, even suicidal, without knowing why.

One of the major objectives of the present book is to show that these innate "social motives" may be as real and as much a part of the human psychology as hunger and thirst. If social and urban planning is to meet human needs effectively, such "innate motives" should be explicitly considered as we develop our public plans. Unfortunately, our present understanding of such motives is extremely vague. Moreover, the complexity of the motives that govern pride and joy makes it most unlikely that a full understanding will be achieved in the near future. What we can expect is that future research will result in a gradual reduction of our uncertainty concerning the relevant factors.

The present state of knowledge leaves the door wide open for consideration of a very wide spectrum of motives that may or may not be real and that may or may not be innate. Until scientific progress can narrow the range of uncertainty, the responsibility will continue to fall where it always has—on the individual, who faces the challenge of trying to understand his own motivations.

THE DECOMPOSITION INTO MOTIVES

It appears that most social motivations are concentrated in just two dimensions of value: joy versus sorrow and pride versus shame. An extremely wide variety of rather stable "social motives" combine to control our emotional state in these two dimensions. This arrangement poses a peculiar methodological problem for research dealing with these "motives."

In principle the innate motives that control the emotions of pride and joy are no different from the factors that govern fear and anger. In both cases, there is a complex relationship between the social or environmental stimuli, and the rise or fall of an emotional value. However, fear and anger motivate a comparatively narrowly defined set of activities, and they respond to a correspondingly restricted set of stimuli. For this reason, it seemed less necessary to decompose the underlying factors into a series of "motives."

The emotions of pride and joy, in contrast, motivate a wide spectrum of activities and respond to a correspondingly wide spectrum of causal factors. The decomposition into a list of "innate motives" represents an effort to provide a verbal description of some of the causal factors underlying changes in the emotions of pride and joy. Although we postulate only *two* relevant dimensions of emotion (pride and joy), a total of *eleven* motives are discussed. We must therefore ask whether it is logically meaningful to talk of eleven or more motives when only two emotions are involved.

In a strictly logical or theoretical sense, it seems that all of the causal factors could be combined into no more than two very complex motives, but as a practical matter these two complex motives would be very difficult to describe. The specific breakdown into eleven separate motives is undoubtedly influenced by the convenience of specific categories in the English language. For example, the English language distinction between "admire" and "like" suggests the need for two separate motives. In some other language this distinction might not be made, and a single word might describe both sentiments. In such a language it would probably seem natural (and it might be just as valid) to combine motives 2 and 3 into a single more complex motive. Logically, the only requirement is that the ensemble of "motives" must provide an adequate explanation of the variations in the emotions of joy and pride. At present, this separation into "simpler" motives is fundamentally a linguistic convenience. As the theory becomes more quantitative, the decomposition into motives may also prove to be a computational convenience. Finally, as we develop a better understanding of the physiology of the brain, we may ultimately identify a natural decomposition which corresponds as closely as possible to the functional roles of specific associated neurological components.

A PRIORI OR INTUITIVE DESIRES

One of the obvious limitations of the human motivation system is that we have difficulty in anticipating the value consequences of our behavior. We discover that certain activities are fun only by trying them. If we are never motivated to experiment, we may never learn. Obviously, the motivation system would be more effective if we could know instinctively (and in advance) what types of activity would be rewarding. Of course, where a specific well-defined action is required, very simple evolutionary mechanisms such as the tropism can automatically generate the behavior. Where some degree of rational control is appropriate, a compromise arrangement is possible, as in the case of facial expression or bodily elimination, where the automatic response can be inhibited by the rational mind. But the practical usefulness of such built-in mechanisms is necessarily limited to very simple automatic responses. To provide effective motivation of more complex behavior (prior to actual experience) it might be helpful if we could have a built-in intuitive knowledge of what we would enjoy.

In artificial decision systems, the value system is always built in by the designer in such a way that it can be used directly for the rational evaluation of future outcomes. In a biological decision system there are two separate valuation functions that must be accomplished by *different* cybernetic components. The values actually experienced are determined by an innate system of emotions (located outside the "mind") which responds automatically to interactions with the environment. In contrast, when we rationally consider alternatives, we consider them inside the "mind" using a mental model of our own value structure. During this hypothetical consideration of alternatives, the innate "value system" does not actually function, so it cannot directly supply the necessary decision criteria. During the fantasy of a dream, we experience emotions much as we would in actual experience. Thus, dreams can serve as a learning experience that allows the individual to gain a better understanding of his own motivations. Although most dreams are not consciously remembered, they may nevertheless be reflected in a subconscious intuition about values or desires that can motivate actual behavior.

This method of learning about our values has obvious limitations. First it is applicable only to those types of situations that actually occur in dreams, and second, the experience within dreams is usually a distorted surrealistic representation of reality. In addition dreams are almost always defective in that the dreamer rarely (if ever) experiences heat or cold or other tactile sensations such as the pain or tactile pleasure that should logically accompany the action of a dream. When a nightmare reaches a point where the dreamer would be hurt, he usually wakes up. He does not continue the dream and experience the pain. Indeed where pain, heat, or cold are actually experienced in a dream it usually turns out, upon awakening, that the sensation is real. Tactile sensations may be missing in dreams simply because the mechanism for producing these sensations lies outside the mind and cannot be easily simulated in a dream.

The absence of appropriate tactile sensations is often apparent in our sexual fantasies. In a dream, a young man may enjoy a naked romp with a young woman in the snow. In real life, the chill of the snow would almost certainly degrade the pleasurable emotional content. Nevertheless, on the basis of dreams that may or may not be remembered, an individual may develop a yearning for activities that have never been experienced.

At least in principle, such an intuitive a priori knowledge of the innate values might also be provided directly by a second genetically defined evaluation system that could be interrogated by the rational mind to help decide how well it might like different alternative outcomes. Subjectively, such a mechanism might be sensed as an intuition about how we would like future alternatives. But the evolutionary development of such a mechanism poses serious difficulties. It requires the *parallel* evolution of *two* independent but matching value-systems! This seems most unlikely. Moreover such value intuitions by themselves would probably not be very effective in motivating behavior after the individual discovered that he did not really enjoy an activity so motivated. From these considerations, it seems probable that evolutionary forces favoring such an intuitive value mechanism should not be very significant until *after* related values had evolved in the basic value system. Thereafter, such intuition might be favored by evolution because it would motivate the individual to experiment with new activities, but one would not expect the intuition to be accurate about the form of values involved. For these reasons it seems unlikely that such built-in intuition about values plays a very important role in the human decision system.

SUMMARY

The illustrative structure of human "Social Values" developed in this chapter only scratches the surface of a very complex and difficult subject. The purpose has not been to provide a comprehensive (or completed) theory, but rather to stimulate future research by demonstrating the feasibility of the value theory as an approach for analyzing human motivation.

To develop the illustrative value structure we used evidence from several sources—human facial expression, the evolution of facial expression from other primates to man, and the major differences between primitive, human, and nonhuman primate societies. This information was combined with McDougall's theory of social psychology to develop an initial structural theory for human social values. The resulting value structure emerges as an integrated "system design" concept for motivating cooperative behavior in primitive human societies. The "system design" uses facial expression as a motivational signaling system, and it is based on complex interactions between the "innate motives," the "sentiments," and the "emotions" of different members of the society. Because this motivational system is based on personal interactions between people who must be recognized as distinct and important individuals, it may not operate satisfactorily within the impersonal environment of a modern urban society.

REFERENCES

1. Arnold, Magda. *Emotion and Personality*, vols. 1 and 2. New York: Columbia University Press, 1960.
2. Arnold, Magda. *Feelings and Emotions*. New York: Academic Press, 1970.
3. Chevalier-Skolnikoff, Suzanne. "Facial Expression of Emotion in Non-Human Primates." In *Darwin and Facial Expression*. New York: Academic Press, 1973.
4. Ekman, Paul, Wallace Friesen, and Phoebe Ellsworth. *Emotion in the Human Face*. New York: Pergamon Press, 1972.
5. Ekman, Paul. *Darwin and Facial Expressions*. New York: Academic Press, 1973.
6. Herzer, Ludwig, and Fritz Löhner. "Yours Is My Heart Alone," from the Operetta *Land of Smiles*. English Version by Harry B. Smith. New York: Shubert Music, 1929.
7. Izard, Carroll E. *The Face of Emotion*. Englewood Cliffs, N.J.: Appleton Century Crofts, 1971.
8. Izard, Carroll E. *Patterns of Emotion: A New Analysis of Anxiety and Depression*. New York: Academic Press, 1972.
9. Lorenz, Konrad. *On Aggression*. New York: Harcourt Brace & World, 1966.
10. McDougall, William. *An Introduction to Social Psychology*. London: Methuen, 1908.
11. Morris, Desmond. *The Naked Ape*. London: Jonathan Cape, 1967.
12. Morris, Desmond. *Intimate Behavior*. New York: Random House, 1971.
13. Osgood, C. E. "Dimensionality of the Semantic Space for Communication via Facial Expressions." *Scandinavian Journal of Psychology* 7 (1966): 1–30.
14. Plutchik, Robert. *The Emotions: Facts, Theories, and A New Model*. New York: Random House, 1962.
15. Service, Elman R. *Profiles in Ethnology*. New York: Harper and Row, 1971.
16. Spitz, Rene Arpad, Ruth W. Washburn, and Florence Goodenough. *Facial Expression in Children*. New York: Arno, 1972.
17. Storr, Anthony. *Human Aggression*. New York: Atheneum, 1968.

18. Tiger, Lionel, and Robin Fox. *The Imperial Animal.* New York: Holt, Rinehart and Winston, 1971.

19. Tomkins, S. S. *Affect, Imagery, Consciousness.* New York: Springer Publishing, 1962, 1963.

20. Van Hooff, J.A.R.A.M. Aspects of Social Behavior and Communication in Human and Higher Nonhuman Primates. Rotterdam: author, 1971.

21. Van Lawick-Goodall, Jane. "A Preliminary Report on Expressive Movements and Communication in the Gambe Stream Chimpanzees." *Primates: Studies in Adaptation and Variability.* New York: Holt, Rinehart and Winston, 1968a.

22. Van Lawick-Goodall, Jane. "The Behavior of Free-living Chimpanzees in the Gambe Stream Reserve." *Animal Behavior Monographs,* 1968b.

Chapter 12

Common Sense
and the World Model

Good sense which only is the gift of Heaven,
And though no science, fairly worth the seven.
ALEXANDER POPE, *Moral Essays*

So FAR we have simply assumed the existence of a world model and have ignored the model-building process. Although this approach was fairly satisfactory for both the "Selfish Values" and the "Social Values," it is not practical for an analysis of the "Intellectual Values." The purpose of the "Intellectual Values" is to guide the individual in the development and refinement of his world model. To understand the intellectual values, we must begin with an analysis of the model-building process itself.

By "model building," we mean the normal mental activities that improve our understanding of the world environment. Each of our interactions with the environment adds to our knowledge. An experience either proves consistent with our understanding, or else it requires us to modify our understanding. It may change our concept of a friend's personality; it may add to our knowledge of a specific machine; or it may change our understanding of a natural law. Each such change in our ideas constitutes a change or refinement in our world model. The world model of the small child is relatively naive and unsophisticated, but as the child grows in experience his model becomes more extensive and sophisticated.

The overall "world model" used by an adult includes numerous interrelated submodels that are specialized for particular types of problems and are used at different levels of detail. For example, the primitive craftsman works with detailed information as he fashions an arrowhead and attaches it to the shaft of an arrow. When he uses the arrow in the hunt such details are forgotten, and the entire arrow becomes just a single element in the hunt. In effect, he has two separate mental models of the arrow. One is appropriate for its construction, another is appropriate for its use.

The efficiency of the human mind in the development of useful "world models" is a consequence of a priceless intellectual asset, common sense, which underlies all human knowledge and wisdom. This chapter is concerned with our understanding of those mental processes that we call "common sense." Our hypothesis is that "commonsense" reasoning reflects an evolutionary "design concept" for a learning machine, and that many of the things we classify as intellectual values originate in the design concept for this learning machine. In accordance with the theory of values, the purpose of the innate Intellectual Values must be to motivate efficient operation of the human mind as a "learning machine."

The world model that is developed by this learning machine includes not only a model of the outside world, but also a model of the valuative characteristics of the world as these are received by the rational mind. Thus the same commonsense learning mechanism is used for both valuative and objective learning. It serves to develop a rational model of the primary value sensations and their relationship to various types of objective experience. It also serves to develop the commonsense network of secondary values and secondary decision criteria. Obviously, therefore, the behavioral efficiency of the decision system is critically dependent on the effectiveness of the learning mechanism.

The theoretical discussion of mental processes in this chapter sets the stage for a discussion of specific human Intellectual Values in the following chapter. Although this discussion of Intellectual Values is necessary to a balanced treatment of human values, it is not essential to the discussion of personal ethics and social policy in Part III. Consequently, some readers may prefer to skip to Part III and return to these chapters at a later time.

The learning processes involved in the creation of the "world model" begin with the genetic design of the brain, and they culminate in the sophisticated theories of our modern cultural heritage. Therefore, to provide a balanced perspective, we must consider the relationship between information that is built into the brain and information that is subsequently acquired.

INHERITED VERSUS LEARNED INFORMATION

The remarkable efficiency of biological systems in acquiring the information needed for their world models reflects the fact that they do not really start from scratch to build the model with each generation. Each species is pre-wired according to its own genetic inheritance so that the essential information that an individual will need either is already present or can be easily obtained. The prewiring reflects the accumulated evolutionary knowledge of the species. In a strictly logical sense this genetically built-in information does not differ from other information that may later be acquired by each individual of the species.

A species can be prewired so that related sensory data is automatically brought together, so that a meaningful interpretation is easier. For example, the corresponding parts of the visual field from *both* the right and left eye are delivered to the *same* region of the brain to facilitate the stereo analysis for a three-dimensional image. The visual prewiring is designed so that edges of objects and especially moving objects tend to be emphasized. This prewired analysis converts the data into a more compact form, which makes meaningful interpretation easier. Thus the prewired processing (which accomplishes the first steps in the "model-building" process) saves a great deal of trial-and-error learning and contributes to efficiency in the model-building process.

Strangely enough, there appears to be no sharp division between information that is genetically inherited and information that is learned. A wide range of intermediate alternatives seems to be possible:

1. Information can be really prewired and built in by the genetic code.
2. The genetic code can specify a special-purpose learning mechanism whose *only* function is to learn to process *specific* types of input information.
3. The genetic code can specify a more general-purpose learning mechanism which is effective at learning certain classes of relationships that can be anticipated in the system design.
4. A completely general-purpose learning mechanism can be specified for dealing with concepts and relationships that cannot be anticipated in the system design.

This list suggests almost a continuum of alternatives, ranging from information that is unambiguously built into the system to information that is unambiguously learned. In this chapter we will be *most* concerned with the more generalized end of the learning spectrum. But the cognitive process incorporates all levels of learning, and the "world model" itself encompasses information obtained at all levels.

PREPLANNED OR SPECIALIZED LEARNING MECHANISMS

Before discussing the design principles for our commonsense model-building process, we will consider briefly some illustrative learning mechanisms that involve specialized or preplanned learning. To illustrate the principles of preplanned learning we will consider two examples from the visual processing system.

Inexpensive cameras with simple lenses produce color fringes around their images because the different wavelengths of light focus at slightly different locations. The human eye is also designed with a very simple lense, so the optical images that it focuses on the retina include similar color fringes. For some years, it was a mystery why we were not conscious of these optical color fringes.

Experiments using prism goggles have shown why (5, 7). The goggles bend different colors by different amounts, thus producing images with severe rainbow-like color fringes. When such goggles are first put on, the individual experiences very troublesome color fringes. But, if he perseveres for several days, the color fringes gradually fade and disappear. If he then removes the goggles, reverse color fringes appear; and these also fade with time, as he abstains from wearing the goggles.

Apparently, the synapses in the optic nerve automatically adjust their interconnections so that the *different* colors in the visual image are brought into the best possible registration. In effect, the optic nerve must be designed to learn from experience, so that it automatically adjusts its wiring to align the different colors in the visual image. This automatic adaptation is not limited to processes as simple as the color fringes. It applies, as well, to the curvature of lines that result from different types of goggles, and even to some complex and variable distortions that *change* as the eye moves relative to the prism.

Recent work with albino cats (6) provides another remarkable example of *automatic* adjustment in the wiring of the optic nerve. Albino cats, as well as albinos of many other species, suffer from a serious genetic defect in the alignment of the stereo image. In normal cats the left field of view from both eyes is processed on one side of the brain, and the right half from both eyes is processed in the other side of the brain. To provide stereo vision, the corresponding half images from both eyes are brought together in a region called the lateral geniculate nucleus. In this region, the stereo image from one eye is aligned with a main image from the other eye to permit three-dimensional analysis of the visual image for that side of the brain. The combined visual information is then transmitted to the visual region of the cerebral cortex.

In albino cats only a small part of the stereo image is correctly aligned with the main image in the lateral geniculate nucleus. The remainder of the stereo images are interchanged between the two sides of the brain! Thus, where the stereo image should be there is substituted a *reversed* image from the same eye, corresponding to the wrong half of the visual field! Clearly unless something were done to correct the problem, the cats would be left with a hopelessly confused visual picture. Remarkably, the cats' brains are able to adapt. The adaptation does not take place between the eye and the lateral geniculate nucleus where the defect occurred, but rather in the optic fibers leading onward from the geniculate nucleus to the visual cortex. Different cats adapt to the same genetic defect in different ways.

In some cases the fibers that would otherwise carry the misplaced stereo image to the visual cortex simply atrophy, so that the bad image is suppressed. Although these cats are left without stereo vision, the expected confusion in the visual cortex does not occur. In other cases a really remarkable thing happens. The optic fibers carrying the misplaced stereo image actually cross each other so that the original reversal of the image is corrected. The fibers are then *redirected* so that the misplaced stereo image does not overlap the normal image in the visual cortex at all. It is added instead (where it belongs) as an almost normal *extension* of the visual image into the other half of the field of view! Thus, each half of the brain receives an almost normal visual image which extends into the other half of the field of view.

This provides a striking example of the way the neurological network can adapt to make sense out of somewhat scrambled incoming sensory data, and it illustrates rather clearly the concept of a very specialized built-in learning mechanism.

In the past, it has generally been assumed that the growth and development of the neural network is controlled by special chemical keys that define the linkage destinations for specific axons. The above examples strongly suggest that the information signals carried by the neurons must play an important role, both in maintaining order in the growing axon network and in determining the preferred locations of synapse linkages. The examples are particularly suggestive because the visual signals themselves seem to be the only possible source of information that could result in such automatic and appropriate realignment of the neural network.

Apparently the growing visual fibers are guided by a principle of "signal coherence" that causes each fiber to seek a path where its signals will be as "consistent" as possible with those carried by its neighbors. Such a mechanism seems to be the only possible explanation of the remarkable rearrangement of fibers carrying the stereo signal, so that they will fit where they belong in the visual pattern. It would also explain the ability of the neural network to realign color images to correct for color fringes, and the tendency of the nerve fibers to atrophy when they fail to find a location where the proper coherence exists. Quite likely groups of neurons are specialized *both* with chemical keys and with different requirements for "signal coherence," so that their growth and interconnections result in different predefined types of synapse linkage. Indeed, the development of the neural network in the embryo may be guided by spontaneous neural signals that serve no purpose except to guide the growth of the network. Whether or not this hypothesis proves to be correct, it provides a concrete example of how specialized learning principles might be built into the design of the brain.

A POSSIBLE GENERALIZED LEARNING MECHANISM

The foregoing examples show that even the physical wiring of the human decision system may reflect the operation of preplanned learning mechanisms that are built into the neuron. It is tempting to think that very similar mechanisms involving the development of new synapse linkages, the enhancement of old linkages, and the atrophy of irrelevant linkages may be an important learning mechanism even in the highest centers of the adult brain. The possibility that such a mechanism, involving growth of new linkages, may be involved in learning is supported by several other pieces of circumstantial evidence:

1. When rats are exposed to an intellectually challenging or enriching environment a significant increase seems to occur in the weight of the brain.
2. Human brains often show a slight enlargement or thickening of the cerebral cortex in the speech region.
3. As individuals mature there seems to be a noticeable increase in the breadth of the skull, as if the skull were making room for an enlargement of the brain. A young man will characteristically have a relatively narrow skull, but as he matures and gains in experience the skull seems to widen. The effect is particularly noticeable when recent pictures are compared with older youthful pictures, and it seems most pronounced in lawyers, politicians, and scientists who work in intensively intellectual activity.

For our present purposes, however, the specific mechanisms of learning are not really important. It is sufficient to note that the boundary between knowledge that is "prewired" and knowledge that is "learned" may be quite ambiguous, even in the physiology of the brain. Fortunately for our present discussion, the distinction between what is "prewired" and what is "learned" is not important. We will simply recognize that each individual is operating at all times with a world model that has reached a certain level of sophistication. Part of the world model is prewired; part of it is generated by automatic learning mechanisms; and part of it is assembled by the most generalized, automatic learning center—the conscious mind.

THE EVOLUTIONARY BASIS OF COMMON SENSE

In the remainder of the chapter, we will be concerned primarily with those principles of cybernetic efficiency that guide the conscious mind as it seeks to update and improve the world model. It is now widely recognized that the principles used in commonsense reasoning (and even to some extent in scientific reasoning) are in large measure a genetically inherited asset. There are, however, rather substantial differences of opinion about the specific principles involved. Those who have approached the problem from the perspective of mathematical logic have tended to describe human learning processes as an imperfect biological approximation to a formal mathematical process known as Bayesian inference.* Those who have approached the problem pragmatically through careful study of the way knowledge actually accumulates have usually maintained that the process is much more complex and that the theory of Bayesian inference cannot provide an adequate basis for understanding the process.† Our purpose here is not to try to resolve such fundamental disagreements about the foundations of knowledge, but rather to provide a pragmatic understanding of the commonsense learning process that will be adequate for discussing the origins of human intellectual values.

Nevertheless, the point of view developed here may provide a step toward resolving the existing differences of opinion. Our analysis suggests that both of the traditional views about the foundations of knowledge are correct in their own way. But they represent different *levels* of explanation, in the same sense that quantum mechanics and organic chemistry represent different levels of explanation for biological phenomena. At the most fundamental theoretical level, it appears that the commonsense learning processes can be explained as an *interaction* between two sets of principles: the principles of *Bayesian inference* and the principles of *cybernetic efficiency*. At the more pragmatic level of explanation, the interaction between these two sets of basic

---

* For example see Carnap (2), (3).
† For example see Popper (10), Campbell (1).

principles produces the rather complex set of (commonsense) information-processing rules analogous to the complex rules of organic chemistry. Obviously it is possible to provide a valid description of the commonsense deductive processes *either* in terms of the fundamental principles of Bayesian inference and cybernetic efficiency *or* in terms of the resulting rules of commonsense information processing.

Our approach will be to begin with the assumption that the commonsense learning processes are based fundamentally on the principles of Bayesian inference. But in order to apply Bayesian inference in a practical way within a finite cybernetic system, it is necessary to elaborate on the basic Bayesian process with a number of tricks and approximations that serve to keep the computational requirements within reasonable limits. As we proceed with such an elaboration of the basic deductive process, we begin to approximate many of the features of commonsense human logic. For the benefit of readers unfamiliar with the principles of Bayesian inference, the following section provides a brief description of the mathematical concept.

THE PRINCIPLE OF BAYESIAN INFERENCE

Because of its close correspondence to common sense, Bayes' theorem appears to be almost intuitively obvious. Fundamentally it says that we can judge the quality of alternative theories (or alternative hypotheses) simply by asking which one provides the most accurate predictions about a system. Relative to a poor hypothesis the behavior of the system will seem unlikely, or even impossible. In terms of a good hypothesis, the actual behavior will seem much more likely. The best hypothesis is simply the one for which the actual system behavior seems the most likely.

Bayes' theorem organizes these commonsense concepts into a formal procedure that can be used to estimate the "likelihood" of alternative hypotheses on the basis of the accuracy of the predictions that they would make about a system. Suppose we have two alternative hypotheses which initially seem equally likely. Each hypothesis makes certain predictions about how a system should behave and assigns a certain probability to various possible outcomes of an experiment. When the experiment is carried out some specific outcome occurs. Suppose this outcome had been assigned a high probability, P_1, by the first hypothesis and a lower probability, P_2, by the second hypothesis. According to Bayes' theorem two hypotheses are no longer equally likely after the experiment. The one which predicted the actual outcome with the highest probability, P_1, is now the more likely hypothesis. To calculate the new relative likelihood of the two hypotheses we simply multiply the previous likelihood for each hypothesis by the probability assigned to the actual outcome of the experiment. If the estimated likelihood of the two hypotheses before the experiment had been H_1 and H_2 then their relative likelihood H_1' and H_2' after the experiment would be given by

$$H_1' = P_1 H_1$$
$$H_2' = P_2 H_2$$

Let us check to be sure that this result seems reasonable. Suppose that for the second hypothesis the observed observation is impossible (i.e., given the hypothesis, H_2, then the probability, P_2, of the observation is zero); then the revised likelihood for hypothesis H_2' is also zero. In simple language, this says that if a hypothesis predicts that an event is impossible, and the event nevertheless occurs, then we must reject the hypothesis. This of course is consistent with "commonsense" logic.

But the greatest power of the Bayesian deductive approach shows in cases that are less extreme. Suppose we have two hypotheses which initially have equal likelihood $(H_1 = H_2)$, and an event occurs that would have a 10 percent probability on the basis of the first hypothesis but a 90 percent probability on the basis of the second hypothesis. Following the event we can say that the second hypothesis is nine times as likely as the first $(90\%/10\% = 9)$. If similar results are obtained in a long succession of experiments, the superiority of the second hypothesis approaches a certainty. For example, if similar results are obtained in a series of ten independent experiments, then the second hypothesis becomes more likely by a factor of 9^{10}.

Thus, Bayesian inference provides a powerful technique for distinguishing between alternative hypotheses. The method operates more efficiently when the probability of the observed events is very different for two alternative hypotheses. Obviously the most efficient situation occurs when the observed event is literally impossible in the context of one of the hypotheses. Mathematicians call such an event a "counterexample." When a counterexample is found, the hypothesis must be modified or rejected. Mathematicians, logicians, and experimental scientists are very fond of counterexamples since they permit rapid rejection of fallacious theories.

In the form previously described, Bayes' theorem does not allow one to make any statements about the absolute likelihood of alternative hypotheses. It only provides information about the *relative* likelihood of those alternatives that are considered. The formal theorem, however, has traditionally been used only in a context where it can be assumed that one and only one of the alternative hypotheses is correct. When this is so, we know that the sum of the likelihoods for all hypotheses must be equal to one. Under these circumstances, Bayes' method can be used to calculate actual as well as relative likelihoods.*

Despite its simplicity and obvious deductive power, Bayes' theorem has remained controversial. It is still not fully accepted by many mathematicians and statisticians. A basic mathematical difficulty with the method is that it seems to depend on being able to assign in advance certain "a priori" likelihoods, H_i. Moreover, regardless of how many observations one may make on a system, the final set of likelihoods, H_i', seems to depend on the initial a priori values that were assigned.

Although this is true in a rigorous mathematical sense, in a practical sense it usually is not. After a few observations the a priori likelihoods are usually dominated by the results of the observations. For example, although different people may disagree by factors as large as 10 to 100 about how to assign a priori likelihoods, a series of experimental observations can produce *astronomical* differences in the relative likelihoods. Thus, the small differences in a priori assumptions tend to be quickly dominated by the actual results of Bayesian inference.

One other problem with the formal Bayesian method is the difficulty of being sure that *all* relevant hypotheses are included in the set, so that we can be sure that the sum of the likelihoods is 1.0. Although in most practical cases it is not possible to be sure that all alternative hypotheses have been included, the method can still be used to distinguish between those alternatives that are being considered.

In the years after Bayes' method was originally introduced, these difficulties were seen as serious defects or flaws in his theorem. This concern was natural at a time when science was supposed to deal with certainties and when mathematics had to be "exact" to be respectable. However, with the present

---

* In this case we know that the sum of the a priori likelihoods, H_i, must be equal to 1.0, specifically

$$\Sigma H_i = 1.0$$

To ensure that the resultant likelihoods *after* an event *also* add to one, the *resulting* relative likelihoods must be corrected by dividing by a factor, R. This gives a revised formula for the final likelihoods.

$$H_i' = P_i H_i / R$$

where

$$R = \Sigma P_i H_i$$

Through the use of this more conventional version of Bayes' formula one can update the *absolute* likelihood for alternative hypotheses. Some mathematicians may disagree semantically with our extension of the formal term "Bayesian inference" to refer to situations when the set of hypotheses is incomplete. However, the principles that apply in such situations follow as an obvious corollary of the standard theorem, so there should be no substantive disagreement.

recognition that science must deal with approximations, there has been a change in the intellectual climate. As the years have passed, and no practical alternatives to Bayes' deductive method have been found, the difficulties have gradually been accepted as fundamental to deductive logic. In our present context the so-called "weaknesses" of the Bayesian method are not defects in the theory at all. They reflect instead some of the fundamental characteristics of the process of deduction and model making.

To develop some basic ideas about the commonsense learning process we will revert to an approach that was used in Part I. The basic principles will be developed first in the context of a computerized cybernetic system. After the principles are developed, we will show how the same principles appear to be utilized in our commonsense mental processes. The reader is once again asked to be patient during the development. The relevance of computerized principles to commonsense logic will not become obvious until after the principles have been developed.

EVERETT'S BAYESIAN MACHINE

Although the Bayesian principle apparently lies at the foundation of human deductive reasoning, it has not been used directly in most computerized learning systems. The obvious impossibility of representing (even in a computer) *all* possible hypotheses for nontrivial issues has discouraged the use of the method. Consequently, in almost all formal applications of Bayes' method more sophisticated statistical techniques have been used that (for special assumptions) can be derived as linear approximations to Bayes' law. Although these techniques are mathematically more complex, they allow the probability distribution (or distribution of likelihoods) for the alternative hypotheses to be represented as a single mathematical function (the normal or Gaussian distribution), and this avoids the need to deal with the large number (or even infinite number) of alternative hypotheses as separate and individual entities. But these traditional statistical methods involve so much mathematical sophistication and are so limited in their areas of application that they hardly seem appropriate for a biological decision system. If the Bayesian concept is to be used by such biological systems, it seems certain that it must be used in a more simple and robust computational mode.

Around 1971 Hugh Everett III proposed and built a prototype computerized learning system which used Bayes' law in its original simple form.* The system performed quite well and demonstrated a number of the practical principles of deductive reasoning. Because of some close parallels to the human deductive mechanism it is worth considering this machine in some detail. This Bayesian machine was originally motivated by problems in the radar "tracking" of aircraft. Obviously the problem of radar tracking is only one very specialized example of a learning process based on sensory information. Although it may seem far removed from the design principles for biological systems, it serves to illustrate some very important general principles.

Traditional statistical techniques have been used for many years to improve the reliability of radar tracking. Similar and even more sophisticated techniques have been used in tracking ballistic missiles and associated reentry objects. One of the most sophisticated methods, called a Kalman filter, provides "optimal" linear tracking. So long as the object being tracked obeys equations of motion that are essentially linear, the performance of the Kalman filter is equal to the best that is theoretically possible.

But aircraft do not follow linear equations of motion. At unpredictable times, they make turns and thus deviate drastically from the most probable predicted course. When this happens, even the best possible Kalman filter does a miserable job of tracking. Using the same radar information any schoolchild would make better predictions.

Everett proposed that, with the large data-processing capacity of modern computers, an artificial "learning system" utilizing Bayesian inference in its original simple form might be practical. If so, it should be possible to construct a system which (in the limit of very large computational resources) would provide an "optimal" *nonlinear* filter! To accomplish this he proposed a computational method in which the usual *continuous* distribution of alternative hypotheses is replaced by a large but finite "statistical sample" of alternatives selected from the distribution.

Although the concept originated in the context of aircraft tracking, the original computerized version was developed to provide better identification of "trends" and "turning points" in financial markets. Later Gary Lucas and I adapted it to provide a prototype system for tracking maneuvering reentry vehicles in the context of a ballistic missile defense (9). In both applications the Bayesian mechanism performed well. Within the statistical error of the sampling method it gave the same results as the traditional statistical techniques when the problem was linear. In the case of very nonlinear problems (such as maneuvering vehicles) it performed better than the linear methods. The

---

* The original system was treated as company proprietary and was not reported in any literature. A subsequent version, however, has been documented (9).

operation of the system is probably easiest to describe in the context of the more familiar aircraft tracking problem.

When a radar detects an aircraft, it does not pinpoint the aircraft exactly. There is always some uncertainty about the exact location. Depending on the type of radar, the uncertainty can be as much as several miles or as little as a few feet. The typical radar antenna rotates about five times a minute, and a new observation of the aircraft position is obtained once on each rotation. As the number of observations increases, the uncertainty about the aircraft position and direction of flight decreases. Nevertheless, there is always some uncertainty and the uncertainty tends to be largest when the aircraft is executing a turn, since one cannot be sure of either the sharpness of the turn or how long the turn will continue.

To deal with this problem, the Bayesian machine considers a very large number of alternative assumptions about the location and flight route of the aircraft. As each observation occurs, the "probability" of the reported radar observation is tested against each of the many alternative hypotheses. Gradually the inappropriate assumptions are eliminated, and only the most probable alternatives remain. Obviously, for any practical application, it is necessary to make the system work with a finite and limited number of alternatives, but this requires some computational tricks that will be introduced later as we begin to address the procedures of commonsense logic. To introduce the concept we will describe the method in a very simplified form, as it might operate if there were no limit on the number of samples.

Let us consider a case where there is only one aircraft in the sky and the radar is suddenly turned on. Before the radar is turned on, the aircraft could be anywhere in the area and could be flying in any direction. The original Bayesian "sample" therefore consists of millions of alternative aircraft "tracks" distributed *uniformly* over the area, each corresponding to a different location and velocity and each with a *very* low probability.

After the first radar observation, the situation changes radically. The first radar observation establishes the location of the aircraft within a rather narrow uncertainty region. Out of all the millions of alternatives only a few thousand alternatives place the aircraft close enough to the reported position to be consistent with the radar observation. The a priori likelihood H_i for all other alternatives is multiplied by very small probabilities, and are effectively eliminated from consideration. Only the few thousand alternatives that place the aircraft sufficiently near the observation survive with significant likelihood. Of course, a single radar observation of aircraft position tells nothing about the direction of flight. Therefore, the surviving alternatives are still flying with equal likelihood in all possible directions and with a wide range of speeds.

After the radar antenna has made a full revolution, a second observation is made of the aircraft position. Since the aircraft is moving, the new reported position is somewhat different. A second application of the Bayesian principle produces another dramatic change in the surviving Bayesian sample. The only samples that will survive the second (as well as the first) observation are those that are flying in the same general direction as the actual aircraft (for these are the only ones that can be sufficiently close to both observations). All other hypotheses are multiplied by very small probabilities and effectively disappear from the sample. At this point we may be left with a few hundred alternatives, all of which correspond roughly to the correct position and velocity of the aircraft. However, our surviving set will include some samples which are turning as well as samples that are flying straight.

When a third observation is made the surviving sample set is further reduced to just a few tens of samples. The surviving alternatives correspond quite closely to the true path in both position and velocity, and they also begin to approximate the actual rate of turn. A fourth radar observation will continue to narrow the uncertainty, but it will also further reduce the surviving sample so that only a very few hypotheses will remain.

This little example illustrates quite accurately both the power of the simple Bayesian approach and its practical computational limitations. The method is extraordinarily efficient in weeding out unacceptable hypotheses. Its practical weakness is that, even if one starts with an astronomical number of samples, the number will be very quickly reduced to a point where only one sample hypothesis will survive. When this happens the deductive power of the method is destroyed. If such a technique is to be used as a practical deductive method, some way must be found to maintain an adequate population of samples.

To achieve this objective, we settled on a sample "regeneration" method very similar to evolution's method of "reproduction" and survival of the fittest. Since the surviving sample of alternatives any time is really only a statistical sample from a much larger distribution of possibilities, it was decided to generate new samples by randomly interpolating *between* the existing samples. In this way, each new sample hypothesis results from a "marriage" or linear combination between randomly chosen but neighboring hypotheses within the existing sample.

When a radar observation occurs, each hypothesis is tested against the observation. The hypotheses "survive" the radar observation with a *probability* which reflects their consistency with the observation. Specifically, a Monte Carlo process is used to select surviving hypotheses, and each hypothesis, H_i, is allowed to survive with a probability proportional to the probability P_i of the observation for that hypothesis. Typically, a large fraction of the original sample will *fail* to survive the observation. To keep the number of samples constant, the creation of new samples continues (each one being tested against the radar observation) until a *surviving* population of alternatives is achieved which is the same size as the prior set of samples.*

In this way the population of sample hypotheses is kept constant, but the structure of the surviving sample reflects all the information obtained in all the prior observations. As this process continues, the surviving set of samples evolves (by a process of multiplication and survival of the fittest) so that it matches quite accurately the actual flight path.

Each sample hypothesis corresponds to a specific possible trajectory for the aircraft. The hypothesis includes a specification of the position, the velocity, and the rate of turn. It also includes a specification of the probabilities for various maneuvers. Therefore, for each such hypothesis the position of the aircraft can be projected forward in time in accordance with the specified maneuvering probabilities and the laws of motion for the aircraft. For all practical purposes, each hypothesis is a simulated possible flight path. As each simulated flight path is projected forward in time, maneuvers are randomly introduced in the flight path as specified by the maneuver probabilities.

Between radar observations the position of each simulated aircraft in the sample is advanced in time to allow for its movement between observations. When this has been done, the system is ready to process the next radar observation. The rules that govern the movement of aircraft in the sample may include a certain probability per unit time that certain types of turns will be initiated. They may also include the probability that an existing turning maneuver will be ended. Obviously the performance of the tracking system will be best if these assumed probabilities correspond rather accurately to the true probabilities for the aircraft.

The behavior of the system is interesting to watch. When the aircraft is flying in a straight line, the radar observations tend to be consistent with the most probable predictions and most of the sample alternatives survive each observation. Under these circumstances each radar observation can be processed with a minimum of calculation. But if the aircraft has just initiated a turn, then only those samples that have *also* initiated a turn will be likely to "survive" the observation. Many thousands of samples may have to be generated to get the required number of survivors. The number of calculations required to process the observation becomes much greater, and the calculations take much longer, almost as if the system were "worrying" about the problem.

This basic Bayesian system has now been used on quite a number of dif-

---

* In practical operation the sample population would probably be maintained at a few hundred to a few thousand samples.

ferent problems. It provides a routine and systematic approach to deductive reasoning. Beginning with almost no information about the flight path, it utilizes a few radar observations to deduce quite accurate information about the position, velocity, and rate of turn of the aircraft.

In principle the Bayesian procedure is not confined to well-structured problems (such as the aircraft tracking problem) where the laws of motion are known in advance. It can also be used to *infer* the laws of motion. For example, we might begin by testing the hypothesis that successive radar returns will usually be separated by less than five miles, and compare this with the opposite hypothesis that they will be separated by more than five miles. A very small number of observations should settle this initial issue, so that more detailed questions could be addressed.

At present, Everett's Bayesian machine is little more than a laboratory curiosity. It is a Model T version of the concept. Nevertheless, on the basis of experience with this system it is possible to provide some general principles concerning the "cybernetic efficiency" of such systems.

PRINCIPLES OF EFFICIENCY IN DEDUCTIVE PROCESSING

In choosing an initial design for the computer system, our objective was to make the design as simple as possible so as to limit the cost of developing the system. Moreover, because we were interested in a system that could approach the theoretical limits of deductive efficiency, we were not prepared to compromise in system performance. Thus, the chosen design was a logically simple approach that depended heavily on the computational speed of the computer. Although it may be feasible for a high-speed computer to process hundreds or even thousands of alternative hypotheses for each event, such an approach is certainly not appropriate for the human mind. If human beings were to apply Bayesian logic in the simple "brute force" way used by Everett's system, they would never develop much useful information.

A deductive procedure for a biological system must have very different goals. It must be compatible with the limited computational speed of the brain, but on the other hand some sacrifice in deductive efficiency is permissible. To provide a practical deductive approach for a biological system, it is essential to keep the number of alternative hypotheses that must be processed as small as possible. We must therefore ask how the design of Everett's system might be modified to decrease the number of hypotheses required for successful operation.

Obviously the requirement for processing vast numbers of hypotheses in Everett's system results from the use of unduly detailed hypotheses, even at the very early stages of the calculation. For example, even for the first observation, before the system knows anything about the position of the aircraft, it is using hypotheses that include detailed assumptions about velocity and rate of turn. This vastly increases the number of hypotheses that must be tested at that time. For *each* position alternative, the system must also consider the whole spectrum of velocity alternatives. Moreover for *each* velocity alternative, it must consider the entire spectrum of alternatives concerning rate of turn. This *simultaneous* processing of all possible *detailed* assumptions ensures that there will be no loss of efficiency in the Bayesian inference, but it imposes an immense computational burden. Most of the computation load can be avoided if one is willing to compromise a bit in deductive efficiency.

To minimize the number of hypotheses that must be processed, one can begin with less detailed hypotheses, and add detail only as the state of knowledge is improved. For example, the initial hypothesis might deal *only* with the position of the aircraft. At this stage, multiple velocity and rate of turn alternatives are not required. When the *position* issue is essentially resolved, a new level of detail can be added dealing with the question of "velocity." The number of alternatives then required for the velocity issue is much less, because there are only a small number of position alternatives to be considered. Finally, when both position and velocity are resolved, alternatives can be introduced concerning the rate of turn. At that time, this does not add an excessive number of alternatives because the position and velocity issues have already been resolved.

A practical human deductive strategy requires a continuous and adaptive adjustment of the level of detail that is being tested. At each stage we must test alternatives that are just a little more detailed than our current state of knowledge. The "commonsense" deductive strategy, therefore, is approximately as follows: We begin with hypotheses that are only a little more detailed than our present state of knowledge. As observations occur, we use them to distinguish between the hypotheses. When the set of alternatives has been reduced so that we have rather high confidence in a single hypothesis (or a narrow range of alternatives) we may add a new level of detail to our model. As we add the new detail, we effectively subdivide the surviving alternative (or alternatives) into a number of more detailed alternatives. The surviving distribution of alternatives from the earlier level of detail defines the "a priori" probability distribution of alternatives to be used at the more detailed level of testing. The deduction thus proceeds from the general to the specific.

In the "commonsense" deductive procedure just outlined, there is some loss in deductive efficiency, because we do not begin to address questions at the next level of detail until questions at lower levels of detail have been resolved. In principle, much of this loss can be recovered (without much increase in computational load) by *reprocessing* some of the *old* observations against the more detailed hypotheses. When this is done we should, of course, return to the a priori probability distribution appropriate to the earlier time. Nevertheless, the additional computational load incurred in this reprocessing can be controlled by limiting the computation to those hypotheses which have been shown to survive the observations at the lower level of detail.

This reprocessing of old observations against a new hypothesis corresponds to standard scientific procedure. When we develop a new theory we begin by testing it against *past* experience. If it survives this test we begin to test it against new experiences. If it survives these tests, we begin to accept the hypothesis as "true," and we begin to use it as a basis for decision. We use it both to limit our selection of new hypotheses and for predicting the consequences of decisions. This "commonsense" human approach to Bayesian inference apparently underlies all scientific knowledge. It seems probable that Bayesian inference provides the *only* source of valid information available to the "rational" mind.

This statement may seem extreme to most readers, because in most of our activities we do not feel that we are dealing with probabilities. For example, when a man's wife walks into the room he may hear footsteps and look up to see her. When this happens, a hypothesis immediately leaps to mind that his wife has walked into the room. Indeed, he has no doubt about it because he has no alternative hypotheses that would be competitive in explaining the sensory information just received.

How does such a hypothesis come so automatically to mind? Presumably the associative memory is at work. The components of the hypothesis are automatically retrieved from memory because their retrieval keys so closely match the sensory stimuli just received. Because the resulting hypothesis provides such a good explanation of the experience, there is no reason to search further. Evolution has provided us with elaborate sensory systems so that almost all such physical events can be interpreted in this way without ambiguity.

However, if the man were to look up and see instead an object of human size covered by a sheet, there might be considerable doubt whether it was his wife or one of the children. He would then begin a typically Bayesian deductive process. Is the size and shape consistent with the hypothesis that it is his wife? Does it match better with another member of the family? Does it move like the wife, or does it move like another member of the family? In short, which alternative hypothesis provides the most likely explanation of the observed phenomenon? In most cases this typical Bayesian approach to uncertainty is highly effective, and the important uncertainties are quickly resolved.

Of course, it would be misleading to imply that the Bayesian deductive process will always operate smoothly. As we all know, scientific investigation is sometimes led into blind alleys. Progress in certain fields sometimes comes to a halt for long periods of time. What explains such difficulties in the process?

Again the experience with Everett's Bayesian machine provides some answers. The system works well as long as it is possible to maintain a significant number of alternatives that are reasonably consistent with the observations. But the process is so powerful that the number of hypotheses about any single issue is usually reduced very quickly to what amounts to one dominant alternative. This is very desirable in a decision system with limited cybernetic resources because it simplifies decision making.

Indeed, the human mind seems to operate *as if* the elimination of ambiguity (in a Bayesian sense) is an innate goal of the system. It is interesting how troubled we can be when significant ambiguities cannot be resolved. Was the boss really scornful of my last report? Or did he just pretend to be scornful to make me work harder? Such ambiguities can keep us in mental turmoil. We worry through the problem time and again in an effort to clarify the ambiguity. The human mind instinctively tries to resolve ambiguities. It seeks to find one dominant explanation which provides the right (i.e., clearly most likely) explanation. When such a hypothesis has been found there is a sense of relief as if an important goal has been reached. But the elimination of alternative explanations also has disadvantages. When all alternatives have been eliminated the deductive process has ended. If the single surviving hypothesis is later found to be inconsistent with the observations, there may be real difficulties in restarting the deductive process. Normally it is possible to prevent premature degeneration into a single hypothesis by using some method for generating new alternatives. This can be done (as in Everett's machine) by interpolating among the alternatives to gain more accuracy in the Bayesian deduction. It can also be done by subdividing the surviving alternatives through the addition of qualitatively new levels of detail.

But although these strategies can help to avoid the degeneration problem, they may not completely eliminate it. As the surviving alternatives converge to a more narrowly defined set, there is always the risk that new types of events may be inconsistent with *all* of the surviving hypotheses. For example, in the aircraft tracking problem this could occur if the aircraft suddenly executed a much more extreme maneuver than had been encountered before. Based on past experience, the estimated probability of such a maneuver might be extremely low, and (within a finite number of sample alternatives) there might be no sample at all corresponding to the maneuver. Under such circumstances, the system could completely lose track of the aircraft and might have to reacquire it, almost as if it had never been previously tracked. This type of situation (which was encountered several times during the testing of the Bayesian tracker) is quite analogous to the case where scientists have been led down a "blind alley."

In the development of scientific theories such surprises are most likely to be encountered when entirely new kinds of experiments are undertaken, for it is possible for a hypothesis to be "true" within a limited experimental context but nevertheless be "false" when the experimental context is broadened. But the previous deductive work is not lost. It served to exclude from consideration a large variety of hypotheses which performed more poorly even in the original context. The new problem, therefore, is not to revive alternatives that were previously rejected, but to generate a new hypothesis that will perform satisfactorily in both contexts.

This kind of problem is typical of the "crisis" situations in the development of science, when an established theory fails to make useful predictions in a new area and no alternatives are available to provide a solution. When this happens there is no easy solution, but the practical approach once again involves the "commonsense" application of the Bayesian deductive method. Hypotheses are formed about the nature of the discrepancy, and these are tested against new experiments. This systematic approach gradually narrows the uncertainties, until it is possible to postulate a new set of relationships that once again permit successful prediction.

Thomas S. Kuhn provides an excellent description of this type of situation in his book *The Structure of Scientific Revolutions* (8). As Kuhn correctly observes, the old hypothesis (or scientific paradigm) is not rejected until a better one becomes available. The transition from an old to a new theory involves a competition between scientists which is ultimately concerned with the predictive capability of the theories. The operation of this informal process is imperfect but nevertheless somewhat analogous to the Bayesian competition between alternative hypotheses in an incomplete hypothesis set.

Successful practical application of the Bayesian method involves a mixed strategy, which has sometimes been described as a mixture of "freedom" and "control." When the deductive process is going smoothly and the range of allowable alternatives is systematically getting narrower, the emphasis should be on control. New hypotheses are formed and tested to add accuracy to the theory and increase the level of detail. This corresponds to the process Kuhn calls "normal science." But when the deductive process has been stalled, and none of the available hypotheses are providing satisfactory predictions, the emphasis should be on freedom—freedom to creatively expand the range of alternatives considered, so that a new hypothesis (or set of hypotheses) can be formed that will survive the tests. The process involved in this search for creative alternatives is not just a random exploration. To be effective it needs to take into account the results of previous deductions that bear on the problem.

It is now possible to summarize some of the most important principles of efficiency in "commonsense" deductive processing:

1. Limit the number of hypotheses that must be tested by choosing hypotheses that are only a little more detailed than the current state of knowledge.
2. Seek to eliminate uncertainty concerning alternatives by looking for counterexamples to a hypothesis or at least by seeking information that will tend to distinguish between alternatives.
3. As earlier uncertainties are resolved, add more detail and accuracy to the hypotheses that are being tested.
4. When a single hypothesis is developed that provides predictions of adequate accuracy, the deductive process usually stops. This is necessary to make cybernetic resources available for other activities.
5. If existing hypotheses cease to provide an adequate explanation, generate alternatives with slight variations and test against the data to see if an improved explanation can be found.
6. If neighboring alternatives fail, then begin a wider search for alternatives that may be more effective.

The foregoing principles correspond closely to the strategy that is used in the intuitive deductive process, and they are basically principles of efficiency.

Obviously the intuitive Bayesian inference is far less exact and efficient than a formal mathematical procedure. Psychological experiments in which intuitive deduction is formally compared with Bayesian inference show that the commonsense deductive process is usually conservative in that it underestimates the effects of new information in changing the likelihood estimates. But such comparative tests must be conducted on formal mathematical problems that are stated in such a way that the available hypotheses are known to constitute a *complete* hypothesis set. The commonsense deductive approach has adapted instead to a situation where the hypothesis set is *incomplete*. Under such circumstances, it is a mistake to assign a very high likelihood to *any* hypothesis without a careful consideration of possible alternatives. Although formal Bayesian inference might conclude that one hypothesis is almost infinitely more likely than the others, a more important practical question concerns the likelihood of the hypothesis compared to others that have not yet been considered. The conservatism of intuitive deduction relative to the formal mathematical estimate appears to be a heuristic adaptation to experience with incomplete sets of hypotheses. This commonsense method of deduction was molded at least in part by evolutionary forces. Although it is unclear to what extent the method is prewired, it seems clear that the brain is designed so that the commonsense deductive method is both natural and easy to learn.

The foregoing deductive method suggests some "Intellectual Values" that might be useful to guide the model-building process. The following list provides some examples:

1. A dislike for ambiguity or uncertainty.
2. A liking for hypotheses that seem to provide adequate unambiguous explanations.
3. A liking for counterexamples that can be used to eliminate inappropriate alternative hypotheses.

These general value concepts will be developed in more detail in the next chapter after some additional principles of the model-building process have been developed.

IMPORTANCE OF SIMPLICITY AND SCOPE

It is apparent that the preceding principles already incorporate a number of the principles of commonsense logic. But to develop some of the remaining principles we must consider the requirements for cybernetic efficiency in some of the other functions of the decision system. The formal Bayesian process is concerned only with the validity, accuracy, and reliability of the world model and of various hypotheses contained within the world model. But the model-building process must be guided by more than the requirement for accuracy and validity. To be useful, the world model (and the components of the world model) must be sufficiently simple so that it doesn't impose an excessive burden to remember or to use it in the prediction of outcomes. Other things being equal, a simple model is preferable to a complex one. In many cases this means that multiple models are needed at different levels of detail. For example, although it might be theoretically possible for an archer to hunt wild game using a detailed mental model of the arrow (including even the structural details of the binding between the shaft and the arrowhead), in practice such details just get in the way. It is more efficient to work with a hierarchy of models at different levels of detail than to try to do all predictions with only one model. It is convenient to have specialized model components that are simplified so that they are particularly easy to use in specific applications.

On the other hand, it is important not to have to deal with too many model components. Thus comprehensiveness or scope is an important consideration when evaluating the usefulness of a model. A model of wide applicability can replace many specialized models and thus contribute to cybernetic efficiency.

These considerations of simplicity and scope are important valuative factors that need to be considered in the choice of model components. The issues are relevant both in the selection of hypotheses for Bayesian inference and in the final use of model components. Other things being equal, we should prefer simple hypotheses over complex ones, and we should prefer hypotheses of broad scope to those of narrow applicability.

It appears that human intellectual activities may in fact be guided by very similar innate values. Ideas of intellectual "elegance" are particularly prized by intellectually creative people. "Elegance" is a quality that is not easy to define, but it seems to denote a combination of simplicity and scope. Theorists intuitively seem to recognize elegance and are much attracted by it, despite the fact that in most cases they have no well-formulated definition of the term.

To be useful, a model must pass the Bayesian tests for validity, but it also must rank high in both simplicity and scope. In somewhat less concrete terms, Smith and Marney (11) recognized these same dual requirements for good models. They described the Bayesian validity requirement as a requirement for "warrantability." Specifically, this involves the definition of a "range of applicability" within which the model has been tested and can be "warranted" or guaranteed. They described the requirements for simplicity and scope as a "serviceability" requirement. Serviceability is fundamentally concerned with the cost-benefit characteristics of the model. If a model is to be "serviceable" the benefits derived from the use of the model must substantially exceed the cybernetic costs involved in its use. Major theoretical advances are usually characterized by an increase in scope and simplicity, as many crude and ad hoc models are replaced by a single "elegant" theory. The development of such a theory usually involves a wide-ranging, creative search for a more widely applicable hypothesis. When good candidates are found they must be tested as always for validity by the methods of Bayesian inference.

It is often alleged that elegance and simplicity in a model (or theory) are in and of themselves an evidence of validity. There is no doubt that these attributes are indicative of potential usefulness if the theory is valid, but are they an indication of validity? There appears to be no a priori basis for expecting the physical world to follow "simple" or "elegant" laws. But it is certainly true that during the advance of science, the most accurate and reliable models have *also* been both simple and elegant. This observation, of course, leads to a natural Bayesian inference that in most cases the true physical laws are likely to be both simple and elegant. Thus, as a practical matter, theorists are probably justified when they use this criterion as an initial validity guide in the search for more generally applicable theories.

GROUP WISDOM AND PERSONAL JUDGMENT

Much of our mental model building is a social or group activity. The words we use to describe our opinions in such discussions are less formal than the Bayesian method might lead us to expect. Even in discussions that are concerned with objective facts (rather than value judgments) individuals will precede their remarks by phrases such as, "In my judgment," "In my opinion," or "My intuition tells me. . . ." It is expected that different individuals will have different judgments and different opinions about many factual issues. It is also expected that an "experienced" individual may have acquired a certain "wisdom" from his experiences that may not be easily explained or justified.

How do these informal concepts of judgment, wisdom, experience, and intuition fit into the concept of Bayesian inference? Superficially we might expect that with everyone using the *same* basic deductive method there should be no reason to have differences in judgment. But with a little reflection, it is obvious that this is not true. Even if two individuals were to observe exactly the same process (or even the same sequence of events) the interpretation of the events would almost certainly be different.

Because of past experience different individuals will begin their deductive approach by formulating *different* sets of hypotheses about the *same* group of events. As the different hypotheses are tested and refined each will develop a unique *personal* interpretation of the events. Ultimately, if the process continues long enough, the observers may independently arrive at a very accurate (and consequently very similar) interpretation. But because the world is usually dynamic and variable, an exact correspondence between independent views is a rare event. When individuals compare their ideas they are usually at an intermediate point in the deductive process. Each has formulated some hypotheses that seem at least moderately successful, but their interpretation of the events may be very different. Moreover, an individual may be using some hypotheses that are not producing really satisfactory predictions, or he may be using several alternative hypotheses that seem only slightly different in likelihood. The "experienced" individual may still lack a satisfactory theory, but has classified situations where his hypotheses fail so that he may have a very valuable "wisdom," without having sure knowledge.

As the individuals pool their understanding, the similarities and the differences in their interpretations can provide new insights. Where they have independently reached similar conclusions the conclusions are reinforced. Where their opinions differ they will seek to reconcile the differences. If they can find a new view which provides a consensus, it is probably more likely to be right than the independent view of either observer.

In many practical situations, it may be inappropriate or even impossible to test the new consensus against the observed events. As a finite cybernetic system, the individual may not even be able to recall the events that led to his opinions. Nevertheless, the individual opinions have value; they represent a distillation of each individual's deductive process. A group consensus, if it seems reasonably consistent with several individual opinions, provides a basis for decision which will typically be more reliable than any individual opinion selected at random. Thus, the consensus provides a practical way of combining individual experience to achieve a "wisdom" which exceeds that of any single individual. It should be apparent by now that many of the characteristics of commonsense logic are almost inevitable consequences of the requirement for cybernetic efficiency in a finite decision system. Although the analysis could be extended to develop additional principles of commonsense logic, we have already gone far enough to provide a framework for discussing some of our innate intellectual values.

WHAT IS TRUTH?

As the deductive analysis proceeds, the world model is constantly updated and revised. Old hypotheses are rejected and new ones are added. But certain hypotheses stand the test of time; counterexamples which might destroy them are not encountered. Moreover they are functional because they permit reliable predictions. Such hypotheses come to be accepted as "true." They are highly valued because they are reliable. They contribute to efficiency in the cybernetic process because they do not have to be repeatedly checked or confirmed.

But it should be emphasized that the "truth" of such a hypothesis is only relative to the context of experience in which it has been tested. No matter how broad the basis of experience, the possibility always exists that in a radically new and untested context the hypothesis may fail. On this basis, it follows that we can never be sure of the truth of our concepts beyond the context within which they have been tested. Practical truth is always relative to a specific experimental context.

Of course, within the context of past experience certain hypotheses can reach astronomical levels of likelihood compared to all other known alternatives. We come to view these things as certainties. For example: Two plus two equals four; or an area on a flat surface bounded by three straight lines (a triangle) will have three corners (or line intersections). Sometimes we think of such "truths" as being *so absolute* that they can be deduced by "pure reason." But this is an illusion. Even "pure reason" begins with axioms. The *relevant* axioms are selected on the basis of experience.

If we were ever to find a context in which either of the preceding truisms consistently failed, we would almost certainly accept the "counterexample" provided by our experience and would deny the conclusion of "pure reason." Either the axioms would fall, or a fault would be found in the logic. This point (which may seem rather extreme) is offered to illustrate how strongly and intuitively our "commonsense" reason adheres to Bayesian logic, in preference to other modes of thought that may seem to be more formal or rigorous. So long as we limit ourselves to an informal or intuitive application of the Bayesian principles (as man almost always has in the past) it seems clear that this pragmatic definition of truth is the *only* realistic meaning. But it is tempting to ask whether we might be able to recognize a more absolute form of "truth" through a more formal and sophisticated application of the Bayesian theorem. For example, the Bayesian method allows us to develop and test any form of hypothesis we may find useful. Is it possible that by selecting and testing some clever combination of hypotheses, we might be able to determine in more absolute terms whether a theory is "true"?

Of course, if a theory were so good that it predicted all outcomes, with zero error and 100 percent reliability, we would have a perfect theory (at least within the range of our experience) and there would be no point in looking for a better theory. But, in practice no theory is perfect. Actual observations always different somewhat from the most likely predictions. The deviations may reflect experimental errors, imperfections in the theory, or statistical fluctuations predicted by the theory (as in quantum mechanics or roulette). A complete theory should define not only the most likely outcome but also a predicted probability distribution of outcomes when all sources of error are considered.

Suppose we have two such theories and wish to determine which is more likely to be true. Proceeding in the standard way, we would ask how likely the *observed* measurements are in terms of each theory. On this basis (unless both theories make exactly the same predictions) we should expect in due course to select a preferred or more probable theory. Having chosen between the alternatives we might wish to ask whether an even more accurate theory is possible, so we might check for internal consistency within the preferred theory. We might therefore ask how the calculated likelihood for the actual observations compares with what the theory itself would predict for the likelihood of a *randomly* chosen set of observations. Do the *actual* deviations from the theoretical mean correspond with the *predicted* deviations? If the observed deviations are substantially larger or smaller than predicted, it indicates that the theory is not internally consistent and a better theory is possible. Indeed a more accurate theory can be produced on the spot, simply by adjusting the predicted size of the deviations.

On the other hand, even if the theory is found to be internally consistent in *all* respects, it does not *guarantee* that no better theory exists. So long as the theory does not make predictions that are 100 percent reliable and 100 percent accurate, then there remains the theoretical possibility of a better theory.

The current status of quantum mechanics corresponds closely to this situation. The theory seems to pass all tests for internal consistency * and experimental observations seem to fully verify the probability distributions predicted by the theory. Nevertheless, there remains at least a theoretical possibility of a better theory. As has been frequently noted, it is possible that by analyzing correlations with certain previously ignored variables, it may be possible to produce predictions that have less uncertainty than the present predictions of quantum mechanics. If such predictions were ever achieved, it would immediately establish that quantum mechanics is not an *ultimate* theory. All that is necessary to eliminate quantum mechanics as an ultimate theory is to produce a predictive method that yields a higher Bayesian likelihood than present quantum mechanics. It is interesting that quantum mechanics (which seems to be an internally consistent theory) might thus be dethroned by an alternative hypothesis that might not even be internally consistent. Although most physicists would agree that such an event seems unlikely, the possibility is interesting because of the insight it provides concerning the ultimate basis of human knowledge.

---

* The internal consistency of the theory is most apparent in the multiworld formulation of the theory (4).

EVOLUTION'S BAYESIAN BOOTSTRAP

As noted earlier, the individuals of a species do not start from scratch when they begin to learn about the world. Prewired learning mechanisms are provided that are specifically designed to learn certain things. In addition, even certain specific pieces of information may be prewired. This raises an interesting theoretical problem. How could such specific information, or specific learning devices, be acquired during the evolutionary process.

The answer should now be quite apparent. The evolutionary process itself operates almost *exactly* like Everett's Bayesian "learning system." Each individual of a species is a manifestation of a specific genetic inheritance. The genetic inheritance is encoded in the sequence of genes within the chromosomes. From a purely mathematical perspective we can think of this total sequence (of almost 10 billion genes) simply as a very long "symbol string." Each such "symbol string" can be thought of as an evolutionary "hypothesis." The "likelihood" of the hypothesis is measured by the ability of the resulting biological organism to survive and multiply.

The mechanism of sexual reproduction with occasional mutations provides an efficient way of exploring neighboring hypotheses, to see if genetic combinations can be found that will produce a higher "likelihood." From this perspective, it should be evident that a large amount of prewired information could easily be developed in our evolutionary inheritance. A symbol string of 10 billion characters can incorporate a great deal of very detailed information. Evolution's Bayesian learning process has been under way for so long that the resulting "hypotheses" can be exceedingly complex and detailed. The principle of Bayesian inference lies at the heart of our commonsense thought processes. It is the bedrock on which human models and theories are built. The search for improved models and theories of course is also guided by the principles of simplicity and scope. With these basic principles as a background we are now ready to address the structure of human "intellectual values."

REFERENCES

1. Campbell, Donald T. "Blind Variation and Selective Retention in Creative Thought as in other Knowledge Processes." *Psychological Review*, 67, no. 6 (1960): 380–400.

2. Carnap, Rudolf. *Logical Foundations of Probability*. Chicago, Ill.: University of Chicago Press, 1962.

3. Carnap, Rudolf. *Logical Structure of the World: Pseudo Problems in Philosophy.* Berkeley, Calif.: University of California Press, 1967.

4. Everett, Hugh, III. *The Many World Interpretation of Quantum Mechanics.* Edited by Bryce DeWitt and Neil Graham. Princeton, N.J.: Princeton University Press, 1973.

5. Gibson, James J. "Adaptation, After-Effect and Contrast in the Perception of Curved Lines." *Journal of Experimental Psychology* 16 (February 1933): 1–31.

6. Guillery, R. W. "Visual Pathways in Albinos." *Scientific American* 230, no. 5 (May 1974): 44–54.

7. Kohler, Ivo. "Experiments with Goggles." *Scientific American,* May 1962.

8. Kuhn, Thomas S. *The Structure of Scientific Revolutions.* University of Chicago Press, 1962; second edition, 1970.

9. Lucas, Gary. "Cassandra, A Prototype Non-linear Bayesian Filter for Re-entry Tracking." *Lambda Report 101,* Nov. 30, 1973.

10. Popper, Karl R. *Conjectures and Refutations.* New York: Basic Books, 1962.

11. Smith, Nicholas M., and Milton C. Marney. *Foundations of the Prescriptive Sciences.* McLean, Va.: Research Analysis Corporation, March 1972.

Chapter 13

Our "Intellectual Values"

How much lies in laughter:
the cipher-key, wherewith
we decipher the whole man.
THOMAS CARLYLE, *Sartor Resartus*

SINCE our mental "model-building" activities occur within the brain, they are difficult to observe or monitor. Consequently, we might expect that it would be impossible to develop anything more than introspective evidence for these values. Fortunately, this is not the case. Man is a social animal, and much of his "model building" is a cooperative social process. Consequently, evidence concerning the "Intellectual Values" can be expected in many aspects of social behavior. We will therefore begin the consideration of Intellectual Values by focusing on those values that seem most obvious in the external behavior of the species.

Evolution has not yet developed Intellectual Values that are perfect in terms of survival of the species. In some cases a motivating value does not coincide precisely with the activities it is intended to motivate. In other cases, the values may motivate by-product activities that may not serve any practical survival function. For example, the enjoyment of art and music seems to be almost an accidental by-product of Intellectual Values that have other very practical survival functions.

Since the concept of instinctive values as motivators of intellectual activity is relatively new, little experimental work has been done on the subject. The concepts developed here are based primarily on personal observations. Although the speculations may be suggestive of the actual structure of our Intellectual Values, careful experimental work will be required to provide a definitive understanding of the value structure.

CURIOSITY

Curiosity is one of the most obvious external symptoms of our Intellectual Values. As in the case of the complex social values, curiosity can be broken down into a number of more fundamental value components. The following list illustrates some of the factors that seem to be involved.

1. *A desire for intellectual activity:* Just as the individual likes to exercise his body and senses a need to do so, so he also wishes to exercise his mind. As with exercise of the body, the desire for mental exercise seems to be an appetitive value. The need can be satiated, and excessive demands on intellectual activity can be unpleasant. Thus the mind seeks an appropriate level of mental exercise. The fireman fills his idle time with games of chess and checkers. A scientist is more likely to seek relaxation in art or physical exercise. Of course, the preferred level of intellectual activity varies widely between individuals.
2. *Selective interest:* This phenomenon is very widely observed. The individual's interest tends to focus on things that are strange, but not on things that are too strange to fit into his world model at that stage of development. When a small child is taken for a ride in a car he shows little interest in the external scenery. The blur of the passing scene is simply too strange to fit into his world model, so it is of no interest to him. The child's interest is focused intead on the inside of the car, on his toys and on the other passengers in the car.

   The Bayesian theory provides an understanding of why this must be so. Efficiency in model building requires that new hypotheses should be concerned with issues that go only a little beyond the range of what is already known. The child's interest coincides with his process of hypothesis formation and testing. For efficiency it must be focused on the boundary region of his existing world model.
3. *Distaste for ambiguity:* There seems to be an instinctive distaste for situations where it is necessary to maintain two or more alternative hypotheses. The desire to get rid of such ambiguity contributes to cybernetic efficiency by reducing the need to work with alternative hypotheses. This desire to eliminate ambiguity undoubtedly explains much of the fascination of mystery stories and puzzles. When an ambiguity is resolved or a problem is solved there can be a very rewarding sense of accomplishment. If it cannot be resolved, frustration and perhaps even anger will ensue. The individual may be motivated to throw the puzzle away or otherwise get it out of his mind.

The importance of curiosity is clearly correlated with intelligence. In general the species with the most highly developed brains seem to be the most curious. Curiosity has obvious survival value since it enables the individual to develop information at leisure that may be needed later in situations of crisis.

When the "Selfish" and "Social" values were discussed, we were able to specify the subjective sensations of value which serve to motivate the "rational" decisions. Should we expect similar subjective sensations of value for our intellectual values? The answer seems to be, "not always." Although some intellectual activity is under deliberate or rational control, other intellectual activities seem to occur automatically or compulsively, and even contrary to our conscious desire. To the extent that such activities are motivated by a value mechanism, the relevant mechanism may lie outside the rational mind. Although values may exist for such processes, we may not be able to identify any subjective "value sensation." Where the mental activity is largely under conscious control we should expect to find subjectively experienced sensations of value, much as with the "Selfish" and "Social" values.

What sensations of value are associated with the three attributes of curiosity just discussed?

1. The desire for mental activity seems to be motivated by an unpleasant sensation of boredom that accompanies inactivity. An excess of mental activity seems to be followed by a sense of mental fatigue, lethargy, and lack of interest that motivates a withdrawal from activity and events which would stimulate mental effort.
2. The selective concentration of hypothesis building in areas which are somewhat strange but not too strange seems to be motivated by the emotion of "interest" which automatically causes these areas to be intellectually attractive. But this is not the only factor that guides interest. Apparently certain genetically defined visual "keys," such as an "attractive" female figure, will also guide or focus "interest."
3. The avoidance of ambiguity seems to be motivated by a rather strong sense of mental unpleasantness associated with ambiguity. No English name seems to be available for this value sensation, but subjectively it seems closely related to frustration. The urge to get rid of ambiguity may also be compulsive, beyond what can be explained by this value sensation. The removal of ambiguity is usually associated with a strong sense of achievement which constitutes the subjective reward.

The combined operation of the above value sensations can generate a strong motivation for mental activity and can make such activity one of life's most rewarding experiences.

HUMOR

Humor is one of the most obvious Intellectual Values. It appears to be a social aspect of the Bayesian model-building process. As our discussion of the Bayesian process demonstrates, the "counterexample" is one of the most effective tools for removing ambiguity and eliminating invalid elements of a world model. An efficient Bayesian machine will find counterexamples to be of great value. Our enjoyment of humor appears to be nature's way of motivating the social sharing of counterexamples!

This interpretation of humor, as a social manifestation of Bayesian deduction, seems to be a new concept so some justification of the idea is needed. First, we must ask whether humor is really based on counterexamples. Second, we will inquire how the enjoyment of humor can motivate a sharing of counterexamples.

What are the common characteristics of jokes, stories, events, or cartoons that make them funny? Probably the most universal attribute of humor is its incongruity or inconsistency with established concepts. Specifically humor is (or at least briefly appears to be) a "counterexample" to established concepts. Let us consider some specific examples:

The classic "humorous event" where a person slips and falls seems to be funny in direct proportion to the status of the one who falls. The more pompous and dignified he is, the more incongruous and funny it seems, because it will seem more like a "counterexample" to preexisting concepts. If we replace the typical falling dignitary with a small toddler who can barely walk, no one laughs. When the little child falls, the event is almost expected; it is not a counterexample and it is not funny.

The art of telling a funny story also involves counterexamples. The skillful story teller uses circumstantial evidence to elaborate a specific (but perhaps erroneous) understanding of the story. In the punch line, he unleashes a bombshell of information which destroys the mental model that he has so carefully nurtured in the listener. The story is funny in proportion to the completeness with which the mental model is developed and the suddenness with which it is destroyed. In short, it is funny in proportion to the impact of the "counterexample" on the listener's existing model.

The same basic principle operates with regard to cartoons. The typical cartoon depicts a situation which is incongruous. It may show an individual of high status in a ludicrous situation, or an individual of low status in a pose or posture that affects great dignity. The caption may be a surprise relative to the picture, the picture itself may be incompatible with prior conceptions, or it may depict a physically impossible situation.

Thus, there is little doubt that humor is closely related to the Bayesian counterexample. It seems likely that humor reflects an innate human value whose basic purpose is to motivate the communication of events that are "counterexamples" to previous ideas. The built-in valuative response associated with humor may not be precisely tuned to the practical value of the information as a counterexample. For example, fictitious funny stories are of little practical value, but they are funny nevertheless. Of course (other things being equal) a humorous story tends to be most funny if it is true and if the persons involved are actually known to the audience.

The practice of telling fictitious funny stories or jokes is apparently a by-product of an intellectual value component that is concerned with more practical objectives. On the other hand, even this by-product behavior may have contributed to survival. There is no doubt that humor is conducive to group solidarity, and this tends to enhance the human motivation for cooperative social behavior.

The development of humor in its present form seems to be a unique achievement of human evolution. It is interesting to speculate on how it may have come about. Certainly, the other primates also utilize the intuitive Bayesian approach to learning. Consequently, we should expect that they also should be particularly interested in events that serve as counterexamples. It seems likely that these animals find such events intellectually rewarding, but without a language for communicating such ideas, there is little practical value in an outward display of this emotion.

With the evolution of man, two important changes occurred that made humor possible. The development of the emotion of joy (as a generalized motivator for cooperative behavior) provided a ready-made facial expression through which "amusement" could be expressed, so that humor could be socially motivated. The development of language provided a mechanism which made the communication of counterexamples possible. With these raw materials at hand it was only a small step to link the existing mechanism for appreciation of Bayesian counterexamples to the emotion of joy. In this way, out of a fortuitous marriage between the emotion of joy and the value of Bayesian inference, humor may have been born.

TRUTH

Logically one might expect that truth itself would be an instinctive value for a Bayesian system, but one must recognize that truth usually cannot be recognized when it is initially encountered. A hypothesis comes to be recognized as true only after alternative hypotheses have been eliminated and the hypothesis has remained stable for quite a while. Consequently, there may be no real need for an innate value associated with truth.

There can be no doubt that as Bayesian thinkers, we inevitably place high value on truth. Truth is valuable not only because it avoids ambiguity directly, but because it provides a foundation for more logical selection of new hypotheses that are more likely to survive the Bayesian tests. But the value associated with truth seems to be more a learned than an instinctive value. Truth is valued less for itself than as a means of avoiding the discomfort of ambiguity. Although it appears to be a secondary rather than a primary intellectual value, it is a very important secondary value.

Individuals who consistently tell the truth are useful to others in their model-building activity, and thus they tend to be socially more acceptable. Individuals who cannot be relied on to tell the truth are likely to be disliked because the resultant ambiguity is distasteful. For this reason, truth telling is an important ethical principle. It contributes to the achievement of the innate objective of social acceptability.

SIMPLICITY, COMPREHENSIVENESS, AND ELEGANCE

As mentioned earlier, the criteria of simplicity, comprehensiveness, and elegance provide important guides to the cybernetic efficiency of a model or hypothesis, but it is not clear whether evolution has provided any corresponding innate or primary values. It is entirely possible that these intellectual values, like the value of truth, are learned almost automatically as a result of experience in Bayesian thinking. On the other hand, there is some evidence that the criteria of simplicity and elegance may actually reflect innate values. Many creative people react to the elegance of a theory much as they would to the beauty of a face. The response is essentially esthetic as opposed to purely rational. This esthetic reaction may reflect the existence of some subtle, but nevertheless innate, Intellectual Values.

ESTHETIC VALUES

Many of the activities that we enjoy involve the operation of multiple, essentially independent instinctive values. The traditional sources of esthetic satisfaction (art, music, poetry, and literature) seem to be of this type. The effectiveness of artistic effort cannot usually be explained in terms of any narrow combination of innate values. The essence of a successful artistic effort is that it combines many elements from our instinctive valuative experience.

A detailed review of the principles of art appreciation would take us far beyond the scope of the present effort. The following paragraphs provide only a brief sketch of some of the instinctive values that seem to influence our appreciation of the arts.

VISUAL ART

Visual art appeals to a surprising number of our innate values. It appeals to our innate sense of "beauty." As a consequence of evolutionary experience, certain shapes, forms, and color combinations are perceived as intrinsically pretty. Others are perceived as intrinsically ugly. In all probability rather specific visual engrams have been defined by evolution which automatically evoke favorable or unfavorable response. One obvious example is the male response to the female figure, but it seems likely that there are many others that are somewhat less obvious. The valuative analysis of the optic image for such engrammic components may occur either in the frontal lobes or in the midbrain. As a consequence of such a mechanism, the creative artist can use his painted image to invoke a wide range of valuative and emotional responses.

Art can appeal to sentiments that are based on past experience. The artist can depict circumstances that tend to recall past emotional experiences and thus generate (by association) similar emotional responses.

Art appeals to our innate sense of visual curiosity. To be most effective, a painting must take account of the law of selective interest. It must not be too simple, but it also must not be so strange that it seems totally random or irrelevant. Really random patterns are of very little interest; but a design that is too simple loses its interest very quickly. Children tend to like art that is less complex than the art that appeals to adults.

Art can also capitalize on subtle response characteristics of the eye to rhythmic variations in pattern and color. Modern op art uses this technique to produce strange responses in the visual cortex. It is possible that some of our esthetic response to pattern and structure may result from similar response peculiarities to patterned images in the frontal lobes or midbrain.

Art can also capitalize on our curiosity about the environment by depicting scenes and situations which appeal to other interests.

The foregoing discussion is a very brief outline, but it illustrates the wide diversity of valuative appeals that are available to the artist. The real trick, of course, is to incorporate a diversity of such appeals to provide maximum interest and appeal within a single work of art.

The artist himself in the creative process experiences a wide range of additional valuative experiences. He creates and tests hypotheses about the esthetic consequences of alternatives. He experiences the instinctive pleasure of contributing to society and of doing something constructive. He also tests on himself the esthetic impact which he hopes his audience will feel.

MUSIC

In order to develop a language, human beings had to be endowed with an intense interest in sounds. They had to have a sensitivity to the rhythm and quality of sounds. Long before language, sounds were a mechanism of emotional communication. To operate in this way, sounds had to be endowed with valuative qualities. The baby's cry was intensely unpleasant. The lover's voice was warm and melodious. Different kinds of cries, screams, and calls conveyed instinctive messages of mood and emotion. Some were pleasant, others were unpleasant.

Music capitalizes on this instinctive language of the ear in much the same way that visual art capitalizes on the innate valuative response of the eye. The analysis of sound for valuative or motivational keys may occur either in the frontal lobes or in the midbrain. Music is based largely on tonality, rhythm, and melody. Each of these elements appeals to a specific component of our instinctive values.

The reaction to harmony versus dissonance seems to be our strongest valuative response to tonality. Evidently nature chose to use this device as a primary valuative key in human response to sounds. The infant's cry includes an exceptionally large amount of dissonance which contributes to its unpleasantness and increases the mother's motivation to quiet the child. The mother's response to the infant's cry is undoubtedly based on other more specific identification keys as well, but the presence of dissonance in the cry is clearly an important motivational factor.

Our innate negative reaction to dissonance can be easily demonstrated. If a person sits at the piano and practices random melodious chords he may be tolerated for quite a while. But if he plays random dissonant chords he will be quickly ordered to stop. Normal conversational speech includes primarily harmonious overtones. A positive response to harmony may be an important factor behind the infant's initial interest in conversation.

It might seem that the instinctive values could just as well have been reversed, making harmony unpleasant and dissonance pleasant. If this design alternative had been chosen, the infant's cry would be harmonious. But from an engineering design perspective, dissonance is more difficult to produce in a single wind instrument like the larynx, so the development of language around dissonant sounds might have been more difficult.

The response to dissonance as opposed to harmony is the most obvious of our instinctive valuative responses to tonality, but it is probably only one of many such responses. For example, the distinction between major and minor chords seems to convey an innate message which parallels the emotions of joy and sorrow. Such relationships may provide a fruitful area for research on the theory of music.

At least a part of our valuative response to rhythm can probably be traced to the human heart beat. The normal pace of musical rhythm corresponds closely to the pace of the normal heart beat. Since a child in the womb is exposed to such a rhythm from the maternal heart, the neural networks in the auditory nerve must almost inevitably develop a sensitivity or innate response to this rhythm. The normal heart rhythm includes a sequence of minor beats that are regularly spaced between the major beats. This combination of a major beat followed by a sequence of minor beats defines a rhythm very similar to that of a primitive drum.

The sensitivity of certain neural networks to such a rhythm was almost inevitable, so it was only a simple evolutionary step to place a valuative interpretation on the rhythm. The basic heart rhythm seems to serve as a valuative key to quiet the fears of an infant. The human infant appears to be instinctively comforted when it hears the maternal heartbeat. Most mothers, whether right-handed or left-handed, unconsciously use this response by holding the child against their left breast when they wish to comfort him. The valuative response to such a rhythm apparently remains throughout life, and it conveys a message of social reassurance.

But our response to rhythm is considerably more complex than just a simple social reassurance. A beat which is becoming more rapid tends to generate excitement, while a beat which is slowing down produces a sense of relief and security. This innate response to changes in rhythm could easily have been learned in the womb. When the mother becomes excited her heart beats faster, and the infant at the same time may be subjected both to increses in adrenalin from the mother's blood and to more rapid movement. Thus the accelerating rhythm becomes naturally associated with excitement.

The human interest in melody is probably a by-product of the instinctive interest in sound patterns that was necessary to motivate the learning of a language. Interest in both melodies and rhythm seems to be influenced by the Bayesian law of curiosity. The simple songs of the nursery may seem trite and uninteresting to a musically sophisticated adult, but they are nevertheless fascinating to the child. Conversely, the music of the sophisticated adult may be uninteresting random noise to the child.

It is often said that as we mature, our taste in music "improves." We come to appreciate "better quality" music. This is a misleading way of stating the situation. It is more accurate to observe that adult music, adolescent music, and a child's music are all different. For each audience there is good music and bad music. Each musical composition is of greatest interest to an audience with a specific range of prior musical experience. For other audiences it may be too simple to be of interest or so strange that it sounds like noise.

The performing musical group can experience the pleasures of creativity much like the visual artist. But in the creation of group music there is a special additional joy, for the creation is a *social* process. It thus invokes the standard Social Values of cooperation, teamwork, and social acceptance.

LITERATURE AND OTHER ARTS

Literature can appeal to our curiosity, to our emotions, or to both. Non-fiction literature and the mystery novel appeal particularly to our curiosity. The suspense story, the mystery, and the novel also appeal to our emotions by allowing us to experience vicariously a variety of emotional situations.

The appeal of any piece of literature is complex, and a great deal has been written on the subject. Since the present theory does not significantly change the traditional interpretations no purpose would be served by a more detailed discussion of the subject here.

Most of the other arts—song, poetry, theater, and dance—involve various combinations of the valuative elements previously considered. As always, the goal of the creative artist is to combine within his work an effective combination of esthetic appeals.

RELEVANCE OF ESTHETIC VALUES IN EDUCATION

In the fields of education and esthetics we can benefit from a better knowledge of the "Intellectual Values." Teachers and parents need to have a better understanding of what interests the child and why. The world view of the child needs time to develop before the adult world can be added. In the past, there has been a tendency to hurry children onward to an appreciation of the "better" adult esthetics. Children are urged not to waste time on "trashy" childish books, or on the simple songs, limericks, and pictures that appeal to childhood. Their attention is directed to "more important" cultural achievements; to classic literature, to symphonies, to modern music, to classic and modern abstract art. The present perspective on esthetic values makes it clear that this is a narrow-minded adult perspective. When we pressure children to participate in adult esthetics we are engaged in a needless form of adult chauvinism.

I recently listened to a "performance" of classical music by an elementary school orchestra. It was clear that most of the children had little appreciation for the music. What they were playing was, for them, an almost random sequence of notes with an almost random rhythm. Of course they played it badly, without spirit and with minimum enjoyment. Undoubtedly the teacher believed that he was teaching the children to appreciate "good music." In fact he may have been teaching the children *not* to appreciate "music." If the same orchestral ensemble had learned some simple melodies that they could enjoy, they would have learned more and played better. This, of course, is only one example of the unnecessary problems we can create by imposing adult values on the children in the educational system. There is a clear need for a better understanding of the development of Intellectual Values as the individual matures.

SUMMARY

We have now completed our survey of the innate human values. No effort has been made to provide a comprehensive or definitive treatment of the subject. The goal has been simply to outline some of the main elements of the instinctive value structure and to provide a starting point for further research.

Although the value concept provides a theoretical framework that seems to be potentially quantifiable, we are far from having the information necessary to permit such a quantitative understanding. Present knowledge seems closest to a quantitative formulation in connection with the "Selfish" values. As we move from "Selfish" through "Social" to "Intellectual" values, the available information is progressively less adequate. The discussion reflects this inadequacy by becoming less detailed in the later chapters, where no attempt was made to describe the values in a quantitative form.

Superficially this may seem incompatible with the objective of providing a quantifiable model of human behavior. In fact, it is only a pragmatic recognition of the tremendous uncertainties in our present knowledge. In accordance with the principles of deduction it is inefficient to develop hypotheses that go too far beyond our present level of knowledge. Consequently most of the speculations go a little beyond, but not very far beyond, our present level of experimental knowledge.

The preceding speculative value framework raises numerous questions that invite systematic research. Until such research can provide reliable answers, each individual must seek his own understanding of his values and those of his associates. As a result of normal scientific methods in which hypotheses are developed and tested, we can expect a gradual narrowing of uncertainties. Obviously, the individual variability in the importance of different values will place practical limits on the accuracy with which formal research can quantify innate values.

In all probability before we develop a full understanding of the human value structure we will find certain inconsistencies that will show that the value concept itself is only an approximation to the human motivation structure. Nevertheless for the present, the concept provides a framework in which questions can be asked and hypotheses can be tested.

PART III

VALUES FOR PERSONAL DECISIONS AND SOCIAL POLICY

Introduction to Part III

For man is man
and master of his fate.
ALFRED LORD TENNYSON,
Idylls of the King

WHEREAS our attention, so far, has been focused on the innate values (or motivations) that are built into the human decision system, the remaining chapters will be concerned with the secondary values that we develop out of experience as an aid to practical decision making. These secondary values include our social and moral principles as well as the routine decision criteria that are used in our daily decision making. Unlike the primary values, these secondary values are products of rational thought and therefore are subject to change on the basis of rational thought.

One of the major objectives of a theory of values is to provide a more scientific foundation for the judgments we must make to adapt these secondary value criteria to a rapidly changing social and technological environment. Our immediate purposes in Part III are twofold:

1. To show how the value-theory perspective relates to existing concepts in related fields such as sociology, psychology, ethics, religion, moral philosophy, and social policy.
2. To explore some of the apparent implications of the theory as it bears on personal decisions and social policy.

Because of the wide scope of these issues, it is impossible to go beyond a very brief treatment, which focuses primarily on how the value theory concepts can be reconciled with existing ideas in related fields.

THE FUNCTIONS OF A VALUE SCIENCE

The study of valuative criteria for human decisions has traditionally been identified as the field of ethics. Consequently, when the decision-theory perspective is applied to these problems it inevitably becomes involved in the subject of personal and social ethics. Although the method of analysis is very different, the subject matter is much the same, so many of the problems and paradoxes of ethical theory must be addressed.

In the past, the field of ethics has been treated almost like an independent discipline that could be developed without much reference to developments in psychology and sociology. Similarly the traditional scientific approach in psychology and sociology has sought to isolate these disciplines from value issues so that research could proceed without emotional bias or prejudice. There is now a growing recognition that the objective and valuative disciplines are intimately related. A mature psychology and sociology ought to provide practical guidance with regard to personal and social decisions. A mature ethical theory should be compatible with a scientific understanding of human behavior and human motivation. The value-theory interpretation of human motivation appears to offer the necessary link which should make it possible to unite sociology and psychology with a scientific ethical theory.

Broadly speaking the functions of such a science of human values can be classified as follows:

1. *Descriptive,* to describe and "explain" the way people actually behave and the criteria that they actually use to make decisions.
2. *Predictive,* to predict the types of decisions that will be made through an understanding of the way these decisions depend on the environment and circumstances.
3. *Prescriptive,* to recommend or help in deciding what criteria *should* be used and what decisions *should* be made.

Obviously these three functions parallel very closely the functions of any "science." If a science is to be valid and useful it must be able to *describe* what is and *predict* what will happen. If a science passes these two basic tests of validity then we can use it as an aid in making decisions. Of course, the most important function of a science is the final one of helping with practical decisions. In the case of a value science it is particularly true that this *prescriptive* function is the most important. But even a value "science" cannot be applied in such a role with much confidence unless it also passes the tests of validity as a descriptive and predictive science.

Traditionally ethical theory has concentrated on the prescriptive function. The fact that most of these theories failed as a descriptive or predictive science was not generally considered a serious defect. It was assumed that if a theory could define what people ought to do, then the problems would be solved because people would naturally want to do what the theory said they "should." Unfortunately this hypothesis has not been confirmed by subsequent experience. Many of the traditional ethical theories have had little practical impact because people did not "want" to do what the theory said they "should." The present theory represents an effort to reconcile the concepts of "should do" and "want to do," in order to achieve a theory of human behavior which can pass the validity test as a descriptive and predictive as well as a prescriptive theory.

Among the traditional ethical theories there have been very few that were intended as descriptive, much less predictive, theories.* The fields of psychology and sociology on the other hand have developed primarily as descriptive sciences. Although they have achieved some limited predictive value, their predictive ability has remained extremely weak.

To find a good example of a theory of human behavior that is truly "predictive" we have to look to the world of science fiction. In his science fiction classic, the *Foundation* series (1), Isaac Asimov explored the consequences of a fictional theory which he called "psycho-history." In a way, psycho-history was the antithesis of a traditional ethical theory. It was descriptive and predictive but *not* prescriptive! As Asimov described the theory, it was "the quintessence of sociology; it was the science of human behavior reduced to mathematical equations."

A theory of human behavior which is truly predictive is likely to be less optimistic than traditional ethical theories, and it may itself pose some serious moral and ethical dilemmas. Probably the best way to illustrate these problems is to quote directly from Asimov's classic.

> The Galactic Empire-was falling.
>
> It was a colossal Empire, stretching across millions of worlds from arm-end to arm-end of the mighty double-spiral that was the Milky Way.
>
> It had been falling for centuries before one man became really aware of that fall. That man was Hari Seldon. . . . He developed and brought to its highest pitch the science of psycho-history.
>
> Psycho-history dealt not with man, but with man-masses. It was the science of mobs; mobs in their billions. It could forecast reactions to stimuli with something of the accuracy that a lesser science could bring to the forecast of a rebound of a billiard ball. The reaction of one man could be forecast by no known mathematics; the reaction of a billion is something else again.

* In some cases, when the theories proved to be unreliable in their correspondence with the author's moral intuition, the theories were described more "conservatively" as descriptive or explanatory of human values, rather than as prescriptive theories.

> Hari Seldon plotted the social and economic trends of the time, sighted along the curves and foresaw the continuing and accelerating fall of civilization and the gap of thirty thousand years that must elapse before a struggling new Empire could emerge from the ruins.
>
> It was too late to stop that fall, but not too late to shorten the gap of barbarism. (1) *

Because the science of "psycho-history" was not prescriptive, Hari Seldon could not recommend a course of action to avoid the inevitable. "Psychohistory" predicted an almost fatalistic evolution of human affairs. Although it might be possible to intervene and make small local changes in the course of affairs, there was no way to reverse major trends that had gathered the momentum of history. On the other hand small changes, introduced sufficiently early, could expand over the long periods of history so that they might ultimately have a major impact on the momentum of human affairs. It was on the basis of such a strategy that Hari Seldon designed his great "plan" for the future of the galaxy. The purpose of the plan was to shorten the inevitable gap of barbarism from the predicted 30,000 years to a more acceptable 1,000 years!

But Hari Seldon was faced with a moral dilemma. He had to avoid the dissemination of any information about his plan. Indeed he had to maintain secrecy about the whole science of psycho-history, because such information would deflect the course of history and might destroy the effectiveness of his equations in predicting the trends.

It therefore might be asked whether the predictive ability of a value theory interpretation of human motivation could introduce ethical issues analogous to Hari Seldon's dilemma. Fortunately, it has not turned out so badly. Our common sense belief that people are likely to make better decisions if they have better information seems to be supported by the theory, so we do not encounter the Seldon dilemma. Of course, there are a number of problem areas where we might find better solutions if human nature were somewhat better than it is, but on the whole, human nature seems remarkably good already. Things could be *much* worse than they are.

Asimov's analysis in the *Foundation* series also raises an important philosophical question: Is it possible, even in principle, for a science of human behavior to be both predictive and prescriptive? It appears that some kind of philosophical "uncertainty principle" may be involved. If a science is perfectly predictive, there is no place for the responsiveness that is needed to benefit from a prescriptive science. If there is too much responsiveness to prescriptive advice, there is little hope for a predictive science. Apparently a science of behavior can be both predictive and prescriptive only if it is quite imperfect in both areas!

* From the Prologue to *Foundation and Empire*.

In fact, the intrinsic inability to really predict human decisions, and the consequent inability to really predict the future, leads to just such a result. The situation is somewhat analogous to the paradox about "free will"; and the resolution of the problem is much the same in both cases. The free-will paradox was resolved by considering the human mind as a decision system. The apparent paradox of psycho-history is resolved because societies in aggregate also operate somewhat like "decision systems." If it becomes really clear that present policies are leading to disaster, the introduction of that information into the system may be just what is needed to achieve a change in policy and thus avert the disaster. For this reason a science of human values with genuine predictive ability can also operate effectively as a prescriptive science. Indeed much of the potential usefulness of the theory in the prescriptive role follows as a direct consequence of its effectiveness as a descriptive and predictive science.

The moral dilemmas of psycho-history arose because it was *too good* as a predictive science and not good enough as a prescriptive science. Intervention to improve the natural course of events had to be accomplished *surreptitiously* by those that understood the science, and this raised the inevitable moral issues that are always encountered when a small group sets out to manipulate the lives of many.

Even if the present theory were fully developed, such manipulation does not seem to be theoretically possible. The predictive power of the theory is not that good. The most effective way to alter the course of events seems to be to provide new information that can help in evaluating descisions. Fortunately, such concepts cannot be used to manipulate events. They will alter the course of events only to the extent that they survive public scrutiny and are accepted by other individuals and the society. Thus the ethical dilemmas that appeared in Asimov's psycho-history appear to be absent from the present theory.

ORGANIZATION OF PART III

Part III is divided into four major chapters. Chapter 14 discusses the relationships between primary and secondary values during the course of biological and cultural evolution. This analysis leads to two value principles that seem to be useful in explaining the biological and cultural evolution of the human social system. One of these is a value criterion governing individual behavior, while the other plays a similar role in the cultural evolution of social norms and social values. Thus, the analysis in Chapter 14 continues in the traditional spirit of a descriptive and predictive science.

The two following chapters, however, explore some of the implications of consciously and deliberately using these *same* evolutionary value principles as a "prescriptive" guide for personal decisions and social policy. It is in these chapters that we encounter some of the basic philosophical problems of traditional ethical theory. Chapter 15 deals with the value principle governing personal behavior and considers some of its implications as a guiding principle for personal ethics and personal philosophy. Chapter 16 deals with the value principle governing social norms and social values and considers its potential usefulness as a formal guide to social policy.

Like almost any new concept, the value-theory interpretation of human motivation raises numerous questions which remain to be explored. Chapter 17 calls attention to some of the unexplored implications of theory and suggests ways in which it may provide a new insight concerning some of the problems of modern society.

REFERENCES

1. Asimov, Isaac. *Foundation* series. New York: Gnome Press: *Foundation,* 1951; *Foundation and Empire,* 1952; *Second Foundation,* 1953. Reprinted 1966 by Avon Books.

Chapter 14

The Behavioral Origins of Social Obligation

> Indeed it is now clear that the further development of the science of ethics waits upon the more thorough clearing up of the evolutionary ideas themselves, and upon their more complete application to biology, psychology, and sociology (including anthropology and certain phases of the history of man) in order to provide the auxiliary sciences necessary for ethical science.
>
> JOHN DEWEY (2)

HAVING COMPLETED our discussion of the values that are built into the human decision system, we are now ready to consider the relationship between these primary human values and the secondary values that are learned from experience. The primary and secondary values seem to be intimately linked in evolution's behavioral plan for the human species. The more recent stages of human evolution undoubtedly took place *within* a human social environment that was already rich in secondary values. Thus, the evolution of the *primary* values was conditioned by *secondary* social or cultural values which were themselves derivative from the primary values. (Of course, this derivative relationship was probably most direct during the preverbal phase of human evolution.) This interlocking relationship, between man and his society, during the later stages of human evolution is described by Burhoe (1) as a symbiotic relationship which, from the perspective of genetic evolution, is analogous to the relationship between two different but cooperative species. Burhoe shows how this type of relationship can lead to a genetic adaptation of man to *his* society. As a consequence of this interlocking relationship between primary and secondary social values during human evolution, it seems certain that the secondary cultural values (at least as they existed in a preverbal society) should be considered as an essential part of evolution's behavioral plan for the human species.

To provide a better understanding of this behavioral design concept it is necessary to consider the way the primary values can give rise to secondary values within a cultural context. Our purpose in this chapter is to extend our previous interpretation of evolution's system design so that it can explicitly include the resulting secondary values. To develop these ideas, we will focus our attention on small cohesive social groups analogous to the primitive band or tribe in which the innate human values originally evolved.

SOCIAL CONDITIONING OF INDIVIDUAL BEHAVIOR

The individual inevitably develops his personal network of secondary values within the context of his own family and social culture. Since the process of valuative deduction is dependent on the environment (i.e., the rules of the game) the resulting value network is inevitably molded by cultural environment. Moreover, because of the specific innate value structure of the human species, the influence of the existing cultural values in the environment is particularly strong. Psychologists and sociologists have long been familiar with the process of "social conditioning" through which the behavior and values of the individual are molded to conform to the values and ideals of the society. The decision-science interpretation of human behavior suggests some specific mechanisms which help to explain *why* this conditioning process is so effective.

From the decision-theory perspective the "conditioning" process operates *both* through the individual's rational value network and through the response characteristics of the innate motivational system. Chapter 8 discussed briefly some of the impact that experience in infancy and early childhood can have on the innate value system. Such environmentally conditioned changes in the response characteristics of the value system appears to modify some of the basic personality characteristics of the individual. Although such subconscious conditioning of the value system plays an important role in the present theory, the role is far less important than in traditional behavioral theories. Within the value-theory interpretation, a large part of the conditioning" process takes place in an essentially rational way, both consciously and subconsciously within the rational mind.

The small child likes to be approved. Within the confines of the family and later as he begins to interact with the external society he gradually comes to understand what types of behavior are "approved." In the beginning he measures all "outcomes" only against his primary value criteria (i.e., his emotional responses, etc.) as they are supplied by his own innate value system; but he soon develops an intuitive set of secondary or judgmental value criteria that he finds useful in making decisions. As he gains confidence in these secondary criteria, he begins to use the secondary criteria themselves as value standards to develop new levels of value criteria that are even further removed from the innate or primary value criteria.

These higher order value criteria can be simultaneously based both on primary and other secondary values. As the value network develops, it even becomes possible to work backwards, adjusting the lower part of the value network on the basis of consistency requirements with higher level value principles. Thus, as the individual grows in wisdom, the distinction of levels within his secondary value network becomes obscured. Although in a fundamental sense the whole value network is derived ultimately from the primary values, the resulting network of secondary values gradually acquires a stability which gives it an almost independent influence over the individual's behavior.

Since the higher levels in the individual's value network are initially developed on the basis of the lower levels, his early experience (which establishes the foundation for this rational value network) can have a profound influence on his subsequent decisions. Although he may on rare occasions be motivated to reexamine and perhaps even radically restructure his fundamental value premises, he does not ordinarily do so. As a matter of cybernetic efficiency (as long as his existing value network seems to be producing fairly satisfactory decisions) he is likely to avoid any major restructuring of his basic value network.

As the individual's *innate* motivation system matures, it begins to define new objectives for him such as social dominance and leadership. Once again, the *way* these innate objectives can be most effectively achieved is determined by the social norms and values of the society. Consequently, the secondary values generated by these new motivations are also conditioned by the "rules of the game" within the social environment.

The network of secondary value criteria produced by this process includes certain very general concepts that we may consciously interpret as values, such as freedom, personal integrity, and self-respect. Such general value concepts cannot be attributed to any single component of the innate value system. They grow instead out of experience with the full spectrum of innate values. Since these rather abstract secondary values seem to provide guidance that is relevant to a wide spectrum of the innate values, we usually attribute to them a validity that is more universal than the specific value structure from which they were derived. On this basis, some readers might comment that many of the secondary values seem to correspond more closely than the primary values to what we usually think of as "ultimate" values. Indeed the development of secondary values of such wide applicability is one of the great achievements and challenges of human intellectual activity.

But the network of secondary values is not limited to such generalized valuative criteria. It also includes a large number of minor and very simple decision criteria that we often refer to informally as "dos" and "don'ts." For example:

> Don't be inconsiderate of others.
> Do brush your teeth after meals.
> Do brush your hair in the morning.
> Don't drink dirty water.

When we consider the way these simple rules are actually used, it is apparent that they are also a *form* of value criteria. Such dos and don'ts are not interpreted as rigid rules. Each is assigned some judgmental importance, and each will be violated if there is sufficient reason to do so. Thus, each of these "rules" really corresponds to a value criterion that is stated in a simple binary form—one value for do and another for don't. Obviously, this simple binary way of representing the secondary value criteria contributes to cybernetic efficiency. Although the actual "value" or importance attached to each rule is rarely verbalized explicitly there is nevertheless an underlying judgmental value associated with each such criterion. The ensemble of such dos and don'ts combined with the other value criteria provide the very complex network of secondary values that we use to make our daily decisions.

THE ORIGIN OF A "HIGHER STANDARD OF BEHAVIOR"

The existence of a "higher standard of behavior" in human society has been one of the most persistent problems of ethical and sociological theory. Any satisfactory theory of human behavior must somehow be compatible with the existence of such a "higher standard." So far, in discussing the behavior of an individual within a society we have described his objectives only in a very abstract form, which involves finding "satisfaction" in terms of his innate values.

Without examining the problem in any greater depth we might naively conclude that the individual's decisions could be made simply by asking which alternative he would "like" the best. But as we all know, we do not make decisions simply on the basis of what we would "like" to do. We are continuously involved in conflicts. We would "like" to take the day off, but we can't because we "should" finish some task at the office. We would "like" to go to a movie, but we can't because we feel "obligated" to stay home with the children. Any satisfactory explanation of human behavior must account somehow for such moral conflicts. When such conflicts arise they can usually be interpreted either (1) as a conflict between long-term and short-term preferences or (2) as a conflict between personal preference and "social obligation."

The conflict between short-term and long-term preferences is fundamentally quite simple. We might have more fun today if we did not go to work; but if we fail to go to work we may lose our job. When all things are considered, we prefer to work today so that we will be able to eat tomorrow. In terms of our concept of a value-driven decision system, this is just a matter of maximizing satisfactions over the whole trajectory, including tomorrow as well as today.

But the conflict between personal preference and social "obligation" is a much more complex issue. In all human societies the values and social norms of the society exert a powerful influence on individual behavior. The individual behaves as if he is obligated to conform to a norm and to pattern his behavior in accordance with the social ideal for his society. This obligation to the social norm often dominates the individual's behavior, even when it is in conflict with his personal "preferences." Although there are wide differences between societies in the structure of their social ideals, the existence of some kind of social norm and social ideal, which defines a "higher standard" of behavior, seems to be a universal property of human society. In effect, the *existence* of a "higher standard" of behavior seems to be an essential part of evolution's behavioral plan for the human species.

As we noted in Chapters 10 and 11, one of the basic objectives of the innate human Social Values is to motivate cooperation within a cohesive social group. To clarify the process by which a "higher standard" of behavior can emerge from the innate human values, we will consider the operation of the innate values within a simple idealized social group. Although the structure of the innate Social Values is complex, the key motives for cooperative behavior are probably the first three listed in Table 11.3 (desire for dominance, desire for approval, desire for social acceptance). To simplify the discussion, we will consolidate these three motives into a single generic motive which we will describe as an innate desire for "social approval." The inventory of human facial expressions—the sneer of scorn, the smile of approval, the frown of disapproval, and the blush of shame—all testify to the central importance of "social approval" as an innate human Social Value.

The small child is strongly motivated by the desire for "approval." In the absence of any other preexisting criteria, the parent is most likely to approve behavior which is helpful to him in whatever he is trying to do. The child soon discovers that he likes to help because the resulting parental "approval" contributes to his sense of pride and personal worth. This is the first step in the development of a cooperative pattern of behavior. In essentially the same way adults learn that they like to help each other. Once a basic pattern of cooperative behavior has been established the approval motive begins to operate in a more refined way. Methods of cooperation that are found to be effective are approved, while ineffective methods are disapproved. Thus to maintain a high "approval" rating the individual must learn to cooperate in an effective and "approved" way. The effective and approved methods of cooperation gradually become formalized as social norms within the society. Although the individual might, on some occasions, "prefer" to depart from these established norms, he knows that he cannot do so without a severe loss in his accumulated approval score. Consequently, he feels "obligated" to comply with the accepted social norms.

The preceding discussion shows how the approval motive within a small primitive group might spontaneously generate social norms, even in a society where no such norms had previously existed. Of course, it should be obvious that the need to generate social norms from scratch would almost never occur in any real society. In any real society, a well-established set of social norms is already in existence.* Consequently, in existing societies, the approval motive serves primarily to reinforce and perpetuate the prevailing social norms. The evolution of such social norms over many generations, within a primitive society, involves a very imperfect process of valuative deduction; so the norms typically include both a large amount of accumulated wisdom and a substantial amount of mythology and error.

The prevailing social norms, when they are first encountered by each generation, appear to be of authoritarian origin. The norms are initially defined by parental authority. The parents recall that the norms were defined for them by their parents. Thus, whenever an effort is made to trace the origin of the "higher standard of behavior" it seems to be of a completely authoritarian origin. For this reason, almost all the early efforts to explain and rationalize the

---

* In larger more complex societies there may be a diversity of norms for different social subgroups. However, for simplicity the present discussion is limited to small cohesive social groups. The problem of subgroups in a larger society is discussed in Chapter 16.

prevailing social norms did so in terms of authoritarian origins. The Chinese credited their ancestors, but most other civilizations felt the need for some mythological first cause.

Although in most societies the prevailing social norms tend to be quite stable over many generations, they are in fact subject to change. Over a period of time they can drift slowly in response to the rational views of social leaders. In a society that exists in a hostile warlike environment, the social ideal will tend to shift so that it places unusually high value on bravery and other militarily beneficial personality traits. In a society that exists in peaceful isolation such military character traits will be deemphasized and perhaps even discouraged in the "social ideal." The wide diversity of moral and ethical standards found in different cultures is indicative of the variability that is possible in these secondary social value criteria.

This interpretation of the human social system makes it clear that the refinement and updating of social norms and social ideals is one of the fundamental responsibilities of the social and moral leadership in any human society. Both the existence of the ideal and the somewhat "rational" *adaptability* of the ideal to the cultural experience of the society appear to be essential parts of evolution's design for human survival. Because these social values are so important to the operation of a human society, most successful societies have been very conservative in making changes. In general the prevailing norms and taboos have been carefully taught to each new generation and have been quite rigidly enforced. One of the most serious dangers now facing our scientific society may be the loss of an effective tradition of social values as a consequence of the prevailing scientific skepticism.

HUMAN SOCIAL VALUES: A SYSTEM-DESIGN INTERPRETATION

Having described informally a simple mechanism which allows the innate values to generate social norms for a society, we are now ready to examine the process from a more formal perspective. Our approach will be to interpret the structure of secondary values from a system-design perspective. Although such a system-design approach often produces oversimplified interpretations, it can nonetheless be very helpful in defining the fundamental interactions within a sociobiological system. Our purpose in the next few sections will be to analyze the conflict between "personal preferences" and "social obligations" from the perspective of an evolutionary system design.

To develop this interpretation we will think of the individual's innate values as divided into two parts: (1) those innate values involved in the broad "social approval motive" (specifically the first three motives in Table 11.3—desire for dominance, desire for approval, desire for social acceptance) and (2) all other innate values. The first set of values defines the individual's motivation for social approval; the second set defines his "personal preferences" as they would exist in the *absence* of the social approval motive.

From the perspective of an individual who wishes to make a decision, these two sets of values are different in a very fundamental way. To the extent that he can estimate probable outcomes for different alternatives, it is usually fairly easy to estimate the relative "desirability" of alternatives with regard to one's own personal preferences. (Previous experience usually provides the individual with a rather reliable guide to his personal preferences.) In contrast, however, there is no way the individual can evaluate the relative "desirability" of the alternatives in terms of the "approval" motive without knowing what types of activity will be approved by others in the society. Consequently he must inevitably consider these two components of his innate values separately. He must weigh the value of social approval against personal "preference" when he makes a decision. The approval motive defines what he "ought" to do; the remaining value structure defines what he would "like" to do. To locate the alternative that will be "best" in terms of his overall objectives, he must search for one that provides a good compromise between the conflicting demands of the two sets of value criteria. Thus, this simple form of "moral conflict" is an inevitable consequence of the structure of the innate human values. The conflict arises because the "social approval" motive requires the individual to explicitly consider the approval of others in addition to his own personal preferences.

Since the approval motive is so important in human social behavior, it is obvious that the criteria that will determine what behavior will be "approved" within a society must also be an important force in the guiding of social behavior. It has become conventional to say that "approval" is governed by the social norms and values of the society; and that these norms and values vary widely from one social culture to another. Although this may seem like an answer, it is almost equivalent to saying that whatever *is* approved in the society *will* be approved. From a short-range perspective, this may be the only possible answer, because the norms and values of a society tend to be self-perpetuating. But over a long period of time, the norms and values of a culture can drift as the innate valuative responses (particularly of the social leaders) interact with the changing environment. Thus, at a deeper level it is necessary to ask about the underlying valuative forces that determine the equilibrium structure of social norms within a society.

Two different kinds of forces appear to be involved. First it is obvious that people will have a natural tendency to approve those types of behavior that are directly useful to them. For example, the bravery of a warrior that contributes to the protection of a tribe will tend to be approved. Similarly the generosity of an individual who provides food for a hungry family will tend to be approved by the family. Evidently, any behavior that helps others in the society to satisfy their innate values will tend to be rationally approved.

But it is also possible that people may have an irrational but genetically defined tendency to approve or disapprove certain types of behavior. We obviously enjoy seeing people smile, so we have a greater tendency to approve the actions of one who smiles. In a very similar way, it may be possible that we may have an innate tendency to disapprove behavior such as cowardice or dishonesty, or an innate tendency, for example, in watching sports events, to approve (or enjoy) skill, bravery, or honesty when we see it in others. Although there is at present no really scientific evidence for such tendencies, there are theoretical reasons for believing that they might exist. Once the innate "approval" motive had been genetically established in human behavior, an innate tendency to approve or disapprove certain kinds of behavior could *also* acquire a real survival value—because it could motivate *others* to behave in ways that will contribute to the survival of the society.

At present it is totally unclear to what extent our tendencies to approve or disapprove may be innate. However, in terms of the basic principles involved, it does not really matter whether we approve because we rationally understand the value to us of certain types of behavior, or whether we approve simply because we instinctively enjoy such behavior. In either case we are approving the behavior because the behavior contributes positively to our own innate values. Thus, it appears that the cultural evolution of what will be approved within a society can also be explained in terms of the innate human value structure. The behavior that is most likely to be "approved" is the behavior that is most effective in actualizing the innate values of others within the society.

Although the act of approving or disapproving a particular kind of behavior may be partly innate or instinctive, it is certainly very strongly influenced by deliberate rational judgments. Within certain limits we can deliberately and rationally decide what kinds of behavior we are going to approve. It has been one of the functions of moral and social leadership within a society to help make judgments on such issues.

Obviously the leaders in any society will have some tendency to approve those types of behavior that seem most favorable to them as leaders. But in a small social group (such as the primitive band or tribe) where leadership is informal, this tendency toward personal selfishness tends to be moderated by a very strong social motivation. Usually the social leaders are the ones that are most "approved"; thus, they also tend to be individuals who have been most strongly motivated by the approval motives. Consequently, in a small cohesive society we can expect that the social leaders would usually be motivated to make judgments that will be approved as "wise" and "just" by others in the society.

It appears that the forces that govern the cultural evolution of social norms in a small informal social group should be closely related to the best interests of the society as a whole. To provide a simplified model of the way cultural norms and values in a primitive society are likely to evolve, we will simply assume that over many generations the judgments made by social and moral leaders will have the effect of moving the social norms and values in a direction that seems to be generally consistent with the interests of the society as a whole.

A MODEL OF HUMAN SOCIAL ETHICS

The preceding discussion has shown that in a primitive society the structure of human social behavior involves an interaction between a number of different value criteria. The motivations of the individual are logically divided into two parts: (1) his motivation for social approval and (2) his own personal preferences as they would exist in the absence of the approval motive. The interests of the society as a whole can be approximated by a quite different value criterion that reflects the innate values of all the individuals within the society. To understand human social ethics in a primitive society we must consider the interactions between these different value criteria.

We will begin by considering the problem faced by an individual within such a society who must make a decision that involves a "moral" judgment. To highlight the moral issues involved we will assume that the individual can estimate the probable outcomes for a large number of different courses of action. Moreover, after projecting the outcomes as far into the future as seems useful, we will assume that it is possible to estimate the "desirability" of the probable outcomes in terms of *each* of the relevant value criteria. The diagram in Figure 14.1 provides an idealized representation of these relationships.

We begin by considering the figure on the left (14.1a). The vertical axis in this figure represents a "utility scale" for the individual's *personal values* ignor-

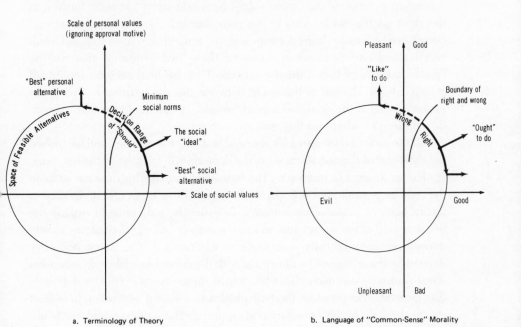

a. Terminology of Theory   b. Language of "Common-Sense" Morality

Figure 14.1
The Theory and Commonsense Morality

ing those personal values that depend on "social approval." In principle any feasible "decision alternative" can be evaluated against this scale of personal values. The horizontal axis represents a "utility scale" for the social consequences of the same set of decision alternatives. It represents total utility of an outcome when the utility is defined to reflect the innate values of all individuals in the society. Each decision alternative that the individual considers can have both personal and social consequences. When any decision alternative is plotted on the chart it will fall at some point defined by these two vector coordinates. To provide a simple example, assume that all possible alternatives have been plotted in this way, and that they all fall somewhere within the circle labeled "space of feasible alternatives." For simplicity we also assume that the alternatives are so numerous that there is some alternative corresponding to any point inside the circle, and that there are no feasible alternatives outside the circle. Consider now the problems the individual will face in making a decision.

If the decision were to be made *solely* on the basis of the "personal utility scale," the individual would simply select the highest point on the circle. This is the point labeled "best personal alternative." If he were to make the decision *solely* in terms of the "social value" he would select the point furthest to the right on the circle. This is the point labeled "best social alternative." Because of his innate desire for approval, the individual wishes to select a balanced alternative (somewhere between these two extremes) that will be "good" in terms of *both* valuative criteria. Thus his final decision should fall somewhere on the arc of the circle between the two extremes. This is the region labeled as the *decision range of "should."* What the individual "should" do will lie somewhere on this arc.

If the individual's desire for approval is strong the decision will fall toward the right end of the arc; if it is weak the decision will fall toward the left. Society has an interest in motivating the individual to make his decision as far to the right as possible; but it is obviously unrealistic to expect the individual to ignore his own personal preferences. To assist the individual in making the decision (and to encourage him to make a socially beneficial decision) society defines certain minimally acceptable social "norms." The moral boundary defined by these "norms" is illustrated with the curved line labeled "minimum social norms." Any decision to the left of these "norms" will be definitely disapproved. This provides the individual with a strong motivation to at least stay to the right of this boundary of disapproval. But society would like to obtain even more favorable decisions from the individual. It therefore tries to define a kind of "ideal" behavior that is illustrated by the arrow labeled "social ideal." This "ideal" offers the individual a model of behavior which will get a much higher "approval" score. An individual who is highly motivated by the "approval" motive will try to model his behavior to correspond closely with this social "ideal." In deciding what to do, each individual must weigh the benefits of social approval against other conflicting benefits in his *innate* value system.

The concepts that are defined by this simple primitive model have a surprisingly close correspondence with the commonsense concepts of morality that prevail even in our present society. The diagram in Figure 14.1b illustrates the way these *same* concepts are incorporated in the language of everyday conversation. The "personal value" vector defines what the individual "would like" to do. Alternatives on this scale are judged by whether they feel "good" or "bad," "pleasant" or "unpleasant." Alternatives are judged relative to the "social value" scale by whether they seem "good" or "evil." The social "ideal" defines what the individual believes he "ought" to do. The minimum social norms define a boundary between "right" and "wrong." Although there are certain contexts in which the distinction between the words "right" and "wrong" seems sharp and reflects this formal boundary, there are other contexts in which the *same* words refer to valuative comparison which includes gradations of *more* wrong or *less* wrong, or alternatively *more* right or *less* right. In this comparative context the concept of "right" and "wrong" seems to refer to the relative position of alternatives along the decision arc.

Of course the word "wrong" is also used to imply a *mistake* or *error*. In this context it refers to a mistake in logic rather than a moral value judgment. If an individual were to select an alternative anywhere in the *interior* of the circle, or *beyond* the limits of the "decision range" of the arc, the choice would be described as "wrong" or "mistaken."

This analysis of the commonsense concept of "right versus wrong" helps to explain a number of *linguistic* paradoxes. For example it helps to explain the apparent conflict between *ethical theory* and *common sense* with regard to whether personal pleasure is "right" or "wrong." For the ethical theorist, personal pleasure is one of the ultimate tests of what is "right"; in the commonsense view, however, personal pleasure is usually associated with what is "wrong." The diagram makes the reason for the disagreement obvious. As a practical matter, there is never any motivation to do "wrong" *unless* there is a conflict between what is "approved" and what one would otherwise decide to do to achieve personal pleasure. In any practical case when an individual deliberately chooses to do "wrong," it is because he has placed personal pleasure above social approval. For this reason, in our everyday language personal pleasure tends to be associated with "wrong," while doing what is socially approved is associated with "right."

The diagram also helps to clarify the commonsense linguistic distinctions between "good versus evil," "good versus bad," and "right versus wrong." Although the linguistic distinctions are far from uniform, the words "good versus evil" are most commonly used to compare alternatives relative to the social utility scale; the words "good versus bad" are most commonly used to compare alternatives relative to the personal utility scale; and finally, the words "right versus wrong" are used to refer to the conflict between the personal and social utility scales and to compare alternatives relative to the "decision arc."

Of course, the definitions of the commonsense words as they are illustrated in Figure 14.1b are considerably narrower and more precise than the typical usage of the words in everyday conversation. The words as defined by the model tend to correspond most closely to the *core* meaning of the words. For example in everyday conversation we often fail to make any real distinction between "should" and "ought," although almost everyone will agree that "ought" conveys more of a sense of social obligation, whereas "should" is more frequently used to refer to the most appropriate course of action when all things are considered. To match the words to the more precise meanings required by a formal theory, we have narrowed the traditional range of the meanings for the words. Nevertheless, the correspondence of the value-theory interpretation to both commonsense ideas of morality and to the commonsense language of morality seems surprisingly good. This correspondence provides additional circumstantial evidence for the basic validity of the value-theory perspective.

The theory differs in one important respect from most traditional ethical theories. It leaves open the possibility that there may be circumstances in which, when all factors are considered, an individual will choose to do what is "wrong." This of course is necessary if the theory is to have a proper correspondence with reality as a "descriptive" and "predictive" theory, for it is obvious that occasions do, in fact, arise when individuals deliberately decide *not* to do what they know is "right." In contrast, since we have identified "ought" with the social ideal, there should never be a situation where one could say that an individual "ought" to do what was wrong.

In commonsense usage, some people also use the word "ought" to refer to an obligation to an *absolute* concept of right as opposed to a social ideal. However, because the decision-science perspective does not encompass the concept of absolute right, this particular meaning is absent in the present model of human social ethics.

THE SYSTEM DYNAMICS OF SOCIAL ETHICS

The schematic representations in Figure 14.1 make it possible to address some of the dynamic interactions in the primitive social system from a "system design" perspective. When the individual is considering a specific decision he must weigh the value of personal pleasure against the value of social approval. If his decision falls to the left of the "minimum" social norms (and he is discovered) he will face severe disapproval. Therefore, if he can find an "approved" option that is not too unpleasant he is likely to take it. But if the minimum social norms have been defined so that they are *too* far from the "best" personal alternative, the individual may deliberately decide that the price of approval is just too high. He may therefore "for good reason" decide to ignore the social norm. If there are too many "good reasons" for ignoring the norms, the social disapproval associated with violation of the norms will be eroded.

The same basic relationship applies with regard to the "social ideal." The individual knows he can earn increased approval as he approaches the social ideal. But if the ideal is *too* incompatible with personal self-interest he may decide that there are "good reasons" for ignoring the social ideal. Therefore, the ideal will be most effective in influencing personal decisions, if it represents a *realistic* balance between the personal and social value systems.

In weighing such decisions, the individual must weigh carefully the long-term as well as short-term consequences. In most cases the value issues that affect personal self-interest are relatively short-term, whereas in a small cohesive social group the acquisition of a good personal reputation is a lifelong process. A decision that places too much weight on short-term objectives is likely to be a "selfish" decision, and it is likely to have adverse long-term consequences. For this reason, the traditional precepts of social ethics tend to emphasize "looking ahead." The shortsighted "yielding to impulse" is associated with moral error, whereas a "farsighted" decision is generally classified as "morally good."

Having described some of the considerations that determine how individuals must respond to decision alternatives, we can now examine the same chart from the perspective of the social leadership. Although the actions of the social leadership affect the same diagram, they affect the diagram in a different way. Generally over many generations the role of the leadership is to adjust the social norms and the social ideal in such a way as to achieve maximum value for the society. Although the social and moral leaders have no direct control over the actual individual decisions, they may be able to influence the decisions by adjusting the structure of the "minimum social norms" (for example, by changing the taboos or the laws) or by changing the structure of the "social ideal." Thus, the "control parameters" available to social leadership are encompassed within the "social norms" and the "social ideal." Over many generations the decisions of the social leaders should tend to adjust these control parameters so that the actual decisions made by individuals within the society will be socially more beneficial.

If the control parameters are moved too close to the horizontal axis the disapproval penalty associated with ignoring them will be degraded, and they will become ineffective because they will be ignored. If they are moved too close to the vertical axis they will probably be obeyed, but the resulting individual decisions will not be very beneficial to society. Somewhere between these two extremes there should be an "optimum" position for the social control parameters that will yield personal decisions that are most beneficial to society. In a primitive environment the societies that are most successful in making these adjustments will have a competitive advantage. They will be the ones that survive and multiply. Societies that are less successful in defining these cultural norms in a satisfactory way will tend to be destroyed.

SOME MODERN IMPLICATIONS

Apparently the search for principles of "social ethics" comes down to a search for principles that will be most *effective* when the *actual* decisions of individuals are measured against the horizontal axis or social value criterion. Although these "control parameters" in the two-dimensional schematic of Figure 14.1 are represented as simple lines or boundaries, they actually involve very complex structures of concepts and principles. In order to make them as effective as possible a number of considerations are important:

1. *Ease of understanding:* Social principles cannot be really effective unless they are easy to understand and apply. For this reason the social ideal should be simple and logically consistent. The social norms should be simple and logically consistent with themselves and with the social ideal.

2. *Reasonableness:* The principles should be reasonable and easily justified as a good balance between the personal and social valuative criteria. The principles should not require unrealistic sacrifice of vital personal values; and *any* sacrifice of personal values should be logically justified in terms of collective values. If a principle is found in practice to be ineffective (as in the case of the prohibition on the sale of alcohol), it should either be changed to make it effective, or it should be abandoned to avoid erosion of the other social norms.

3. *Sharpness or clarity:* Particularly in the case of the minimum social norms, it is important to have an unambiguous boundary between "right" and "wrong." The definitions of "lying" and "stealing" provide good examples of such sharp moral boundaries.

Many of our present laws, particularly the tax laws and the laws which regulate business practices, can be seriously criticized because they are neither simple nor unambiguous. The disregard of the fundamental ethical principles of clarity and simplicity within our present legal structure tends to create a general disregard for the law, much to the disadvantage of society. The problem is not that the laws are unreasonable, but that they are too complex to be easily understood. There is a critical need for a simplification and clarification of this legal structure.

The modern social planner needs to give consideration to one other important factor. Much of the strength of the traditional human social ethics is founded in the *long-term* consequences of a violation of social norms. This long-term damage to personal reputation tends to outweigh the short-term pleasure that might come from yielding to impulse. Unfortunately, as the society becomes more mobile and transient, these long-term social penalties tend to be degraded. If we are to create a more satisfactory modern social system we may need to consider methods for restoring the traditional long-term penalties for the violation of social norms.

THE METHODS OF TRADITIONAL ETHICS

The schematic representation of Figure 14.1 also helps to clarify the wide diversity of approaches that have been used historically to deal with the subject of ethics. Authors who have attempted a scientific perspective with regard to personal motivation (for example, Thomas Hobbes) or philosophers such as Epicurius who have asked, "What is good for the individual?" have tended to develop hedonistic theories which emphasize the personal axis in the figure. Authors such as Jeremy Bentham who have been concerned with what is "good" or "right" for society have tended to focus on the social axis, or the "utilitarian" principle, as the foundation of social morality. Most classical ethical thinkers were really concerned either with the "social norms" or the "social ideal." These seem to fall broadly into two schools. Those who based their analysis on an appeal to authority tended to focus primarily on the subject of "minimum social norms." This mode of reasoning provides the foundation for both religious and secular law. Others such as Plato, Christ, John Locke, or Immanuel Kant have been more concerned with consistency with certain moral ideals and have tended to focus on the "social ideal" rather than on "minimum social norms." Apparently all of these approaches are relevant approaches within a more comprehensive theory of ethics.

Historically most of the work in the field of ethics has been motivated by the desire to provide a more consistent or rational justification for the precepts of social ethics. Thus, most authors have been interested primarily in the body of principles which define the "social ideal" and the "minimum social norms." In the present theory, these principles do not appear as simple or direct consequences of any fundamental or absolute ethical principles. They are more nearly analogous to an engineering compromise, in a very complex "system design" problem. The real justification for the social principles is that they define a practical and socially beneficial balance between the self-interest of the individual and the interests of the society as a whole.

In practice both the social norms and the social ideal draw their strength from social tradition, and from the weight of "authoritative" opinion. The appeal to "authoritative" principles was the earliest and remains the most widely used method for addressing problems in practical ethics. This method of reasoning (by appeal to authority and precedent) is called "casuistry." Unfortunately "casuistry" has also acquired a negative implication as sophisticated but specious logic. In the present context no such negative connotation is intended.

Casuistry is a valid and very useful approach for dealing with practical problems in social ethics. The casuist asks questions such as: "What are the authoritative rules and precedents?" "Which authoritative opinions are most relevant to the specific present issue?" "Are the opinions really properly applied in this context?" "Have the definitions of words been distorted, or are the analogies sufficiently close that they do not distort the original logic?"

When such logic is combined with commonsense flexibility it provides a mechanism by which the practical precepts of social ethics can be refined and made more consistent. When specific situations arise where common sense demands changes in the old precepts, that possibility should be considered; but in the absence of compelling reasons, it is better to adhere to precedent. In this way the precepts of social ethics can evolve gradually, and individuals are able to judge their conduct against predictable established principles. During the course of cultural evolution the casuist principles gradually accumulate a real wisdom about the principles of social ethics. Precedents that fail to operate successfully are rejected or changed. Precedents that operate successfully continue to be applied, and in the process they acquire more prestige and authority. The casuist logic provides a mechanism much like Bayesian inference through which the experience of many generations can be applied to ethical problems.

The extremes of casuist logic have been very properly criticized by reformers and revolutionaries. When the appeal to authority is used as a *sole* method of gaining ethical wisdom it can become a roadblock to progress; but when used in conjunction with current judgment and experience it provides a systematic mechanism by which the principles of social ethics can evolve to more accurately serve the needs of a society.

REFERENCES

1. Burhoe, Ralph Wendell. "The Source of Civilization in the Natural Selection of Coadapted Information in Genes and Culture." *Zygon: Journal of Religion and Science* 11, no. 3: September, 1976.
2. Dewey, John. "Ethics." *Encyclopedia Americana*, 1938.

Chapter 15

Values and

Personal Decisions

> . . . three things last forever:
> faith, hope, and love;
> but the greatest
> of these is love.
>
> 1 CORINTHIANS 13:13

THE INDIVIDUAL is motivated by an innate value structure that includes the desire for social approval as well as a wide range of other innate values that define his personal preferences. According to the decision-theory perspective, the individual's objective is to achieve maximum satisfaction in terms of this *total* complex of innate values. As a consequence he is inevitably involved in a conflict between the objective of social approval and his personal preferences. According to the concepts developed in the preceding chapter, the social ideals of a society emerge from a pattern of cooperative behavior which, over many generations, adapts the social norms and ideals so that they are reasonably efficient in serving the innate values of the society as a whole. Of course the process is far from perfect (it undoubtedly includes a bias favoring the interests of social leadership, and tends to accumulate a certain amount of error as well as wisdom), but within a small band or tribal society it operates as a reasonably efficient adaptive mechanism that contributes to the welfare and survival of the society.

THE TRANSITION TO A PRESCRIPTIVE THEORY

Our approach in the remainder of Part III will be to pursue the logical consequences of deliberately accepting the basic value criteria that were developed in the preceding chapter as ultimate valuative standards for all human decisions. This chapter considers the individual's innate human values as a guide for personal ethics; and the following chapter explores some of the consequences of accepting the innate values of all individuals in the society as an ultimate valuative standard for social policy.

From a decision-science perspective, this transition to a prescriptive theory is a perfectly natural (and indeed almost inevitable) extension of the theoretical framework previously developed. But with this extension, the character of the theory shifts from a passive intellectual construct to a concept that can have an active impact on both personal and social decisions. For this reason this extension of the theory will undoubtedly be more controversial than the concepts developed in the previous chapters.

The logical validity of this extension of the theory depends on whether or not one accepts a purely scientific philosophical perspective. For this reason, some philosophers will disagree fundamentally with the approach. Without attempting to argue the complex issues that separate the alternative philosophic schools of thought, we will simply pursue the logical implications of a scientific perspective that is compatible with the foregoing decision-science concepts.

RELEVANCE TO HUMAN DECISIONS

Regardless of whether one accepts the premises of such an extension of the theory, it seems clear that the ideas as developed so far have a substantial relevance to the human decision process. The decision-making process, whether in our personal lives or in social policy, can be described in essentially the same way. We consider alternative courses of action and project probable "outcomes." We then evaluate the projected consequences to decide which of the alternative "outcomes" we would prefer. Our ability to make good decisions, therefore, depends both on our ability to project outcomes and on our ability to evaluate the "desirability" of the projected outcomes.

The present theoretical framework seems relevant to *both* of these processes. By providing a more fundamental understanding of human motivations and behavior, the theory should help in predicting the behavior of others, and thus should contribute to a better prediction of "outcomes" for personal decisions. From a social-policy perspective an improved understanding of innate human motivations should help in defining what types of societies are likely to be feasible and thus, in the long run, it should contribute to a more realistic consideration of social policy alternatives. From a decision-science perspective, however, the contribution of the theory to the *valuative* aspect of the decision process should be at least as important. A better understanding of the fundamental and innate human values should allow us to do a better job of evaluating which of the alternative outcomes we would really "prefer." Because the rational mind does not have an innate or built-in understanding of the innate human value system, we must learn about this value structure from experience. Just as the scientific approach has proved helpful in understanding the outside environment, it seems likely that a more scientific understanding of human values should provide a similar improvement in our understanding of our own innate values. Thus a more scientific approach to human values and motivation should contribute to better decisions both by improving our predictions of the objective "outcomes" and by improving our evaluation of which outcomes we would "prefer."

OBJECTIVES FOR A PRESCRIPTIVE THEORY

Because of our accumulated cultural wisdom concerning valuative issues it would be unrealistic to expect an embryonic scientific approach to add much in those areas where the cultural tradition seems relevant and applicable. Within the near future, a scientific approach to value theory seems most likely to make useful contributions in those areas where the present cultural tradition leaves a void of valuative or ethical guidance. There appear to be at least three important areas where the present valuative tradition does not provide effective or relevant guidance:

1. Scientific advances, particularly in medicine, genetics, and biology, are raising ethical issues that have no counterpart in our cultural tradition. Consequently, the traditional guidance on these issues appears to be either nonexistent or irrelevant to the real problems.
2. Rapid changes in our physical and social environment resulting from technological development (for example the huge size of urban society and the impact of industry on the environment) are creating social and ethical problems that

have no counterpart in our cultural tradition. Once again the traditional guidance on these issues appears to be either nonexistent or irrelevant.

3. As a result of the obvious success of the scientific method and the emphasis in our educational system on "objective" scientific thinking, we have developed a large segment in the society that has no faith in *any* source of knowledge unless it is "scientific." But until very recently, science has been completely silent with regard to valuative issues, so this segment of the society has been left almost completely without valuative or ethical guidance.

The material in the remaining chapters is deliberately focused on these specific voids in our present valuative and ethical knowledge. Probably the most important objective is to show how the value theory approach can be used to define a human value framework that is not dependent on any religious or philosophical commitments—except those that can be derived from scientific knowledge. Although this approach may seem strange to those with deep religious or philosophical commitments, we must remember that almost half of the society has no such commitments. If we are to develop realistic approaches to social policy, we must be prepared with a theoretical perspective that can provide guidance that seems appropriate and credible to this portion of society. This does not mean that there needs to be any conflict between the new approach and the more traditional points of view. Indeed, as we hope to show, the new approach seems to be generally consistent with, and complementary to, the more traditional ideas concerning human values.

When the innate human motivations are combined with a strictly scientific "world view" they impose rather rigid constraints on the types of ethical principles that can be effective in influencing human behavior. In effect, the innate human motivations become design constraints for the development of a viable ethical theory. Using this conceptual approach, we can ask: What types of social principles can operate successfully for a society that requires scientific justification for its religious and philosophic commitments? What types of social ideals would be applicable? What types of social norms and legal constraints would be required? How are the requirements for legal constraints and sanctions for such a society likely to be affected by the impersonal nature of the modern urban environment?

No claim is made that the resulting ethical principles will necessarily be any better than those that might be derived from other religious or philosophical foundations. It is my personal hope, however, that the value-theory approach will be able to generate a scientific foundation of ethical principles that will be much better than the present absence of ethical guidance for much of the society.

The extent to which a more scientific approach to value issues may in the long run be able to improve on the traditional commonsense approach remains an unanswered question. In the early stages of any "science" it can be danger-ous to rely too heavily on the scientific findings, especially in those areas where they seem to be in conflict with commonsense judgment. As the science matures and we gain more confidence in the findings, we inevitably give them greater weight in our decision activities. In practice, the issue about when to trust scientific results in preference to common sense will usually never arise. What happens instead is that as the scientific ideas are confirmed and become more familiar, we begin to incorporate them as a part of our com-monsense world model.

We can probably anticipate a very similar process in the development of a science of values. In the beginning it will be wise to rely primarily on the traditional sources of valuative wisdom. However, as the scientific perspective begins to be confirmed both by detailed experiments and by our common-sense experience it will seem natural to give it greater weight in our daily ac-tivities.

In order to make progress in the development of a science of values, it will be necessary to formulate and test hypotheses dealing with valuative is-sues. The material that follows is presented in the spirit of such hypothesis testing. One of the first objectives is to test the validity of the value-theory hypothesis by exploring its compatibility with existing knowledge in related areas. Because of the present embryonic state of the scientific foundation, we are limited to commonsense reasoning based on the illustrative value struc-ture defined in Part II.

From the perspective of our rich cultural tradition, most of the observa-tions will seem trite and unduly simplistic. What is new in most cases is not the ethical concepts themselves but the *way* the concepts are developed through commonsense reasoning, using the preceding theory of human mo-tivation as a starting point. Perhaps the most impressive result of the analysis is the close correspondence between principles developed in this way and the practical ethical principles reflected in our cultural tradition. This corre-spondence provides additional circumstantial evidence in support of the basic value-theory perspective, and it may offer new insight concerning the behav-ioral origins of our cultural traditions. From a religious perspective, the corre-spondence to traditional religious principles may provide evidence of divine guidance in the structuring of our innate human values.

SOME PHILOSOPHICAL PROBLEMS

Our discussion, at the beginning of this chapter, of the relevance of the value theory to human decisions was carefully worded to avoid some critical philosophical issues. It deliberately avoided any explicit statement that our innate human value system can be used as a criterion to decide what we "should" do or what we "ought" to do. The section stated only that the value system can be used to decide what outcomes we would "prefer." If we had tried to reach conclusions about what we "should" do or what we "ought" to do, we would have run into serious trouble with many readers who would ask why we "should" do what we "prefer." Even when we define the concept of "prefer" so that it includes an appropriate concern for a higher standard of behavior, many readers will still insist that such a criterion of preference is not an adequate basis for deciding what we "should" do or what we "ought" to do.

In most cases such concerns are based on a philosophical perspective which, either consciously or unconsciously, includes an absolute or authoritarian concept of "right." Since the validity of such concerns depends rather fundamentally on one's philosophical perspective, it is appropriate to review briefly the traditional philosophical approaches to ethical theory.

The traditional methods used by philosophers to develop criteria for human decisions can be classified into three broad categories:

1. Authoritarian
2. Absolutist
3. Naturalistic

The *authoritarian* tradition has looked to absolute authority or divine guidance as the only valid source of information, particularly for the higher decision criteria. Practical efforts to develop ethical principles on this basis, however, have encountered serious problems because of the difficulty of validating the divine authority for any specific ethical principle.

The *absolutist* tradition has sought to avoid this problem by developing absolute or universal ethical principles based on purely rational or logical considerations. This approach also has encountered serious difficulties because of what might be called the "absolutist fallacy." Specifically, one cannot derive decision principles in a vacuum. Logic and reason can be used to derive decision principles *only* if there are certain fundamental goals or objectives that one is willing to define axiomatically as the objectives of the decision process.

The *naturalistic* tradition has sought to avoid this axiomatic problem by looking for value principles or objectives that are innate to human nature. The naturalistic theories, however, have been vulnerable to what is called the "naturalistic fallacy." This is usually stated in an epistemological form which asserts that one cannot use "is" type statements (i.e., statements about what is or what now exists) to draw conclusions about what "should" or "ought" to be.

Both the "naturalistic fallacy" and the "absolutist fallacy" are concerned with logical requirements for drawing conclusions about values. When the two "fallacies" are placed side by side, they form a paradox which seems to exclude all except the authoritarian approach to ethical theory. Since the present theory identifies the genetically inherited *innate* human value structure as the *ultimate* source of human value criteria, it obviously falls within the naturalistic tradition. It seems appropriate, therefore, to check the validity of the "fallacies" against both mathematical logic and commonsense experience.

THE NATURALIST FALLACY AND VALUATIVE DEDUCTION

Our everyday experience in valuative deduction makes it clear that, in fact, it is possible to generate valuative conclusions from objective facts—so long as the facts include either preexisting valuative information or some other selection criteria that can be used to define certain outcomes as somehow superior to others. The present theory postulates that human values have been generated through a natural process of valuative deduction that originates in the evolutionary process of genetic selection.

There is no logical fallacy in such a process. In the normal practice of decision science we are routinely involved in very similar valuative deduction. The conclusions from such a process always take the form of an "if . . . then . . ." statement. Starting with a particular valuative objective, we can draw conclusions about secondary value criteria (within the context of a known factual environment where alternatives and outcomes can be calculated). Such conclusions are always conditional on the postulated objectives and on the factual definition of the environment.

On the other hand, from our decision-science experience in valuative deduction, it appears that the absolutist fallacy is a real barrier to the development of a logically consistent absolute ethical theory, for there seems to be no rational way to generate valuative criteria without using some a priori valuative criterion or selection principle as a starting point or axiom for the theory.

The game of chess provides an excellent example of this process. Suppose that all rules governing moves and captures in the game had been specified but that no object had been specified for the game. Obviously, without an objective it would be impossible to define decision principles for the game or to generate "values" for the chessmen. Once the object of the game has been defined, however, it becomes possible to generate decision principles and to deduce values for the chessmen as a result of experience in the play of the game. The values that are produced of course are only approximate heuristic surrogates for the ultimate objective, and their validity is conditional on both the rules of the game and the specified objective.

From the foregoing considerations, it appears that *either* the "authoritarian" or the "naturalist" perspective might provide an internally consistent valuative framework. Indeed even a mixed source of values seems possible. For example, we might ascribe the higher standards of behavior to a higher or authoritarian source, while attributing the lower value criteria to natural sources. This perspective corresponds rather closely with many traditional religions. As mentioned earlier, the main problem with this approach lies in the difficulty of validating the divine authority for any particular ethical principle. Thus, although the authoritarian approach is quite satisfactory for those who feel they have access to divine guidance, it is of little practical value for others. Our approach in the present analysis, therefore, is aimed at providing practical valuative guidance without invoking *any* contribution from divine authority.

In the present theory, we postulate that the evolutionary development of human values can be explained by a chain of primary-secondary valuative deductions. In the first step, the primary human values were created by a process in which genetic survival defined the "object of the game" and the natural environment determined the "rules" of survival. The innate human values that resulted from this process bear a relation to the evolutionary objective of survival which is essentially the same as the relationship between the values of the chessmen and the objective of winning the game. In the second step, the resulting innate human values define the "object of the game" for each individual human decision system. Based on experience within a particular social and cultural environment the individual uses these primary value criteria to develop his complex network of secondary value criteria.

The real philosophical problems, however, are not concerned with the validity of this process of valuative deduction. The *real* problem arises when we ask whether the resulting human values are "right" relative to some unstated but absolute or authoritarian definition of right. Obviously from an absolute or authoritarian perspective there is no reason to believe that the resulting value criteria would be "right." From a scientific perspective, however, if we were asked whether the resulting values are the "right" criteria for guiding human decisions, we would have to respond with another question, "Right, relative to what criteria?" Obviously the logical validity of the "naturalist fallacy" depends on whether or not one accepts the philosophical concept of an absolute definition of "right." If one does not believe in an absolute or authoritarian definition of right, then it is meaningless to ask whether a valuative deduction is valid—except within the context of some *specified* primary valuative criterion.

Many questions can be raised concerning the theoretical "validity" of the innate human value system. The deduction of surrogate values within a complex environment is always an imperfect process. There is no reason to believe that the present innate human values are anything more than an imperfect surrogate for the evolutionary objective of species survival. Moreover, because the process of valuative deduction depends on the accumulation of experience, it is inevitably a slow process. The evolutionary deductive process took a very long time. Consequently, there is every reason to believe that the resulting innate value structure must be out of date. It undoubtedly defines a motivational structure more appropriate to a very primitive human society than to a modern urban society.

Although these logical complaints about the theoretical "validity" of the innate values are undoubtedly correct, they are also quite irrelevant to the development of a scientific theory of human values. If the existing motivation system is a fact of life, then any practical system of ethics must somehow be compatible with this innate human motivational system.

From a value-theory perspective, it hardly makes sense to ask whether the individual "should" be motivated by these innate values. It would be like asking whether two plus two "should" equal four or whether the sun "should" rise in the morning. Although such questions may be interesting intellectual curiosities, they do not correspond to any real world choices.

Although the innate motivational system must be accepted essentially as a constraint in the development of a scientific ethical theory, this is *not* true of the social values and ethical principles of the society itself. These social values in each society are fundamentally products of the rational mind. It is the responsibility of each generation to ensure that these principles are refined and adapted to deal effectively with changes in the human environment and to insure that a valid and workable set of principles is passed on to each new generation.

In a very fundamental sense this is what this book is about. In order to intelligently select appropriate norms and values for a modern society, we need a better understanding of the role that these criteria play in the operation of society, and we need some rational criterion for making a choice between alternative forms of the social norms and values. We cannot be effective in transmitting our social ideals unless we have some ideals to which we honestly and sincerely subscribe. For those who already have deep religious or philosophical commitments, this is no problem. But for those who demand an objective scientific foundation for belief, it can be a very difficult problem. Thus our objective in the remainder of the book is to try to lay the foundation for a more scientifically based moral and ethical perspective.

A NATURAL THEORY OF PERSONAL ETHICS *

The analysis in Part II showed that the individual is motivated by a complex set of innate values. Some of the values serve the individual's physical welfare; others serve the welfare of society. As seen by the individual, some values correspond to his physical needs; others correspond to his social and psychological needs. The achievement of a satisfying life-style requires a degree of success in meeting all the innate needs—*both* the physical and the psychological.

Obviously the individual's success in satisfying these needs will depend on the society in which he lives. Because many of the innate psychological drives are reflected in the individual's need to be socially useful, society can serve these needs by providing opportunities for the individual to serve society. The widespread failure of urban society to provide such opportunities in a form that permits the individual to achieve psychological satisfaction may be one of modern society's most serious failings. Whatever the failings of the society, the individual has no choice but to do the best he can, within the constraints imposed by his environment. If he cannot find socially useful ways to satisfy his innate psychological needs, he will strive to satisfy the needs in ways that may not be socially beneficial. This dilemma of the individual is fundamental to many of the problems of urban society.

According to our decision-theory concepts, the individual will strive to satisfy his innate values as efficiently as he can. This mode of behavior is inevitable because that is how the individual is designed as a value-driven system. Society can help or hinder this effort, but it cannot alter the individual's primary objectives. It should be a goal of social policy to provide a social structure in which the innate social motivations of the individual are harnessed to

---

* The word "natural" is used to describe the theory, in part because the theory is a "naturalistic" one in the traditional sense, but more importantly because it rests very specifically on the natural human values that are the evolutionary inheritance of the human decision system.

the welfare of the society and in which the individual can achieve appropriate psychological satisfaction from his efforts to serve society.

The field of personal ethics is concerned with what the individual "should" do. But how relevant are such ethical precepts within a theory that treats our innate values as the ultimate criteria for human decisions? It is tempting to assume that this formulation makes the problem of personal ethics trivial. Nothing could be further from the truth. Although the individual's ultimate objectives are a product of his genetic inheritance, the development of really effective decision rules for realizing those objectives is a very difficult problem. The individual may miscalculate, or he may be misled. What he should try to do is to avoid miscalculation and avoid being misled. A satisfactory science of ethics should provide the individual with practical principles of behavior that will help him avoid such mistakes in his interactions with his social and cultural environment, and thus assist him in his search for personal satisfaction. The development of such a scientific body of ethical knowledge is an important long-term goal, but it would be unrealistic to expect our initial applications of the value theory to yield such an understanding, or even to make any substantial improvements over existing commonsense knowledge.

The individual faces many difficult problems in developing his personal network of secondary goals and ethical principles. He can benefit greatly from informed guidance. The innate motivations shift as the individual matures; so the individual needs to understand and anticipate these changes. In pure self-interest, the decisions of the child and the youth should take into account the values and goals of the adult. This anticipation of future values is difficult at best. It cannot be accomplished at all without external guidance.

Even if the values did not change with age they would still pose many difficult problems for the individual. Most individuals quickly learn to understand and respond to the simple "Selfish Values," such as hunger, thirst, and taste, but their understanding of the "Social" and "Intellectual" values van be very inadequate. People need to be taught what to expect from their innate value system, particularly in circumstances that they have not yet experienced. In the context of a natural theory of ethics, individuals should be encouraged to experiment, so that they can confirm (from personal subjective experience) the teachings of the ethical theory. But they also need to be warned about areas where experimentation is dangerous physically or psychologically to themselves or to others.

Individuals need to be taught about the established norms of society, and about the way these norms contribute to everyone's benefit. They also need to be taught the probable consequences of violation of the norms, in terms of their own health and safety, their own chances for social acceptance, and the risk of legal sanctions. The fact that personal ethical principles are built around the ultimately selfish goal of a satisfying personal life does not exclude a concern for others and for the opinions of others. Indeed, the structure of the innate human value-system, as developed in Chapters 10 and 11, ensures that these factors *must* be considered. The individual's happiness and self-respect depend critically on his interactions with others in society.

Scientific ethical principles must be honestly and accurately founded on enlightened self-interest. When the individual asks why he should behave in a certain way, the ultimate reason should always be "because if you do, you will be happier," or "have more self-respect," or whatever similar reasons are honest and accurate. In a scientifically oriented society, the answer cannot involve an appeal to absolute "right," or to unsubstantiated authority. Such realistic moral and ethical advice should be welcomed by the individual because it should help him in achieving the objective of personal satisfaction. If the society is basically healthy, such honest ethical advice should also cause the individual to contribute more efficiently to society.

One important contribution that we might expect from a scientific ethics is a better understanding of the *innate* motivation for socially *altruistic* behavior. When these issues are discussed in traditional ethical theories, they usually seem vague and mystical, so it appears that they fail to pass the test of objective rationality. As we develop a better understanding of the innate social motivations we may be able to provide young people with more effective and convincing ethical guidance.

In a purely theoretical sense, this "natural" ethics is almost indistinguishable from the hedonist philosophy. In application, however, because of the altruistic or "Social" component of the innate values, it is likely to be very different. The hedonist view has been systematically opposed by many people of good will, because it did not seem to provide an adequate basis for altruistic or socially beneficial behavior. This apparent failing of the hedonist philosophy has been a deep concern to the absolutists, who sought to remedy the defect by appealing to divine authority or to some abstract or mystical conception of "right."

The structure of human values as developed in Chapters 10 and 11 shows that this concern is largely unjustified. The innate values include a great deal of individual motivation for altruistic and socially beneficial activities. To bring these altruistic behavioral characteristics to fruition, society needs only to harness these motivations. Far from being a philosophy of cynicism and despair, the "natural" ethics can be a philosophy of realism and hope. It is an expression of faith in the tremendous power of what is good in human nature.

CULTURAL NORMS AND SANCTIONS

An important objection that can be raised to this naturalistic concept is that it seems to leave open the possibility of devious or dishonest behavior. If socially altruistic behavior is motivated *only* by enlightened self-interest, then what is to discourage antisocial behavior when the individual feels he cannot be discovered? It is obvious that this is a very real problem. Fortunately evolution has provided some safeguards. As a protection against such behavior, the human species has developed a high sensitivity to insincere or dishonest behavior. An individual that is perceived to be insincere or devious is instinctively disliked and distrusted. Thus as a practical matter, in most cases, an honest and sincere behavioral strategy is the most effective in realizing the innate human social goals.

In a small town as well as in a small primitive band or tribe, this evolutionary deterrent to deceptive behavior is particularly effective because each individual is so well known to the others. In such a society the individual quickly learns that such behavior will usually fail. Unfortunately, in the comparatively impersonal environment of an urban society these evolutionary deterrents to antisocial behavior are far less effective. Moreover in the impersonal urban society, many of the incentives provided by evolution to motivate socially altruistic behavior operate much less effectively. For example, it is almost impossible to achieve a position of respect or approval, much less dominance, in an impersonal and anonymous society. Thus, within the impersonal urban society the problem of antisocial behavior has become acute.

One thing that a scientific ethics cannot do is prevent antisocial behavior by obligating the individual to either a supernatural or an absolute standard of ethical behavior. If the social environment fails to harness the innate human social motivations, then in fact we must expect antisocial behavior. From the value-theory perspective the real fault in such situations lies not with the individual, but rather with a social environment that fails to provide psychological rewards and penalties to which the innate value system can respond.

From our decision-science perspective, the problem is fundamentally a problem in motivations. To decrease the frequency with which antisocial behavior is likely to occur many obvious and familiar social policy alternatives are available. For example:

1. Increase the probability that such behavior will be detected.
2. Increase the psychological and legal penalties associated with being detected.
3. Decrease the potential rewards or benefits that can be obtained by such behavior.

4. Ensure that legal sanctions are more swiftly and reliably applied.
5. Provide better moral and ethical education so that individuals will understand the social norms and the social ideal and be more aware of the social risks and penalties involved.
6. Provide better education concerning the personal social and psychological benefits that are to be derived from a life-style that contributes positively to society.
7. Modify the social or physical structure of the society so that innate altruistic motivations can operate more effectively.
8. Provide better social conditioning experience in early childhood so that such antisocial behavior will be foreign to the normal pattern of behavior.
9. Use jails (and perhaps in extreme cases, capital punishment) to remove chronic offenders from society.

The foregoing list is by no means special to the present theory. The same list could be generated simply on the basis of commonsense and experience. Many measures suggested above could also be justified within the behaviorist doctrine as appropriate techniques of "behavioral modification." However, the interpretation of their effectiveness is quite different within the present theory. According to the behaviorist doctrine prompt and reliable punishment can reduce the frequency of crime because it provides an aversive or negative reinforcer for antisocial behavior. According to the present theory, the prospect of prompt and reliable punishment makes crime seem like a less attractive option. Whereas in strict behavioral theory, one would not expect such punishment to be effective until after it had been experienced, within the present theory the *prospect* of punishment is important because it serves as a rational deterrent to crime.

The foregoing list of course does not include certain traditional alternatives such as appeal to divine or absolute authority that could not be substantiated by scientific methods. Although this omission may be seen by some as a fault, it has some distinct advantages which should allow such a strictly scientific ethical theory to serve a useful role in the society. When the existing incentives in the society fail to discourage antisocial behavior, the traditional ethical or religious philosopher may attempt to solve the problem by developing a more convincing sermon or a more elegant and compelling theory about how people "ought" to behave. But the scientific ethical theorist has no such options. He must look for solutions through changes in the legal, social, and psychological incentives provided by the society. If during the development of a scientific ethical theory it becomes apparent that certain kinds of antisocial behavior are not sufficiently deterred, then it is obvious that changes are needed in the social environment to correct the problem.

Although effective cultural norms and sanctions are absolutely essential for satisfactory operation of a society, the choice of an appropriate set of norms and sanctions is far too complex a problem to be addressed here. The following chapter deals with some valuative principles that appear to be relevant in the development of such principles, but no attempt will be made to convert the general principles into specific social ideals, social norms, or formal sanctions.

In small tightly knit societies it was possible to depend much more on informal ideals and moral suasion, but as the society becomes more anonymous the dependence on formal legal sanctions tends to increase. The formal sanctions can help to deter antisocial activities that may not be adequately discouraged by moral suasion in an impersonal society. Where such sanctions fail to deter, the imprisonment of offenders at least protects society against really antisocial individuals.

From this point of view, most types of criminal behavior are treated as an essentially normal and rational behavior which is likely to occur in any situation where the apparent advantages to be gained by such behavior are perceived to be greater than the probable risks and penalties. One of the objectives of social policy should be to ensure that such situations do not occur too frequently. Obviously the legal sanctions cannot be effective unless they are actually enforced. Consequently when individuals deliberately decide to violate the law, they must be held accountable for their act. Of course, it may not really be the fault of the individual that he found himself in circumstances where the illegal behavior seemed attractive or the legal sanctions seemed inadequate. Nevertheless, if the legal sanctions are to continue to operate as an effective deterrent to crime, it is absolutely essential that the individual must be held responsible.

Legal doctrine in the United States holds that under certain circumstances (for example, if the individual is insane, or under extreme emotional pressure) his degree of "responsibility" for the crime may be less. Within the present theory, this concept has a certain logical justification. Since the primary purpose of the sanctions is to deter crime, there may be little advantage in actually imposing sanctions for behavior that is so irrational or emotional that it could not realistically be deterred by any sanction no matter how severe. In such cases, the real issue is whether or not the individual poses a continuing threat to society.

The establishment of really appropriate legal sanctions is complicated by the fact that there are wide variations in individual motivation. In any society there will be individuals who do not respond normally to moral suasion. Their innate motivations may fall well outside the normal range either because of abnormal experience in infancy or because of their particular genetic inheritance. Although such individuals are not responsible for their innate motivations, society must, for the common good, take appropriate steps in protecting itself against the threat to social stability they pose. Insofar as possible, these steps must be chosen so as to interfere as little as possible with the individual's freedom of choice and action.

The spectrum of cultural norms and sanctions can be divided broadly into two groups:

1. The rather concrete and objective incentives and sanctions that tend to be favored by the conservatives. These range from fines and imprisonment on the negative side to monetary incentives (e.g., no work, no pay) on the positive side.
2. The less clearly defined social and psychological incentives that tend to be favored by the liberal members of society.

In terms of the analysis of personality types in Chapter 8, it is noteworthy that each group tends to advocate those types of incentives that are most important for their own personality type. Within the present framework, both sets of incentives are important.

It appears that the value-theory perspective may be most useful, however, in providing a more concrete understanding of the social and psychological incentives. It also appears that both liberals and conservatives may have underestimated the importance of the negative social incentives such as scorn and disdain for individuals that fail to measure up to the social ideal. Because cultural norms and sanctions are so important to the operation of a society, there is a critical need for general principles that can be used to decide what specific norms and sanctions are "right."

In the present "natural" approach to ethics, the innate personal decision criteria are accepted, together with any of their faults and limitations, exactly as they are determined by the human genetic inheritance. This *existing* structure of the innate human values defines a fundamental problem in social system design: specifically, how to structure society so as to minimize the problems created by the unavoidable discrepancy between personal and social objectives.

Fortunately, because of the way evolution has defined our innate human values, the problem is not nearly as difficult as it otherwise might be. The innate goals of the *individual* have evolved so that they include many evolutionary surrogates for the welfare and survival of the society. For this reason the actual discrepancy between the objectives of the individual and the objectives of society is surprisingly small.

ETHICS AND THE WISDOM OF THE AGES

One of the most important tests of a new theory is its correspondence with "common sense." This is especially true in the field of value analysis—where (even in principle) human judgment is the ultimate guide. The innate human values as developed in Part II contain a very important component designed to motivate socially altruistic behavior. Thus they can provide a rational foundation for many altruistic ethical precepts. Although it is premature to make any broad judgment, our preliminary analysis suggests that there should be a remarkable compatibility between the natural ethics and the traditional ethics of the major religions.

One way to exhibit the compatibility is to show how some of the traditional commonsense ethical precepts might look when restated to conform to the terminology of a natural ethical theory. Of course, if we were to search very far we could find familiar ethical precepts on both sides of almost any ethical issue. However, there are a few precepts that seem to be so widely accepted that we might be inclined to reject a theory if it disagreed with the precepts. In the following paragraphs some of these very familiar precepts have been recast to reflect our preliminary understanding of the human social values (as developed in Chapters 10 and 11). The paragraphs are intended only to show that some of these most familiar commonsense ethical precepts seem very reasonable and plausible when stated within the context of the theory. There is, however, no intent to imply that these precepts have been independently or rigorously "derived" from the theory.

1. *Know Thyself.* Each individual is different. Each is endowed with a profile of innate values that is his private inheritance. Each must seek satisfaction in terms of his own personal values. Each must experience life to learn his values. Only by experience do we learn what we like to do, and what we do well. Life is an adventure. Not the least of the adventure is learning who and what we are. In knowing oneself is the beginning of wisdom.
2. *To Thine Own Self Be True.* Be honest with yourself. Recognize your strengths and weaknesses. Act in accordance with your own objectives and in terms of your own values. Satisfaction in life must be measured by your personal values, not by the goals of others.
3. *Be Moderate in All Things.* The innate human values represent a careful balance of priorities. When any activity is carried to excess, the enjoyment of the activity tends to be destroyed. This built-in response of our innate value system protects us physically and psychologically from the abuse of excess. Full enjoyment of life requires a variety of interests and activities, and moderation in the pursuit of any individual activity.
4. *Love Thy Neighbor.* Man is a social animal. The most intense and valuable

emotional experiences are achieved in interactions with others. To achieve such experiences we must really care about other people. We must come to know and respect them as individuals. When such relationships are carefully nurtured, love blossoms. It is in the combination of fellowship and love that mankind experiences his own greatest value. (The choice of the "neighbor" as one to love was good practical advice in a less mobile society. It may be somewhat less feasible in our present transient society. Nevertheless, it is obviously still desirable to develop close human relations with one's neighbors.)

5. *Honor Thy Father and Mother.* Our innate human values are a priceless *family* inheritance. Because our innate value structure is genetically inherited, there will usually be a close correspondence between our own innate values as we mature and the values that our parents now express. For this reason parents can provide a unique wisdom that cannot be obtained from any other source. Honor the advice of parents, for they know and care.

The foregoing brief list of ethical precepts is, of course, only illustrative. As in earlier parts of the book, our objective has been to define a theoretical framework on which future concepts can be developed. Although no attempt has been made to provide a complete ethical theory, there does not at present appear to be any obvious reason why a natural or scientific ethics cannot reproduce most of the major conclusions of traditional practical ethics.

In any great civilization the principles of ethics and practical morality must be compatible with the world view of society. The natural ethics is especially compatible with a scientific world view. As new knowledge becomes available, there should be a corresponding growth in the value and usefulness of the theory. Unlike traditional ethical commitments, the natural ethics promises to grow with the expansion of scientific knowledge.

Logically, the precepts of a natural personal ethics fall into two main categories: the "general" precepts and the "specific" precepts. The "general" precepts, such as those just listed, seem to follow logically as an almost obvious consequence of the structure of the innate values. Because they do not depend on specific social norms or the technological environment, they should be essentially the same in any human society. "Specific" ethical precepts are concerned with how the individual should interact with the specific customs, institutions, laws, and social norms of a particular society. These can vary widely from one society to another.

GOD'S WILL OR MAN'S?

Like any scientific theory, the theory of human values is mute with regard to ultimate theological questions. It does not provide any answer concerning the existence or nonexistence of God. Although the development of a scientific and nontheological foundation for a practical ethics is one of the major contributions of the theory, the theory itself is nevertheless quite consistent with a theological view. If we accept evolution as the work of God, then man's innate values are also the work of God. They are a biological manifestation of God's will for man. For this reason it can be argued that the innate human values ought to be acceptable, even within the context of traditional theistic religions, as a fundamental criterion for human behavior.

Obviously, this view will be most convincing if the resulting "principles of behavior" in fact seem to be consistent with what has traditionally been recognized as the "will of God." So far in the analysis there has been a surprising degree of compatibility. But there is one potentially serious problem that has not been addressed.

The idea that our genetically inherited values could be considered as a manifestation of God's will for man, or even as an ultimate ethical criterion may seem naive to anthropologists who are familiar with some of the cruel and manifestly evil practices of many primitive societies. Some obvious examples of such practices include the primitive human sacrifice, deliberate mutilation of the body to produce "decorative" patterns of scars, and the cruel initiation rites that are sometimes applied to young boys as they reach maturity. How can we explain the existence of such cruel and evil behavior?

The presence of such practices in primitive human society is particularly disturbing because it occurs *within* the species, and indeed within what otherwise *seems* to be a cooperative social group. Although the infliction of pain is not uncommon in the rest of nature, it usually has a functional purpose to provide food for the species or to protect a social group from outside intruders. Deliberate cruelty within a cooperative social group does not appear to serve any evolutionary purpose, and such behavior is quite rare in other species. From a broad evolutionary perspective such practices seem counterproductive—human sacrifice reduces the productive population, mutilation of the body risks infection and may otherwise interfere with the adaptability of the body. Why should evolution's design for the innate human values lead to such strange behavior in primitive human societies? The fact that these practices seem so manifestly evil, in terms of our modern moral intuition, suggests that they must be in conflict with the innate human values. It is almost as if something had gone wrong with evolution's design; and so it is natural to ask how this could have come about.

Upon closer examination, we find that such behavior almost always serves a ceremonial function. The behavior is "justified" and serves a practical "function" within the context of a complex and superstitious world view. Thus the cruel behavior may actually be a consequence of a distorted world view, rather than a rational consequence of the innate human values. The cultural inheritance of a primitive society includes a world view which incorporates many supernatural forces. This culturally sustained, superstitious world view is possible *only* because of the availability of a language that can communicate abstract concepts. Without the superstitions it seems likely that most of the cruel behavior would disappear.

There is rather strong evidence that human "cultural evolution" accelerated spectacularly with the appearance of modern man about 40,000 years ago. It seems very likely that it was an *improved* linguistic capability that initiated this *cultural* revolution. If we think of *modern* language as an evolutionary invention that was superimposed rather recently on an older, well-established human motivation system, then many of the apparent paradoxes are removed. Although the improved linguistic capability was on balance a very advantageous genetic mutation, it carried with it a number of side effects that were nonadaptive or bad. The cultural potential for cruel and evil practices within human society may have been among the nonadaptive side effects. It seems unlikely that such superstitious behavior could have been prevalent in the preverbal societies. Because such societies cannot develop a complex or sophisticated world view, the behavior norms in such societies should be a relatively direct consequence of the genetically inherited innate values.

With a little knowledge, man is likely to be led astray. For this reason there is a continuing need in human civilizations to reassess behavior norms against basic human values. In traditional religious terms, there is a need to abandon "the errors of man" and return to "the way of the Lord." This traditional ethical insight is symbolized in the story of Eve and the apple. In the biblical version, Adam and Eve ate of the "tree of knowledge" and were forced to leave the Garden of Eden. The present theory suggests a modernized version of the old parable. Having partaken of the genetic "tree of language" they were irrevocably committed to the path of "cultural evolution," with all the potential for evil that comes from partial and incomplete knowledge.

This point of view suggests that the cruel primitive practices are not a logical consequence of the innate human values, but that they arise as a consequence of a superstitious world view. From the evolutionary perspective, they were not a part of the evolutionary "intent" but were an unfortunate by-product of the introduction of language. It seems likely that as we gain a better understanding both of the world and of our own innate values, such cultural aberrations should become progressively less common. Thus the ethical goal of perfecting human behavior in terms of our innate human values may, in the long run, prove to be quite compatible with the traditional religious goal of implementing God's will for man.

THE ROLE OF ETHICAL LEADERSHIP

With the general decline in religious authority there has been a corresponding loss of faith in the ethical teachings of the church. When ethics was founded on authority, a loss of faith in the authority was able to destroy the subject of ethics as a serious topic for study. Consequently, religious schools have been robbed of a relevant modern curriculum. The loss appears to be largely unnecessary. There is a great deal that is critically relevant to modern living. But to reestablish confidence it is important to place the subject on a sound scientific basis.

There is a critical need for new research. For example, human social behavior is motivated primarily by the dual emotions of joy versus sorrow and pride versus shame. Individuals could benefit immensely if they had a better understanding of how these emotions respond to different types of social activity. Basic research aimed at understanding these relationships should be a high-priority project in the field of ethics. Obviously, unless the experts themselves know the answers, they cannot provide really informed leadership.

Of course, moral and ethical leadership requires the ability to present the ideas in a way that inspires trust and confidence. This is far from an easy task, but it is one that needs to be faced by ethical and moral leaders. It is clear that the public is hungry for realistic and convincing moral and ethical leadership. Consider the popularity of magazines dealing with marriage and sex counseling, or with parental problems. Consider the success of the newspaper columns dealing with personal advice for readers. The public is looking for reliable practical guidance in their private lives. They will not settle for advice that may have been relevant at the dawn of recorded history.

RELATION TO HUMANIST AND
EXISTENTIAL PHILOSOPHY

The theory of values provides a natural alternative which seems to reconcile some of the most important differences between the humanist and the existential philsophers. The humanist perspective is based on an essentially scientific philosophy which recognizes the evolutionary origin of the human species. The humanist typically believes in the dignity and worth of man and emphasizes the capacity to improve the human condition and to achieve self-realization through the proper application of rational thought. The present theory with its emphasis on the evolutionary origin of human behavior is a natural descendant of the humanist point of view. But the new perspective is also compatible with more recent thinking which tends to de-emphasize rational thought. The existentialist view and many of the present counterculture movements largely reject rationality as a fundamental human directive. The present approach accommodates this rejection by recognizing that fundamental value commitments are not under rational control. Yet unlike some existentialist views, it explicitly retains rationality—as the servant of fundamental values.

The natural ethics involves a formal recognition of innate human values, or the human spirit, as the ultimate guide for all human behavior. Reason is assigned tactical responsibility for finding ways to satisfy the instinctive values, but it is not assigned the strategic responsibility for selecting the primary goals or objectives. Because different individuals may genetically inherit differing value structures, the resulting "rational" conclusions may differ between individuals. Thus, the present point of view recognizes (as the absolutist versions of humanism did not) that basic personal values cannot be derived by rational thought.

The similarity of the natural ethics to existential philosophy is most apparent in the resulting personal philosophy of life. According to the existentialist, it is up to each individual to *decide* what he is, and what he is going to be. According to the natural ethics, it is up to each individual to *discover* what he is, and *decide* what he is going to be. Although there is a real enough difference in the words and in the point of view, it is doubtful that there is any practical difference in what might *actually* be done by an individual following either philosophy.

The existentialist view also corresponds to the present theory in its emphasis on learning by doing. According to the existentialist, we can achieve goodness by being active in life and by experiencing what it is to be good. We cannot achieve goodness by being taught or by rational analysis. According to the value theory we can learn the structure of our personal innate values (and learn what is good in terms of those values) by being active in life and by engaging in those activities which are defined as "good" within our own personal value system. Although a natural ethics would attach much greater importance than existentialism to what can be learned from the experience and knowledge of others, there is basic agreement that the ultimate criterion of personal "goodness" lies within.

Both points of view agree that we are ultimately responsible for our own decisions and that we must find the basis for our decisions within ourselves. However, some existentialists, particularly the Christian school, would insist on a somewhat more comprehensive form of personal "responsibility." The Christian existentialist might insist on responsibility for creating even his own primary values, and he would probably refuse to blame those values on a genetic inheritance. From the value-theory perspective the individual has a responsibility for discovering and applying his primary values, but he does not have a responsibility for creating them.

A surprising number of modern intellectuals have found the existentialist view to be a congenial personal philosophy. The most serious weakness of the existentialist doctrine has been its inability to provide practical criteria for social action or to offer the hope of constructive social change. This failure of the existential philosophy seems to stem from two key features of the existential view which are in conflict not only with humanism but also with the value-theory approach:

1. *Excessive emphasis on personal individuality:* The existentialist seems to disregard or ignore the very broad similarities between the innate values of individuals. In his refusal to recognize the essential similarity between individuals, the existentialist loses the ability to make generalizations about human behavior or to apply those generalizations to the solution of social problems.
2. *Insufficient emphasis on the role of reason:* All problem solving depends ultimately on the use of rational faculties. To the humanist, this application of reason provided man's ultimate hope for humane social improvements. In the deemphasis of reason, the existentialist tends to destroy the hope and expectation for constructive progress.

The value-theory perspective makes it possible to retain most of the attractive features of existentialism as a personal philosophy, but at the same time it preserves the role of reason and the hope for constructive social progress that characterized the humanist philosophy.

A HISTORICAL PERSPECTIVE

Our society is experiencing extremely rapid change. Traditional ethical guidance, which once seemed dynamic and relevant, no longer seems applicable to our lives. It appears that at least once before in human history, there was a similar crisis in the value tradition. The crisis occurred as mankind emerged from a primitive agricultural and hunting society and began to develop a civilization with stable cities and towns. With this fundamental transition in lifestyle, the instinctive primitive approach to ethical issues was no longer adequate. The innate human values did not function satisfactorily in the urban setting without formal ethical guidance. This crisis of values led to what has sometimes been called the axial period in religious history. Around the sixth century B.C. there appeared, independently and in widely scattered areas, such great spiritual leaders as Buddha, Confucius, Zoroaster, and the prophets of Israel. Although the teachings of these leaders were quite varied, there was a substantial core to their message that was shared by the greatest leaders.

This ethical core apparently filled a critical human need in the unfamiliar urban setting. It appears that this message may have served to strengthen vital interpersonal human relations that were being weakened in the urban environment. It enabled the human animal to extend into an urban setting the close personal ties on which he had depended in more primitive societies.

From time to time philosophers have tried to provide formal or logical explanations for these simple ethical precepts. Without a real understanding of the innate human values, the efforts have not been very successful. It seems probable that the philosophers failed to achieve popular acceptance because they did not understand man's genetically inherited value system, so they failed to appreciate the impossibility of superseding the innate human values with a philosophically "rational" value system.

Conversely, the success of most of the world's great religions, as well as the intuitively developed Oriental philosophies, can be attributed to the harmony between their teachings and man's genetically inherited value system. For this reason, the religions and intuitive philosophies have had immense emotional appeal—and have achieved popular influence despite any weakness in their logical or rational foundations. Philosophers have had more popular success when, like Confucius, Karl Marx, or Jean-Paul Sartre, they have paid less attention to formal logic and more attention to the economic, social, and emotional needs of man in his environment.

From the point of view of the theory of values, the religious and the intuitive philosophies were basically right, for an individual of any species can achieve satisfaction *only* in terms of his *own* value system. Humans can find "satisfaction" and (that elusive goal) "self-fulfillment" only in terms of the primary human value scale. By explicitly recognizing man's emotional and "spiritual" needs as reflections of a nonrational primary value system, the natural theory offers the possibility of an ethical framework that might have emotional appeal comparable to traditional religions. Thus, it offers the possibility of a gradual reconciliation of the religious, philosophical, and scientific points of view.

Chapter 16

Decision Science and Social Policy

Let us raise a standard to which
the wise and honest can repair;
the rest is in the hands of God.
                    GEORGE WASHINGTON *

THE PRESENT CHAPTER is concerned with the social norms and social objectives of a society. How should the individual decide what types of behavior to approve or what social changes to support? How should social planners decide what changes to propose? Is there a "natural" principle that can be used for evaluating social norms and social policy proposals in terms of the innate human values? In Chapter 14 we observed that the cultural evolution of social norms seems to be governed by valuative forces which correspond roughly to the best interests of the society at large. Since an objective (strictly scientific) criterion for evaluating social norms and social policy could be of considerable importance, this chapter explores the possibility of using this same basic valuative concept as a formal criterion for social decisions.

Once again there is a philosophical problem. Even if such a principle seemed to be useful as a guide to social policy, philosophers would have to ask why individuals in their own personal interest "should" propose or advocate policy alternatives compatible with such a criterion. For our present purposes, however, it is not necessary to address this issue in any depth. It is obvious that individuals will *not* always find it in their personal interest to advocate specific social policies that are in the best public interest. But it is equally ob-

---

* Speech to Constitutional Convention, 1787.

vious that when we are engaged in cooperative activities it can be very helpful to have a decision principle that is recognized as "fair" that can be used to help resolve differences and select a practical course of action. This chapter develops a social policy principle which might provide such a basis for compromise within the context of cooperative social activities. Regardless of whether we are working at the city, state, federal, or community level, situations often arise where such a fundamental valuative criterion would be useful in defining appropriate goals for social policy.

Our approach will be first to develop the valuative criterion and then pursue the logical implications of the concept, testing the results against common sense valuative intuition. If the criterion fails to correspond with common sense, then it will undoubtedly not be used and will not be useful. However, if it appears that it usually provides a good criterion for social policy decisions, then it is possible that the use of the criterion might itself develop into a social norm which could have a very real effect on the way decisions are made.

A VALUATIVE CRITERION FOR SOCIAL POLICY

If our ultimate personal objective is to achieve satisfaction in terms of our own innate values, it follows that we should prefer those forms of social policy that are most favorable to that objective. We should prefer a social structure that will allow us to be as successful as possible in satisfying our own innate values. This fundamental preference of the individual provides a "natural" valuative criterion for choosing between social policy alternatives.

Obviously social policies cannot be tailored to the desires for any specific individual, so we must be willing to settle for a less personalized social objective which seeks equally the common good of all. This simple concept, which is sometimes stated as "the greatest good for the greatest number," provides a familiar commonsense compromise between the conflicting interests of individuals. The same basic principle of social compromise will be applied within the present theory.

Since each individual should prefer a social structure in which he can be most successful in satisfying his innate values, it follows that an equitable social compromise should permit the greatest success of this type for *all* individuals in the society. To be most beneficial for the common good, the society should be structured so that *when individuals act* (as they must) *to achieve personal satisfaction in terms of their own innate values they will be as successful as possible.* This is the *basic* valuative principle for selecting social norms and social policy within a "natural" social ethics. When the objective of social policy is explicitly stated in terms of *innate* human values, it highlights the real importance of the Social and Intellectual values (as well as purely physical needs) as criteria for social policy.

The preceding "natural" social principle rests logically on two separate postulates:

1. The recognition of the genetically inherited innate values as the ultimate source of all human values.
2. The acceptance, at least in principle, of a simple method of valuative compromise which provides a commonsense balance between the conflicting objectives of different individuals.

The postulate which *distinguishes* the present theory from earlier ethical and social concepts is the first one. Our main objective, therefore, will be to clarify the consequences of this new valuative postulate.

The second postulate parallels almost exactly the method of compromise developed in the 1700's by Jeremy Bentham which became the foundation for "utilitarian ethics." Consequently, we will sometimes refer to the "natural" social principle as a "natural utilitarian principle." But the traditional "utilitarian" compromise has been subjected to a number of serious objections which must be addressed; otherwise, readers may assume that the proposed "natural" social principle would be vulnerable to the same form of criticism. In order to show how the new social principle avoids most of the pitfalls of utilitarian ethics, we will give particular attention to some of the features that distinguish the "natural" social principle from traditional utilitarian ethics.

RELATION TO UTILITARIAN ETHICS

When Jeremy Bentham originally proposed the utilitarian principle, he was concerned with some very practical issues in British law and public policy, and he was looking for an objective criterion of social policy that would expose the obvious evils of existing laws. His early publications include two notable assertions which he used to justify his "utilitarian" principle:

1. "Nature has placed mankind under the goverance of two sovereign masters, *pain* and *pleasure*" (4).
2. ". . . it is in the greatest happiness of the greatest number that is the measure of right and wrong" (3).

These two assertions are of particular interest because they summarize the fundamental assumptions of Bentham's approach and make it easy to identify the major similarities and the differences of the present approach.

With just a small change in terminology, it is possible to bring Bentham's two principles into almost exact correspondence with the present theory. Bentham's first assumption would correspond with our "natural" principle of personal ethics—if his valuative criterion "pain and pleasure" were interpreted to include *all* the innate human valuative sensations. It seems rather clear that this, in fact, is just what Bentham intended. To avoid logical inconsistency in his analysis Bentham denied that there are any qualitative differences among the pleasures. In his view all pain and pleasure should be compared quantitatively in terms of its intensity and duration. With this bold denial of the qualitative differences between valuative sensations, Bentham achieved the same objective that is achieved in the present theory by comparing all valuative sensations against a common scale of "utility." The "utility" concept of modern decision theory makes it possible to place Bentham's first assumption on a solid theoretical foundation and thus avoids one of the inconsistencies that motivated criticism by later writers.

With a similar change in terminology Bentham's second assumption can be made to correspond to our "natural" principle of social ethics. If we interpret "happiness" as a measure of the total value actualized by each individual in terms of his own innate values and if we interpret "right" or "wrong" to mean socially good or socially evil (in the sense shown in figure 14.1b), then we can achieve an excellent correspondence with the present theory. Although the foregoing changes in terminology may seem trivial, they are of great importance in avoiding some of the inconsistencies that have subsequently plagued "utilitarian" ethics.

Traditional utilitarian ethics has tried to use the principle of "greatest good for the greatest number" not just as a measure of "good" or "evil" for social policy, but also as a principle of personal ethics to decide what the individual "should" do. From our present perspective this utilitarian approach to personal ethics seems very unrealistic because there appears to be no rational reason why an individual should use such an altruistic criterion for making his personal decisions. Even as a basis for a social ideal, the principle seems naive, since to be most effective even the social ideal should give reasonable consideration to personal preferences. The present theory avoids these pitfalls by limiting the application of the utilitarian principle *only* to social policy. The "natural" principle of personal ethics, which corresponds very closely to Bentham's first assumption, continues to be used to decide what the individual "should" do.

But the utilitarian principle has been severely criticized, even as a criterion for social policy. The most serious criticisms have been based not on the logic of the theory, but rather on certain conclusions of the theory which seemed inconsistent with intuitive concepts of equity and justice. These apparent "flaws" in the "utilitarian" principle were difficult to reconcile when the value criterion was defined only vaguely in terms of a balance of "pain and pleasure." A number of writers have attempted to avoid these difficulties by "extending" the valuative criterion to include "higher pleasures" (6) or a wider range of "primary goods" such as esthetic enjoyment, knowledge, and personal affection (7). But this approach did not solve the problem because the suggested "higher pleasures" did not really correspond to the structure of the *innate* human values. The problem has become so serious that in 1969 H. J. McCloskey wrote ". . . if consistently and fearlessly worked out, utilitarianism leads to morally shocking conclusions which involve inhumanity, injustice, and dishonesty" (5).

The remainder of the chapter is intended to show that these conflicts with moral intuition developed because the critics were basing their arguments on "self-evident" but incorrect assumptions about the structure of the pertinent human values. While we do not yet have an accurate knowledge of that value structure, it is already apparent that the agreement of utilitarian theory with valuative intuition seems to improve as the representation of the innate value structure becomes more accurate and more complete. To provide a proper valuative criterion for either personal or social ethics it is obviously essential to include a proper representation of *all* the dimensions of the innate human value structure.

Many of the apparent problems with the utilitarian principle arose because the relevant human values were described as "primary goods," which (by an unstated analogy with a fixed inventory of "physical goods") might be "distributed" among the individuals of the society. This description of the problem gives a completely erroneous impression of the real ethical issues. The structure of the innate "social values" (as developed in Part II) makes it obvious that the ultimate human values do not follow any conservation law and that they cannot be "distributed." When a gift is given *both* the giver and the receiver benefit. The gift itself may be almost irrelevant. The real transaction is in the shared pleasure and pride that come from mutual concern. The traditional language places too much emphasis on the issue of "distribution" of benefits and too little emphasis on the real issue, which is the *generation* of human values.

One of the most persistent criticisms of the utilitarian principle is that it does not show any preference for a more equal as opposed to a less equal "distribution" of the "primary goods." But this is not a valid criticism of the present "natural utilitarian principle." If different distributions of tangible assets were evaluated against the "natural utilitarian principle," the principle itself would show a strong preference for more equal rather than less equal distributions. This preference for equity in the distribution of resources arises quite naturally within the theory as a consequence of the extremely nonlinear "innate human values" that are used within the "natural utilitarian principle" to evaluate alternatives.

One trivial example will serve to illustrate the point. If we do not have enough to eat there is a very real discomfort (a large negative value), but there is no corresponding increase in comfort if we have more than enough to eat. Thus a large increase in the total social benefit can be achieved by redistributing excess food from those that have more than enough to those that do not have enough. This is only one, almost trivial, example of the way the nonlinearities in the human values produce strong preferences for more equal rather than less equal distributions of real resources. As will be shown later our strong intuitive preferences for equality, equity, and justice are direct natural consequences of the nonlinearities that evolution has built into the innate human value system. There is no need to compound these natural preferences by superimposing an ad hoc "equity preference" within the utilitarian principle itself.

RATIONALE FOR A SOCIAL VALUE CRITERION

The present theory differs from most traditional ethical theories in that it explicitly includes two different valuative principles. The natural principle of *personal* ethics provides the fundamental valuative criterion for personal decisions, while the natural principle of *social* ethics provides a similar valuative criterion for social policy. Because there are two separate valuative principles involved, it is necessary to ask when we should use the personal criterion and when we should use the social criterion.

Fundamentally, of course, all the decisions that we make are personal decisions. From this point of view we should always be guided by the personal decision criterion. If the theory is to be consistent, we should be able to use the principle of personal ethics to decide when and how to apply the principle of social ethics. In effect, the social principle must play the role of a "secondary" value that can be derived (at least approximately) from the "primary" principle of personal ethics.

To clarify the need for such a social policy principle, consider the general role of social values in a society. Most of the cooperative functions of a society can be accomplished only if an agreement or consensus can be reached on the actions to be taken. For example, the behavior "norms" of a society reflect an informal "consensus" about what types of behavior are to be "approved." The government structure and the laws reflect a "consensus" that is reached by more formal procedures.

In the absence of an authoritarian source for a cooperative social principle, we can ask whether there is any natural principle which would be most likely to produce decisions compatible with a group consensus. Is there any unique "principle of consensus" which would normally be the most advantageous for the individuals within the group? These questions lead almost automatically to the "natural" social principle. The logic is as follows: Each individual in the group operates as a decision system. His personal decisions are guided by his own personal value system. There are certain areas, however, where the group itself must operate as a "decision system." The question arises, What value principle will yield the most appropriate decisions for the group as a whole? The answer seems obvious. The most appropriate value structure for the group as a whole can be obtained simply by *averaging* or (equivalently) *summing* over the value structures of all the individuals within the group. This simple valuative concept makes it possible, at least in principle, to use the innate values of the individuals to derive a valuative criterion for decisions by the group.

If group decisions are systematically and correctly made in terms of this criterion it will automatically maximize the average level of satisfaction of the individuals within the group. On this basis, it appears that this is the valuative criterion that would be the most advantageous for the group to adopt. More specifically, if the members of a group had to choose such a "social principle" in advance (without knowing either their own particular place in the group or the specific issues to which the principle would be applied) this is the principle that they "should" choose, because application of this principle (and this principle alone) will maximize their "expected utility" when they take their (previously unknown) places in the group.

Of course, the preceding argument is really only a theoretical justification for the proposed principle of social ethics. It will not be a practical principle unless it actually works in practice. In the sections that follow, we hope to show that the principle is practical, in the the sense that it leads to conclusions that seem generally consistent with "commonsense" valuative intuition.

JUSTICE AND NATURAL LAW

One serious complaint that is often leveled against the traditional "utilitarian" principle is that it does not provide a foundation for justice, equity, or fairness. Because the utilitarian value principle appears to be concerned only with the summation of the total value and not with how the value is "distributed," there has been widespread concern that such a principle could be used to rationalize grossly unjust social structures which discriminate against a few individuals "for the greater benefit of all."

But this objection to the utilitarian principle once again arises because of the failure to consider the actual structure of the innate human values. The innate human values operate in such a way that they automatically generate an intuitive sense of "fairness" and "justice." This remarkable property of the innate value structure is incorporated automatically in the natural social principle.

The concepts of justice and fairness that exist in any mature society are likely to be quite detailed and sophisticated. Obviously these detailed concepts of justice are a product of the cultural experience of the society and they vary widely from one society to another. However, certain fundamental or elementary concepts of justice seem to exist in some form in all human societies. It seems likely, therefore, that the modern sophisticated concepts of justice are really cultural elaborations of certain basic Social Values that are included in our genetic inheritance.

The human "sense of justice" matures surprisingly early. Even a very young child will respond to simple injustices with an indignant "That's not *fair!!*" Although the child's specific concept of fairness may be culturally conditioned, his response to unfairness includes an emotional component that is almost certainly innate. The child's reaction is not just a rational observation that the situation is unfair. The child is responding emotionally; he is angry, frustrated, and indignant. The experience for the child is one of intense unpleasantness (i.e., very large negative value). The negative value associated with the experience is far in excess of whatever reward may have been denied. What distresses the child is not the simple loss of the reward, but the unfairness of the denial.

Evidently the human emotion of "anger" operates in such a way that injustice itself generates a very large negative value. Anger is automatically generated when one individual (or group of individuals) is singled out for adverse discriminatory treatment. It also arises when an individual has labored in full expectation of a routine reward and the reward is arbitrarily denied.

The anger that such events generate is an intense negative value which can be relieved by a direct attack on the source of the "injustice." When circumstances prohibit such an attack, the anger lingers and the sense of distress is acute. Reflecting on my own own experience, I believe the most unpleasant experiences that I can remember were occasions when I believed I had been "wronged" but there was no practical recourse.

It seems probable that our elementary human understanding of justice and fair play is a natural consequence of our social experiences with the emotions of anger and frustration. Because "justice" avoids anger, it follows that justice is intrinsically "good." The emotion of anger is probably the ultimate source of the legal concept of "natural law."

This relationship between "justice" and "anger" is rather analogous to the relationship that was developed in Chapter 14 between the "approval motive" and the cultural norms of cooperative behavior. In both areas, any mature society will exhibit a rather elaborate set of cultural norms. However, there is reason to believe that these norms are really complex cultural elaborations of more fundamental and innate human values. Thus the basic concept of justice seems to be an almost inevitable social response to the emotion of anger.

If we were to use the proposed social principle to evaluate social policy alternatives, the contribution of "negative value" by the emotion of "anger" would produce a very strong tendency to select just or equitable alternatives. For example, if we were to evaluate alternatives which included the institution of slavery, we would have to include within the resulting values a very strong negative value reflecting the anger and resentment of the slaves to their "unjust treatment." It is most unlikely that there would be any comparably important positive *innate* value for the rest of the population which could justify such a policy.

But the emotion of anger is a much more effective incentive for social justice than is implied just by its direct negative impact on the values of the natural utilitarian principle. When a substantial segment of society is angry the rest of society is in real physical danger. Thus, the avoidance of anger is a high-priority objective. In a human social system it must override almost all other objectives. Once again the actual structure of the innate human values eliminates what seemed to be serious objections to the traditional "utilitarian" principle.

The origin of justice and natural law suggested here may help to provide a logical foundation for a modern "theory of justice." In the absence of a theory of human values, legal scholars have found it extremely difficult to produce a consistent theory. The difficulties are most apparent in a highly respected treatise by John Rawls called *Theory of Justice* (8). In an effort to provide a theoretical "foundation" for his theory, Rawls was forced to rely on some distorted logic which has been criticized by subsequent writers (2).

Rawls' argument is based on the thesis that a "just" society is one which would be selected by rational individuals who did not know in advance what position they might later occupy in the society. His assumptions are such that if the logic had been correctly pursued, individuals "should" prefer a form of society that maximizes the "expected utility" of each individual. Thus, if Rawls had accepted the obvious consequences of his postulates, he inevitably would have produced a "utilitarian principle." If he had then adopted the innate human values as the ultimate basis of the human "utility scale," he would have produced a theory of justice in exact correspondence with the one defined here.

In the absence of a theory of human values to provide the necessary *nonlinearity* in the utility scales he was forced into a much more difficult approach. His task was to show that in the "original position" rational individuals would *not* have chosen to maximize "expected utility," but instead would have preferred a social structure that would maximize the utility of the least fortunate individuals in the society. This is a difficult proposition to demonstrate, because it is in direct conflict with the mathematical definition of "utility." The intense concern for fairness and equity that Rawls tried so hard to incorporate in his theory arises quite naturally in the present approach because of the nonlinearities of the innate human values.

INDIVIDUAL FREEDOM

It is unlikely that popular support could be obtained for any social concept which did not include individual freedom. Fortunately, the preference for individual freedom also appears to be a natural consequence of the "natural" social principle. Each individual has his own innate values. The individual is by far the best judge of his own values. In order to maximize the total social value each individual must be free to actualize his own innate values. Interference with this freedom is appropriate *only* when it is necessary to protect the value opportunities of other individuals or of society as a whole.

This formulation differs slightly from the traditional definition of freedom as an "inalienable right." When freedom is treated as a "right" it is hard to explain why freedom must sometimes be limited, but we all know that such occasions arise. In the present formulation the need for limitations on individual freedom to protect the general welfare follows automatically as a consequence of the "natural" social principle. Moreover, at least in principle, the social principle tells us when such interference is justified and when it is not.

THE PROBLEM OF COALITION AND CONSPIRACY

So far we have been addressing the problems of social ethics as if each individual were a member of one and only one homogeneous society. In fact, the individual may be a member of many different formal and informal social groups. Each such group has certain norms and social ideals which apply to the membership; so the individual may be subject to multiple and possibly conflicting loyalties.

To address the value principles involved in the formation of social subgroups it is helpful to begin with a simple homogeneous society and ask how such a society "should" view the formation of coalitions within the larger society. In a simple homogeneous society, as each individual balances personal self-interest against desire for approval, the individuals will behave as if they are working partly for themselves and partly for the larger society. But many situations can arise where individuals will be able to serve their personal self-interest more effectively if they band together into smaller cooperative groups. Similarly some individuals may find that they can serve the interests of the society more effectively if they organize into small cooperative groups.

Such cooperative groups can increase the total value actualized within the society, so it appears that they "should" be encouraged. To decide what specific projects to undertake, each cooperative group must act as a separate "decision system," and it needs a valuative principle for making its decisions. The natural utilitarian principle, applied over just the membership of the group, is the most natural valuative principle for the group. When a group operates as a separate decision system within the society, we will call it a "coalition." In this broad meaning of the word, a city is a coalition, a company is a coalition, and a labor union is a coalition. Modern society owes much of its effectiveness to the operation of coalitions. As we might expect, the coalition or cooperative group is generally considered to be a "good" or a desirable thing.

But the existence of coalitions also poses a serious threat to the society. A "coalition" can become a "conspiracy," which operates for the personal self-interest of the members in a way that is clearly detrimental to the larger society.

The innate human motivation for cooperative behavior and social approval operates effectively in the face-to-face activities of a small group, but it generally does not operate effectively beyond this range. An individual who is a member of a conspiratorial group can achieve social acceptance and approval within the group. Because he can satisfy his innate social needs within the group, he may not be much concerned about the approval or disapproval of the larger society. This is the *fundamental* threat of the "conspiracy." It helps to explain the very bad moral connotation of the word "conspiracy." It also explains the vigilance that societies exercise to protect against "conspiracy" and the vigor with which conspirators tend to be prosecuted. It may also help to explain the weaker "rules of evidence" that are applied in conspiracy trials.

Conspiracy becomes a real threat within any society that is large enough to include many self-contained face-to-face social groups. It is easy to think of examples: the Mafia, the pirates, the urban street gangs, the nineteenth-century "robber barons" of Wall Street, and the "power conspiracy" of the Nixon administration. The "conspiracy" threat becomes most severe when there is a general loss of faith in the validity of the ethical principles of the larger society. Under such circumstances it is particularly easy for coalitions to develop in which the ethics and norms of the larger society are not a part of the "social ideal" or "social norms" of the coalition.

One of the fundamental problems of social policy is to design the social ideal and the social norms of the larger society so that they will be effective in moderating and controlling the behavior of coalitions.

THE SOCIAL BENEFIT FUNCTION

Although the theoretical structure of the natural social principle seems to be superficially identical with the traditional utilitarian concept, there are a number of very important differences. These differences arise because the underlying values are interpreted not as "goods" but as "decision criteria." Because of this change in interpretation, the mathematical relationships are different and many of the remaining inconsistencies of the utilitarian principle are removed.

To clarify these differences, it is helpful to define a formal mathematical value function that could be used within a decision-science analysis. We will define the total social value, V, for any policy alternative to be equal to the summation over all individuals of the expected valuative satisfaction v_i that the alternative should yield (now and in the future) for each individual, i. If we are not sure that all individuals should be treated equally, we can use a weighting coefficient α_i to allow us to assign greater importance to some individuals than others. The criterion might then be written mathematically as follows:

$$V = \sum_i \alpha_i \, v_i$$

In this form, it states simply that the total social value, V, of any alternative is equal to the summation of the value v_i for all individuals i, weighted (or multiplied) by a factor α_i which reflects the "importance" of each individual. Although superficially it may seem obvious that all individuals should be weighted equally regardless of their "importance," it is not so obvious whether the weighting should be independent of age—particularly for the very young or the unborn fetus. The inclusion of the weighting coefficient α_i makes the issue of weighting explicit, and allows us to consider differences in weighting depending on the maturity or age of individuals. The formula defines a "social benefit function" which is simply a more mathematical representation of the "natural utilitarian principle." Theoretically, to use such a formula as a decision criterion, we should simply select the social policy alternative with the highest estimated value of V.

Of course, in practice we cannot hope to get the specific numbers needed to evaluate V for the benefit function. Nevertheless, it will clarify some of the basic issues if we pretend that we might actually get the numbers in order to consider the consequences of using such a benefit function as a formal criterion for social policy. From a mathematical point of view, this "social benefit function" is generated simply by *adding* together the valuative criteria used by *all* members of the society. In principle, therefore, when it is summed over all the members of a society, it automatically represents a consensus of the objectives of the society.

But there are some potential ambiguities and issues of equity that need to be discussed. How does the criterion provide for the interests of future generations? Should the weighting α_i depend in any way on the age or importance of an individual in the society? To acquire the popular appeal necessary for a social norm, a social objective must be compatible with our commonsense ideas of equity. How does the social benefit function survive such a test?

THE ISSUE OF EQUALITY

Let us consider first the choice of the weighting coefficients. Is it appropriate to assign an equal weight to the interests of all citizens, regardless of their importance or contribution to the society? To address this question we must distinguish carefully between the weighting of "wisdom" and the weighting of "values." Obviously there will be some individuals within a society who are recognized as wiser or more expert in certain areas, and it is in the best interests of the group to make use of the most reliable mental models available. Consequently the technical opinions of the experts should be given greater weight. But it is not in the best interest of the society to allow the *valuative* goals of the experts to guide their decisions. For this reason, professional ethics encourages the expert to present his recommendations in a form where the valuative assumptions are explicit: "*If* this is the objective, *then* this is the preferred course of action."

However, we are not concerned here with the weighting of wisdom, but with the weighting of values or objectives. In terms of values it seems intuitively clear that the objectives of all persons should be treated equally. There are several reasons why equality in the weighting seems to be the only practical choice.

First, it is compatible with our democratic tradition of equality before the law, as well as our intuitive concepts of fairness and equity. Second, equality of treatment provides a simpler criterion, which is more likely to achieve popular support. Third, the assignment of equal importance to all citizens directs attention away from the destructive question of how to divide the pie and toward the more constructive question of how to make the pie bigger for everyone.

The equality of weighting in the traditional "utilitarian principle" has been attacked from two directions. On the one side, there are the conservatives who are concerned that an equal weighting of the interests of all individuals might yield too uniform a distribution of goods and benefits and thus not provide adequate incentives for the more productive members of society. On the other side, there are the liberals who are concerned that a decision criterion using an equality of weighting would reflect only the total benefits without regard to the distribution of benefits, and might thus fail to show any real preference for more equal rather than less equal distributions of wealth and resources. In the context of the present valuative formulation neither of these concerns seems justified.

As mentioned in an earlier section, the emotion of anger combined with the other nonlinearities of the innate human value structure causes the "social benefit function" to show strong preferences for more equal rather than less equal distributions of all tangible resources.

The equal weighting in the benefit function is also entirely consistent with the concept of individual incentives. When the benefit function is applied to choose between policy alternatives, it will of course select those alternatives which produce the highest *total* benefit. Practical experience has shown that in order to make the total benefits as large as possible, it is desirable to provide incentives which reward those who contribute substantially to society. This universal rule applies just as well in a socialist or communist economy as it does in capitalist economy. Because of this practical need for incentives to encourage the generation of social benefits, the equal weighting in the "social benefit function" should also be consistent with just and equitable incentives.

On the basis of the foregoing arguments, neither the issue of justice nor the problem of incentives seems to provide any compelling reason for departing from the simple rule of equality in the weighting of values within the "social benefit function." There is, of course, another reason why this seems like the only practical choice. If the principle of equality were not followed, that in itself would probably generate "anger" and resentment in the individuals who are discriminated against and thus reduce the total social value that could be realized. Therefore (with the possible exception of an age dependence, which will be considered later), we will assume that all individuals should be treated equally within the social benefit function.

WHO TO INCLUDE IN THE BENEFIT FUNCTION

When the natural utilitarian principle is formalized as a benefit function, it is immediately apparent that we must decide which individuals should be represented in the benefit function. When the function is summed over the individuals within a particular social group, it provides a consensus of the objectives of the group. It is obviously not realistic to extend the summation to include outsiders, because the resulting function might not correspond to an acceptable objective for those inside the group. To decide what individuals to include within the function we need to know for what group we want to get a consensus. Do we want a social benefit function appropriate to a social club, a professional society, a city, a state, or a nation?

Up to this point, the decision about whose interests to include within the function seems straightforward, but there are some other issues that seem less obvious. Although it seems clear that all individuals should be weighted equally regardless of importance or contribution to society, it does not necessarily follow that all individuals should be weighted the same regardless of age. How about the young infant, the unborn fetus, or future generations not yet conceived? Where and how do we draw the line?

When the classical utilitarian principle was formulated in terms of "primary goods," questions of this type proved to be particularly troublesome. But when it is clear we are dealing with a valuative *decision criterion*, the solution to the problem seems obvious. The innate values of the individual were designed by evolution to provide for survival of the species. Any actions needed for future generations had to be motivated through the values of the *living* generation, for there was no other way that evolution could provide for the future. It appears that the interests of future generations should be already represented, at least to some extent, in the innate values of the present generation. Evidently logical consistency within the theory requires that the "social benefit function" be summed only over the living; for from an evolutionary perspective, it is only the living that decide.

From this strict evolutionary perspective we should weight the values differently depending on age. The consensus for a group decision should include only those capable of participating in the decision. The values of an infant should be given very low weight because its potential influence on decisions is small. On the other hand, if the evolutionary view is correct the real interests of the child are quite adequately represented in the innate values of the parents.

From a formal theoretical perspective this means that when a social policy decision is being considered, the appropriate value criterion should include only the innate values of the living, and these values should be projected as far into the future as seems relevant. But to avoid double counting, the values of unborn generations should be included only as those interests are represented in the values of the living.

An adequate validation of the social benefit function as a valuative principle would require a much more comprehensive analysis than is possible here. Our objective in the following sections is primarily to show that the social benefit function avoids some of the most obvious pitfalls that have caused trouble for previous ethical theories. It remains to be seen whether it can deal satisfactorily with a wider range of more complex valuative issues.

CONCERN FOR FUTURE GENERATIONS

Because the benefit function does not explicitly include the values of future generations, it is of considerable importance to see how these concerns for the future are incorporated within the theory. Although some cynics may defiantly assert that they do not care in the least what happens to the world after they are dead, I believe that few, if any, really feel that way. We have a natural concern for the welfare of our children and our children's children. We have a desire to have our work and our objectives carried forward. We have a desire to be well remembered, even after death. All these natural desires seem to be manifestations of the innate sense of responsibility of a normal primate adult.

This natural concern for future generations is not at all surprising. The innate human values are ultimately derived from the simple evolutionary goal of survival of the species; consequently we should expect a close correspondence between the ensemble of human values and the basic evolutionary goal. On this basis, it would be most surprising if we did not feel a personal desire to have our species survive.

If we were suddenly to learn that some calamity would cause the extinction of human life on earth within the next hundred years, most people would feel an acute loss of meaning and purpose in their lives. Indeed, it would probably become a project of high moral purpose to find ways to avoid extinction of the species in the calamity. Like all other innate human values, this concern for future generations cannot be rationally explained. It is a natural consequence of the built-in human value system.

Of course, the degree of concern varies widely among individuals, but it seems clear that if the real interests of the citizens could be measured, the concern for the future would prove to be a significant factor in the innate values of the society. What weight we should actually give to the future, relative to the present, is a legitimate subject for debate, but as a practical matter the issue can only be resolved in terms of the innate values of the living.

THE VALUE OF HUMAN LIFE

In this section we examine the implications of the "social benefit function" with regard to policy on some life and death issues. These issues are important both because of their current interest and because the predictions of traditional utilitarian theory in some of these areas has been a source of serious criticism. Whether the conclusions of the "natural" utilitarian principle are consistent with our valuative intuition is obviously something that each individual must decide on a personal basis.

LIMITS ON POPULATION GROWTH

One of the arguments against the simple "utilitarian principle" was that it seemed to lead to almost unrestrained population growth. In the traditional formulation one could conclude that the benefit function would always favor an increase in population, because this would increase the number of terms in the sum! To avoid this problem, a number of theorists have suggested that the "benefit function" should be divided by the population, so that it would reflect the average benefit per person rather than the total. Although this ad hoc correction removed the population explosion problem, it still left the possibility that current decisions could be dominated by the interests of generations as yet unborn.

The decision-theory perspective, however, provides a simple and logically consistent solution. The benefit function that is used to evaluate decisions at any point in time should include only the innate decision criteria of the living. To do otherwise would be equivalent to assuming that the innate decision criteria of the unborn (represented by the new terms in the sum) are relevant to decisions affecting the birth rate, even though such decisions must be taken before the new individuals are present in the social system. This does not mean that the welfare of future generations will be ignored. The welfare of future generations will be considered precisely to the extent that it is included within the innate values and aspirations of the living.

In the past, theoretical ethical views concerning population growth have reflected different points of view about what values to include in a "social benefit function." The extreme altruistic view would include the values of all future generations, whereas the pragmatic or selfish view would include only the values of the living. In theory the "natural" ethics parallels the selfish or pragmatic view, but it incorporates some of the objectives of the altruistic view, because the interests of unborn generations are represented, at least in some degree, in the innate values of the living.

Because this rationale may not be fully satisfying to those who hold to the altrustic view, a few comments on the practical similarity of the two views may be in order. In its extreme form, the altruistic view is represented by a literal interpretation of the phrase "the greatest good for the greatest number." It is easy to assume that this goal requires one to maximize both the amount of good per person and the number of persons.

A little reflection will show that this naive interpretation of the altruistic decision criterion must in fact be in error. The evolutionary goal is survival of the species, and the evolutionary time perspective is very long. As world resources are stretched to the limit, the risk of error or misjudgment that could lead to destruction of a species is increased. Even from the strictly "altruistic" view it is foolhardy to risk the nearly infinite future for a small increase in population in the finite present. Thus when we look carefully at the problem of long-term human survival, even the extreme altruistic criterion leads to essentially the same conclusion. Select a cautious strategy. Do not press the world resources too hard at any time, because resulting damage to the environment could cause extinction of various species of life (including the human species) and might thereby completely foreclose future opportunities.

BIRTH CONTROL

Social policy on birth control seems straightforward. Among those living within the society, a birth has by far its greatest valuative impact within the family. Therefore, the prospective parents should be primarily responsible for the decision. They should judge the desirability on the basis of their perception of the satisfactions and responsibilities of parenthood. The state should make birth control methods available so that the family will have a choice.

Of course, as noted earlier, the society has an interest in the birth rate because of the impact of population growth on the society at large. Social policy to influence the birth rate should be evaluated primarily in terms of the effects of population changes on the quality of life and the effects that such changes will have on the total valuative satisfactions of the existing population.

EUTHANASIA

Euthanasia (or mercy killing) has little direct effect on the satisfactions available to the rest of society. It is of intense interest to the individual and to his family. If an individual is so hopelessly ill that his life is a burden both to himself and to his friends and family, the present theory suggests that he should be permitted to choose death. To require him to continue to live would contribute only negative values for all concerned. Of course, the society has a responsibility to avoid abuses. Procedures should be established to insure that mercy killing does not occur unless it is clearly approved both by the victim and his family.

AGE AND THE VALUE OF LIFE

In many traditional ethical doctrines the value of human life is treated as absolute. All lives are of equal value, and the value is essentially infinite. But the way people actually behave does not seem consistent with this view. In time of famine, the elderly in many societies will sacrifice themselves to save food for younger members of the family. In some societies infants will be deliberately sacrificed for the welfare of the society. Intuitively we are more concerned about the death of an individual in the prime of life than we are about the death of an infant or the very aged. Does this intuition have any foundation within the present theory of values?

The question is of more than academic interest. Many policy decisions particularly in the field of medicine ultimately involve judgments about the value of different lives. The doctor must decide whether to save the mother or the baby. Resources for medical research are limited, and research priorities for different kinds of diseases must be established.

The present theory provides rather clear support for the view that the value of a life depends on the age of the individual. The declining value of life with old age follows naturally from the "social benefit function." As the individual becomes older his valuative satisfactions are more in the past and less in the future. Thus, the anticipated value of the remaining life-span becomes progressively less.

Of course, it is impossible to specify exactly how the value changes with age. Even at the same age, a healthy individual can expect a more favorable balance of values in his future than an individual who is burdened with pain and illness. Indeed, in cases of extreme pain and hopelessness (as was mentioned in the discussion of euthanasia) the value of the remaining life-span may actually be negative.

It is important to emphasize that there is a very important distinction between "the value of the remaining life" and "the weighting of an individual's innate values within the social benefit function." The decrease in the "value of life" with old age and illness arises despite the fact that the values of the elderly are given their full weight within the benefit function.

But how do we explain the lesser importance that is usually attached to the death of an infant? If the physician is forced to make a choice, he will almost always choose to save the mother rather than the infant. Superficially we might conclude that because the infant has its entire life in the future, maximum social value should be attached to the infant life. To deal with this apparent paradox we need to look once again at the theoretical basis of the "social benefit function." The social benefit function is designed to reflect the collective interests of those who might be involved in social decisions. Therefore, in accordance with evolution's design principles, the values of the newborn infant should not be separately included in the social benefit function. Evolution's concern for the well-being of the infant is represented at this stage in the desires and motivations of the parents. As the infant matures and acquires an independent decision capability and an independent foresight, there should be a gradual change so that the personal values of the child begin to be independently represented in the "social benefit function." Probably the child should be fully represented only when he becomes truly independent.

From a decision-theory perspective, this gradual introduction of the values of the child is consistent with the principle of not including the values of future generations. The values of the child should not be explicitly included in the "benefit function" beyond the child's own ability to anticipate the future. The infant's future interests are not ignored in the benefit function; they are represented in the values of the adults, particularly the parents. This valuative adjustment which brings the independent interests of the growing child into the social benefit function on a gradual basis is also consistent with long-standing human traditions. The parent is normally granted wide latitude in deciding how to deal with the child. Indeed, it is generally assumed that society has no obligation to intervene except in cases of obvious disregard of the interests of the child, such as neglect or active child abuse.

The concept also makes sense from the evolutionary perspective of survival of the species. The fetus, the infant, and the very young child are basically a burden to the species. During this time, society is making an investment in the child. The life of the child acquires social value as the investment is made. It takes on full value only when the investment is complete. The fetus is fundamentally the responsibility of the parents, and it is represented indirectly in the social benefit function through the values of the parents.

From the foregoing considerations there emerges a generally consistent theory about the value profile of a human life. At conception, the life begins with a zero value. The value increases steadily until it reaches a maximum level in the young adult. This increase in value parallels the investment by society in the individual. As the individual ages, the value of the remaining life gradually declines. The decline reflects the smaller remaining life expectancy, and it parallels the transfer of the social investment from the individual to the next generation.

While the value of life is lower both for the very young and the very old, the reasons for the reduced value are different in the two cases. The values of the very old are fully represented in the social benefit function until the day of their death; for them, the reduced value of life reflects the limited pleasures of old age combined with a lower life expectancy. The very young have a full life expectancy, but the value assigned to their life is lower, because they are not yet fully represented as an independent decision system in the social benefit function. The investment by society in their life is not yet complete.

THE PROBLEM OF PERSONAL SACRIFICE

One of the most difficult ethical questions concerns those occasions when there is a need for an individual to sacrifice himself for the welfare of others.

There are obviously many cases where a mother or father will deliberately sacrifice his or her life in the interest of the rest of the family. Our evolutionary perspective makes it clear that when circumstances become so critical that the individual values the welfare or survival of his family above his own life, such self-sacrifice may well be appropriate. It is undoubtedly consistent with the evolutionary objective.

But a much more difficult problem arises if we ask whether any individual should be *required* by society to make a sacrifice for the benefit of society. In this case there is a clear conflict between the objective of equity and fairness and the general welfare. Traditionally societies have often required such self-sacrifice despite the manifest "unfairness." Military conscription and subsequent assignment to a very hazardous (if not suicidal) mission provides an obvious example. Certain ethical theories that would place equity and justice categorically above all other considerations (as well as others that would place life above all other values) would have to conclude that this is necessarily "wrong." Within the present theory, the issue is essentially a problem of balancing values. The extreme loss of value by the single individual must be weighed against the gains for the rest of society. It seems clear that the sacrifice could be justified only in very exceptional circumstances. But it is also apparent that in certain circumstances such sacrifices may be necessary. In this respect the present theory seems more consistent with the way societies actually behave.

When such sacrifices are necessary the theory indicates that society should try to achieve the sacrifice in a way that minimizes unfairness. The use of a draft lottery, as well as the military tradition that troops will draw straws for a dangerous assignment, shows that this principle seems to be followed in our commonsense behavior.

DELIBERATE MODIFICATIONS OF INNATE VALUES

Modern medicine and genetic research have focused attention on some difficult ethical issues which involve deliberate modification of our genetically inherited innate value structure. Under this general heading, we can include issues such as the use of drugs to produce altered states of consciousness; the implantation of electrodes in the brain to artificially stimulate valuative signals such as joy, contentment, or sexual pleasure; the use of drugs or neurosurgery to block pain signals; and even deliberate genetic redesign of future generations through selective breeding or test tube "genetic engineering."

A first intuitive reaction to these problems is that because they involve changes in our ultimate value criteria, they cannot be addressed within a theory that uses innate human values as an ultimate valuative criterion. However, a few practical examples will show that we routinely make just such decisions (involving modification of innate value structure) in terms of value criteria that are obviously determined by the value struture itself!

When we visit a dentist, we do not hesitate to use novocaine to block the pain. The novocaine block is a deliberate (although temporary) modification of our innate value structure. Once we have made the decision to proceed with a tooth repair, we prefer to avoid experiencing intense and nonfunctional pain. The decision to block the pain is directly motivated by the innate value structure itself.

In contrast, if a normal healthy person were asked to decide for or against an operation that would permanently eliminate all sensations of pain, he would almost certainly decide against the operation. The rare individuals who are born with birth defects such that they do not experience pain are in serious danger. They may inadvertently tear the flesh from their fingers even while doing rather routine jobs.

Our preferred solutions to the contraception problem provide another example of deliberate choices concerning changes in the innate value structure. An obvious alternative solution to contraception would have been to change the innate value structure so that we would not experience sexual pleasure or sexual desire (for example, by castration or by hormone treatments). Our decision to solve the problem by retaining sexual pleasure but blocking the fertilization process is a deliberate choice motivated by the innate value structure itself.

From these examples, it is clear that we do, in fact, make decisions about modifying our innate value structure based on valuative criteria that originate within the value structure itself. Obviously, the above examples were deliberately chosen to illustrate cases where both the decision and the reason for the decision seem obvious. Many issues involving modification of innate values are much less obvious. And they pose more difficult ethical issues.

The problem of drug abuse provides a practical example. The problem poses potentially difficult dilemmas for both individual decision and social policy. It is conceivable that a drug might someday be developed that would be very attractive to individuals and very detrimental to society. For example, one can envision a "happiness pill" that would produce a sensation of continual euphoria. Although such a drug might be very attractive to many individuals, it might be socially destructive because it would short-circuit the innate motivation system so that persons using the drug would become a burden on the society. Apparently social policy would have to be directed to discourage or prohibit the use of drugs that might be socially destructive.

Drugs can also pose some difficult decisions even from the perspective of a single individual. For example, the short-term pleasure or relaxation that may be offered by alcohol or tobacco must be weighed against the long-term hazards to health and future enjoyment. Addictive drugs such as heroin produce long-term changes in the innate value structure such that the individual may be literally trapped by his need for the drug. When such long-term risks are weighed against the short-term pleasure of the drug, the prudent individual will decide to forego the short-term pleasure.

Some doctors, however, have begun to deliberately prescribe certain addictive drugs to relieve the suffering of terminally ill patients. Within the present framework, such decisions seem to be quite justifiable because the patient will not live long enough to face the long-term risks of the drug, and the short-term benefits may allow an extension of a pleasurable existence.

Decisions concerning genetic redesign of future generations pose a different type of ethical problem. Such decisions will have their largest effect on individuals that cannot participate in the decision. Nevertheless, according to the present theory, such decisions must necessarily be made in terms of the goals and objectives of those now living. One of the least controversial forms of genetic planning might be a voluntary birth control program designed to minimize genetic defects in future generations. Such a program would probably receive general support because it would reduce the social and economic burden of genetic deformities. At the other extreme, one can visualize compulsory genetic planning and perhaps even test tube genetic redesign of new generations. The potential benefits of such policies should obviously be weighed carefully against the immediate cost in human freedom and dignity.

Moreover, it is obvious that great caution should be exercised in any attempt to genetically redesign the innate human value system. Although the present motivational structure may not be perfect, its social implications have been tested over thousands of generations. Any substantial changes in this motivational system could have unforeseeable consequences for the human social structure. Given our present limited knowledge, it seems far more prudent to adapt the social environment to the requirements of human motivation rather than to try to redesign human motivation to fit the needs of a modern social environment.

As with all types of human decisions, some choices are obvious, some are difficult. In some cases the balance of risks may lead to obvious decisions; in other cases the balance may be close and the decision may be difficult.

Although our present innate human value structure might not provide an optimum value structure for an abstract society on a new planet, it does contain sufficient valuative information to provide guidance concerning the incremental decisions we are likely to face on this planet. Of course, the human race may nevertheless fail to make the right decisions and may even ultimately destroy itself. But if this should happen, it seems more likely that the disaster will occur because of a failure to anticipate or properly predict the *outcome* of decisions, rather than because of any fundamental failure of the primary human value system to select desirable long-term outcomes.

If the results discussed in the foregoing sections seem to be generally consistent with valuative intuition, we might be willing to use the natural social principle at least on a preliminary basis for other applications. However, this would not mean that social decisions would become routine or even that the same decisions would be appropriate for all societies.

SOCIETY AS A WORK OF ART

The nations of the world represent an extremely varied pattern of social structures. History tells us that human society can exist in an even wider diversity of forms. Each society must follow its own sense of balance and harmony. What is right for one society may be wrong for another.

In a sense, each society is like a work of art. Each artist begins with essentially the same canvas and paints as raw material, but each uses the materials differently. As a work of art develops it acquires a unique and individual character. Elements that are appropriate and proper in the context of one work of art would be inappropriate in another.

Similarly, different civilizations can begin with the same raw material and develop in very different ways. Each civilization begins with human beings that have essentially the same innate values. But as the societies develop, specific social and political concepts begin to emerge, and the societies begin to acquire a unique and individual character. The individuals within the society are a product both of their innate inheritance and their specific experience within the society.

GUIDE POSTS FOR SOCIAL POLICY

Specific social principles must grow out of the environment and context of ideas that constitute a specific civilization. The natural theory adds some new guide posts that should help in defining such concepts. The natural utilitarian principle provides at least a theoretical criterion for evaluating the desirability of specific social alternatives. The theory of personal behavior provides a scientific framework which should be useful in anticipating the response of individuals to specific proposed social changes. As we develop a better scientific understanding of the innate human values, we should be better able to judge which social changes will seem beneficial.

Most of the innate personal values can be realized only as a result of initiatives taken by the individual. In such areas, society can be most effective by simply allowing the individual the freedom to undertake such initiatives. But there are many areas where the individual has irretrievably lost control of his destiny. This is particularly true with regard to those aspects of human behavior that affect the life environment, the economic system, the use of natural resources, and the prospects for survival of the species now and in the long-term future. For better or for worse these environmental and survival issues are now in the hands of large governmental organizations. When changes in social policy are being evaluated the following considerations should enter the analysis:

1. The natural utilitarian principle to evaluate the policy alternatives as they affect the physical, social, and economic environment of the individual.
2. The context of the existing social structure in which policy changes must operate and be evaluated.
3. The long-term objectives involving resource conservation, the welfare of future generations, and survival of the species.

In principle the long-term considerations should be treated as a part of the innate human values and included within the natural utilitarian principle. But as a practical matter it may be easier to treat these as a separate issue, and then omit the related long-term innate human values in the application of the utilitarian principle. In this way the consideration of long-term social welfare is substituted as a more concrete surrogate for the individual's vague concern about his posterity and their future.

Over the long run, one of the most important functions of a scientific social ethics should be to provide better guidance in the choice of the laws and cultural sanctions of the society. However, these issues are far too complex to be effectively addressed within this book or within the present limitations of a very embryonic theory of human values.

THE ESSENTIAL ROLE OF JUDGMENT

Whether we are considering issues of personal behavior in a specific social environment or issues of public policy dealing with that environment, we must necessarily consider some very complex interactions. Theoretically, an evaluation of such alternatives requires a detailed understanding of all interactions in the society. In practice, such a complete understanding cannot be achieved, so we cannot expect to produce results concerning specific personal decisions or policy issues by methods that would survive critical scrutiny by a mathematician or logician. Such conclusions always require the application of judgment beyond what can be achieved by formal or rigorous logic.

The role of judgment in social decisions is rather analogous to the role of secondary values in the game of chess. Although the appropriate "values" of the pieces are a consequence of the rules and the objectives of the game, there is no formal mathematical procedure that can be used to derive the "values" from the structure of the game.

Fundamentally, we develop the "values" from our experience in playing the game. Using our "commonsense" version of Bayesian inference, we ask what value criteria will usually produce the "best" decisions within the game. The development of such "judgment" is essential to an effective game of chess. It is also of critical importance in the evaluation of social policy alternatives. The need to have information from past experience in order to test hypotheses concerning policy alternatives helps to explain the importance of historical information as an aid to policy decisions.

Of course, we cannot expect to develop principles of social policy that will lead to the best possible decisions. Even the best estimates of the "values" in chess do not produce a *perfect* game of chess. However, as social principles are developed and refined by experience they should gradually help us to make better decisions.

Although the innate human values are a genetic inheritance, the development of social norms, social values, and social policy is a human responsibility. The development of a just and humane society that can survive in an uncertain future is a human responsibility. The future is not fatalistically determined by our genes; it is not even theoretically predictable. Although we may be able to improve our judgment through the application of scientific information, our decisions must ultimately be based on the best available human judgment.

COLLECTIVE CHOICE AND SOCIAL WELFARE

The development of a decision criterion such as the "social benefit function" should not be interpreted as a solution to social problems. It defines an objective or a criterion of decision, but it does not specify how best to achieve any objectives. It is more like a clearer statement of the problem than a solution to the problem.

The fact that the "social benefit function" is stated in formal mathematical form should not mislead us into thinking that practical solutions will be mathematical. It is almost certain that they will not be mathematical. There are two important reasons for this:

First, the actual human value structure is not yet known in enough detail to permit any real quantification of the values.

Second, even if the value structure were known, the calculation of the consequences of different policy alternatives for the individuals in society would hardly be practical because of the complex interactions involved.

If practical policy decisions are to effectively serve social objectives such as those defined by the social benefit function it will be necessary to use public opinion and public judgment as a guide to the valuative consequences. At present in the United States this is accomplished through the use of the ballot together with public opinion polls.

This raises an interesting theoretical question. What form of practical voting and election procedures will result in actual policies that come closest to maximizing the social benefit function? Some excellent pioneering work (1, 9) has served mainly to show that there is no practical method of voting that will give ideal results. What now seems to be needed is an analysis of plausible voting and election alternatives to provide an understanding of how much improvement is practical relative to our present procedures.

In the past, the theoretical work on this subject has ignored some problems which seem to be of even greater importance, specifically: How can society effectively use expert knowledge without becoming a victim of the valuative objectives of the experts? How can we gain the benefits of leadership without an undue bias in our valuative decision criteria favoring the leadership group? Can we devise voting procedures which allow the voters to define the valuative objectives and at the same time use the knowledge of experts to identify the most effective ways of realizing objectives defined by the voting public?

While the subject of voting procedures is of considerable importance, it is a technical issue that we will not pursue further. Interested readers are referred to the books by Arrow (1) and Sen (9). Although an improvement in voting procedures undoubtedly would permit somewhat better social decisions, the most serious inequities in our society are not a consequence of voting procedures. More often, they reflect our failure to understand human values and to take them into account. These problems cannot be corrected by an improvement in voting procedures. What is needed is a broader understanding of the structure and relevance of the basic human values.

REFERENCES

1. Arrow, Kenneth J. *Social Choice and Individual Values*. New York: Wiley, 1951.
2. Barry, Brian M. *The Liberal Theory of Justice*. Oxford: Clarendon Press, 1973.
3. Bentham, Jeremy. *A Fragment on Government*. First published in 1776; reprinted in 1891 by Oxford University Press.
4. Bentham, Jeremy. *An Introduction to the Principles of Morals and Legislation*. First published in 1789; 1823 edition reprinted in 1907 by Oxford University Press.
5. McCloskey, Henry John. *Meta-Ethics and Normative Ethics*. The Hague: Martinus Nijhof, 1969.
6. Mill, John Stuart. *Utilitarianism*. First published in 1863; reprinted in 1910 by J. M. Dent & Sons, Everyman's Library.
7. Moore, G. E. *Principia Ethica*. Cambridge University Press, 1903.
8. Rawls, John. *A Theory of Justice*. Cambridge, Mass.: Belknap Press of Harvard University Press, 1971.
9. Sen, Amartya Kamar. *Collective Choice and Social Welfare*. San Francisco: Holden Day, 1970.

Chapter 17

Humanizing the

Social Environment

No wonder the fellow
is fast turning mad
with gloom and frustration
and doubt.
It must be unbearable
being so sad
with nothing to be it
about.

PIET HEIN *

THE TWO previous chapters have been concerned primarily with a form
of hypothesis testing in which the implications of the theory with regard to
human value issues are tested against cultural tradition and commonsense
judgment. Although that kind of analysis is essential as a part of theory valida-
tion, it can convey a very misleading impression that the cultural tradition is
necessarily "right," or that the value-theory approach is nothing more than a
way of justifying the status quo. Moreover, since the principles of evolution
have often been misused in the past as a justification or rationale for existing
oppression and social injustice, such a concern might appear to be justified.

The concern, however, is not really justified within the context of the
present theory. The decision theory interpretation of human values is not a
status quo theory. It is more like a blueprint for moral and social reform.
Throughout recorded history, the evolutionary design for human social behav-
ior has worked *very* imperfectly. With the development of a better under-

* From *Grooks IV* by Piet Hein. PHI Aps, Copenhagen, Denmark; Doubleday and Co.,
Inc., U.S.A. All rights reserved. Used by permission of the author.

standing of human goals and objectives we should have a new opportunity to make society function more effectively in the service of our fundamental human values.

The most basic social problems of modern society arise from a fundamental mismatch between the large size of the present social environment and the limited range of human altruistic social motivations. This mismatch seems to be largely responsible for the rapid increase in antisocial behavior in the society, for the rising problems with mental disorders, and for the international dangers that threaten human survival. What appears to be needed is a basic restructuring of the existing social system so that it can more effectively harness our innate altruistic motivations within the context of a social environment that is orders of magnitude larger than the small social groups for which the innate human values were originally designed.

The development of organizing principles for such a society should be a high-priority objective for a new social science. At present it is clear that we do not have any real answers. In fact, we are only beginning to understand the nature of the problem. But a more scientific understanding of the driving forces of human behavior should help to open our eyes to existing evils within the present social structure. Once the problems are identified, the theory offers a principle of value consensus that may help in obtaining agreement on necessary changes in social policy. The objective of the theory is *not* to justify existing evils but to provide a spotlight for identifying the evils—and a social procedure for correcting them.

The purpose of this chapter is to show how the value-theory perspective can help both in focusing attention on existing social problems and in providing a better understanding of the fundamental causes. The chapter deliberately avoids suggesting any solutions. But as we gain a better understanding both of the problems and of the human objectives, it seems likely that we will also recognize what types of social changes are needed.

Because of the essential role of judgment in issues of this type, there is no way to avoid personal bias in the discussion. Both the selection of issues to be discussed and the choice of approaches for discussing the issues will necessarily reflect personal judgments about value priorities and the structure of society. Fortunately, however, this should not interfere with the real purpose of the chapter, which is to stimulate independent thinking through the use of examples that show how the new perspective can be related to modern social problems.

THE EVIL OF URBAN SIZE

Some of our most serious personal and social problems result from the faceless social environment of large urban centers. In large impersonal metropolitan areas our innate "social values" operate very poorly. With the loss of these innate social controls, the crime rate increases and society encounters a new economic burden because of the losses from criminal activity and the costs of law enforcement. Figure 17.1 illustrates the way the crime rate in the United States depends on city size. The crime rate remains relatively constant for towns and cities up to about 20,000 population. Within such communities most activities take place on a personal basis, people know each other as individuals, and the innate social motivations can operate much as evolution intended. Even in cities as small as 100,000 the general crime rate is almost twice this normal level and the incidence of armed robbery is more than three times the normal level. But the largest increase in the crime rate occurs between populations of 100,000 and 1 million. The robbery rate leaps to ten times the normal level while murder, rape, and auto theft rise to four or five times the normal level. In cities of a million or more the style of life is very different from that in the smaller communities. Most activities are on a very impersonal basis, and there are many self-contained special interest groups that have little concern for the overall welfare of the community. In such cities the problem of social "conspiracy" which was discussed theoretically in Chapter 15 becomes an acute practical problem.

The individual typically has much more difficulty in developing a meaningful life-style in the impersonal environment of large urban centers. This problem of personal isolation shows up in an increased incidence both of real mental illness and borderline mental depression. These mental disorders swell the welfare rolls and impose an economic burden in the form of care for the mentally ill.

The available statistics on mental illness show striking differences between rural and highly urbanized states. On a nationwide basis in 1965 there were an average of about 250 resident mental patients for each 100,000 persons, but nine states had a resident mental illness rate more than 300 per 100,000 population, and eleven states had a rate below 150. The contrast between these two sets of nontypical states is impressive. Table 17.1 summarizes the statistics for these states.

In most cases the states with the highest mental illness rates tend to be urbanized industrial areas. The average population density of the nine states with the highest reported mental illness rates is about 260 persons per square mile, whereas the population density in the states with the lowest mental illness rates is about 22 persons per square mile. The facts seem clear: As the size of the cities becomes larger the crime rate increases, and there is a corresponding increase in the amount of mental illness.

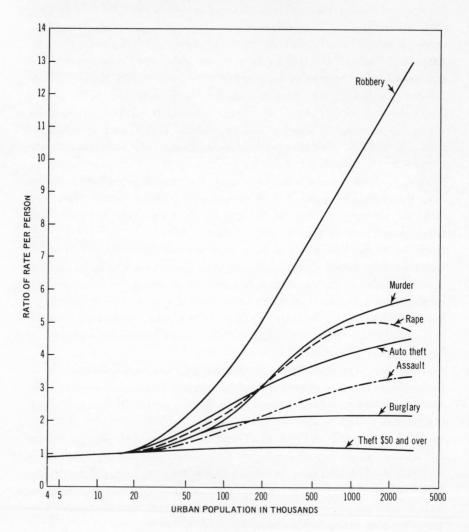

FIGURE 17.1
1972 Crime Rates versus Urban Size
(Offenses Known to the Police)

Plotted as a ratio to corresponding crime rate in cities of 15,000.

Source: F.B.I. Uniform Crime Reports 1972, Table 10.

TABLE 17.1

States with Unusually High or Low Rates of Mental Illness *

ABOVE 300 PER 100,000 POPULATION		BELOW 150 PER 100,000 POPULATION	
N.Y.	489	Tex.	149
R.I.	359	Alaska	148
Del.	356	Hawaii	138
N.H.	354	Kans.	121
Ill.	348	Nev.	115
Mass.	330	Idaho	108
Wis.	326	Ariz.	107
Pa.	317	N.Mex.	96
Vt.	313	Iowa	87
		Utah	60

* Data from *Health, Education, and Welfare Trends*, 1966–67 edition (Page S–15).

As cities become too large they can become almost impossible to govern. Individuals begin to operate ruthlessly and selfishly either for their personal interests or for the interests of a narrow economic group. Labor disputes and strikes become the rule rather than the exception. Extortion, bribery, and corruption become common, and the theoretical efficiency of the specialized urban economic system is lost.

The underlying problem is that the altruistic component of the innate human value system does not operate effectively in an impersonal social environment. Human social values depend on being able to recognize others as individuals and being able to treat them accordingly. Our innate values were not designed to operate in a massive impersonal society, and the evidence is overwhelming that in fact they do not operate effectively in such an environment.

There is a growing recognition that the essential character of human society depends critically on the size of the social group. For example, prison officials are beginning to recognize that small institutions can be more successful in returning offenders to a productive role in society. It seems very likely that some of the problems of discipline in the large urban public schools may also be a consequence of the excessive size of these institutions.

The problem of the impersonal urban society appears to be one of the most serious problems in modern social policy. The search for practical solutions to the problem should be one of the high-priority tasks of social planning and political leadership. Although the basic cause of the problem now seems fairly clear, it is far from obvious how we can most effectively deal with the problem.

At least in theory, one way to extend the effective range of the altruistic human social motivations is through the use of a hierarchical structure in which face-to-face human relationships are established at each level in the hierarchy. Obviously this procedure was widely used in historical societies. Unfortunately it also tended to create authoritarian and often oppressive regimes. However, with a clearer understanding of the objectives it might be possible to develop a modern form of social structure which would provide the altruistic organizing advantages of a hierarchical structure without jeopardizing the benefits of freedom and equality of opportunity that are the great achievements of modern democratic society.

The Chinese appear to have had considerable success in controlling antisocial behavior through the use of local political action groups. Unfortunately it is unclear to what extent their success is simply the consequence of a coercive totalitarian state and to what extent it may reflect the creation of a real community spirit through the local action groups.

Of course, the most obvious approach to the problem is to encourage and perhaps even subsidize the development of smaller urban centers and to try to encourage the development of *stable* and cohesive communities *within* the larger urban centers. The development of smaller cohesive communities might also contribute to the solution of a number of other problems.

The economic specialization in the urban areas has led to an unfortunate specialization in housing patterns. Neighborhoods tend to be segregated not only by ethnic background but also by age and income level. A great deal of attention has been focused on the issue of ethnic isolation, but in terms of basic human values it is not even clear that ethnic isolation is bad. It seems probable that the isolation of age and income groups is much more damaging to basic human values.

Such isolation tends to remove many of the opportunities for human kindness that can make life worthwhile. It separates those who may need help from those who are able to provide it. It deprives the elderly of the joy of encouraging the new generation. It deprives youth of the opportunity to care for the old and infirm. It deprives the well-to-do of the opportunity to personally assist the impoverished. It makes it harder for the young to foresee and understand the future roles they will play in society. When this type of specialization in housing patterns exists, charitable activities which could be an enriching personal experience become an impersonal chore for professional social workers. The development of smaller cohesive communities structured around a more balanced population could restore the opportunities for meaningful personal contact.

It could also do a great deal to restore the dignity of the elderly. In our present society respect for the elderly has seriously deteriorated. The pace of change has been so rapid that wisdom based on experience in a different time hardly seems relevant; but in addition, because of the high mobility of the society, no one remembers the contributions that were made by the elderly to the community. At present, the average urban man or woman can look forward only to an old age in which they are likely to be isolated and forgotten in a state of useless retirement. In a more stable community, personal contributions to the community would seem to be more worthwhile. Service to the community would be a way of building a personal reputation, and it would help to ensure a climate of dignity and respect in old age.

REORIENTING OUR SOCIAL PRIORITIES

From the perspective of traditional social policy, it is natural to focus on crime and mental illness as the key evils of an impersonal urban society. These problems involve direct financial cost and they drain resources that otherwise could fill human needs such as food and housing. However, from the perspective of the present theory, these problems are really only symptoms of a more profound human problem that involves an even greater cost in human values. The impersonal society poses a very serious problem not only for social policy, but also for many individuals for whom it poses a serious barrier to the achievement of a really satisfying personal life-style. The shift in priorities which focuses on this problem as a serious social concern is a direct consequence of a fundamental change in perspective concerning human social and psychological "needs."

Within the present theory, the traditional physical and psychological "needs" are interpreted as a set of innate human "values" which provide the basic valuative criterion for personal decisions. The traditional characterization of these innate values as basic "needs" tended to focus attention on *minimum* requirements. What is the minimum requirement of vitamins to maintain health? How much social and intellectual stimulation is required to maintain psychological health? This traditional psychological perspective tends to place human aspirations at the lowest tolerable level. If basic "needs" are "met," what goals remain?

When the innate human motivations are redefined as primary human values the perspective is completely reversed. The focus shifts from the maintenance of minimum requirements to the attainment of the best possible life-style. How can we achieve *maximum* satisfaction and fulfillment in terms of the basic human values? What is the very best of which we are capable? This perspective encourages a new and critical look both at our lives and our society. It highlights defects and limitations which previously were only vaguely or intuitively perceived.

THE URBAN HERD

From the value-theory perspective the evils of the impersonal urban society go far beyond the problems of crime and mental health. Even for those who are not directly affected by such obvious problems, the urban society takes a severe toll in human values.

The impersonal anonymous form of urban society might be more compatible with the innate values of herd animals such as cattle or sheep. It is difficult for the very social human animal to adjust to an environment where individuality is lost. The issue of how to deal with this basic dilemma constitutes one of the most difficult problems of modern personal ethics.

The essence of the urban corporate society is that individual personality differences are suppressed. Our routine interactions with others consist essentially of "role playing" or contractual relations. A lawyer meets a client. A doctor meets a patient. A customer discusses auto repairs with a shop foreman. In each case there is a set of roles and a contractual relation which defines the essence of the interaction. The system achieves its "efficiency" *because* it is impersonal. But when it is impersonal, the human meaning is lost.

The theory of human social values as developed in Chapters 10 and 11 provides a new understanding of the impact this impersonal environment has on the innate human value system. The innate human goals of social dominance, social acceptance, and social approval are *meaningless concepts* in an impersonal faceless society. It is impossible to achieve a position of respect and dignity when personal interactions are on an impersonal basis. When the personal distinctions are destroyed, the innate human value system does not function. The human meaning of the interactions is lost and the individual experiences a "meaningless" existence.

Human society probably first had to face this dilemma of impersonal interactions during the development of the first urban societies. Even then, in the much smaller urban centers, the problem apparently produced acute psychological problems. This was the axial period of great spiritual leaders. As useful solutions emerged, the solutions began to solidify as religious and spiritual doctrine. What was really important about these doctrines were the precepts concerning interpersonal relations, which served to produce a sense of meaning within the urban neighborhoods.

But modern man is now confronted with a convergence of circumstances that is tending to destroy this carefully woven social fabric.

The increasing *transience* of the urban population interferes with development of really meaningful personal relations. A friendship only begins to develop when a move disrupts the relationship.

Because of the *oppressive size* of large urban centers, we cannot afford to make personalized contact with even a tiny fraction of the people we meet. Our contacts are trivialized, and the pressure of the trivial drives out the nontrivial.

The influence of established *religion* is losing its effectiveness for much of the populations. The ethical precepts of the church, which provided a useful social crutch, no longer seem relevant in the new scientific environment.

In earlier urban societies the sense of *individuality* could be retained, at least with co-workers at the place of business. But now the businesses have become so large and the transience of employees is so great that no real refuge is offered.

The lack of individuality in our acquaintances is exaggerated by the uniformity of mass-produced homes, apartments, automobiles, clothes, and appliances. Not only are people interchangeable in their roles, but they seem interchangeable in their homes, clothing, and possessions. This total interchangeability is an affront to human individuality and an insult to the innate personal objectives of dignity and self-respect.

Technology has conspired to produce an unprecedented degree of *personal isolation*. Air conditioning keeps us inside away from neighbors in the summer, central heating keeps us isolated in the winter. When we move about we are socially isolated by the automobile, or we are afflicted with a crowd environment such as the subway or elevator which is incompatible with intimacy. Television produces an easy but inadequate social substitute which destroys the incentive to visit with friends. While the benefits of these technological developments are undeniable they also have an undesirable social impact.

These combined influences have created an environment in which a large fraction of the population seem unable to find a significance of meaning in their lives.

THE QUEST FOR MEANING

The failure to find meaning in life has become one of the most common complaints encountered by psychologists and psychiatrists. Piet Hein's "grook" about the frustrated young man typifies the problem. In its extreme form the sickness involves a continuing inability to believe in the truth, importance, usefulness, or interest of any of the things one does or can imagine doing. There is a general absence of emotions (pleasant or unpleasant) with the exception of boredom. This humdrum cycle of apathy, boredom, and indifference is broken only by occasional states of acute depression. Salvatore R. Maddi, a psychologist at the University of Chicago, has studied this type of existential illness in some depth (5).*

In addition to the basic syndrome of meaninglessness, Maddi has identified two others—nihilism and crusadism—which he believes are alternative manifestations of the same basic problem. He points out that each of these syndromes occurs with such frequency today that they are not fully recognized as psychological disorders. In effect the nihilist finds meaning and purpose only in his despair and in the angry defiant pleasure he obtains from his efforts to destroy meaning. The crusader is driven to follow causes in order to keep one step ahead of meaninglessness, apathy, and aimlessness. As soon as one cause is fulfilled he must quickly find another so as not to fall apart existentially. Such individuals may contribute a great deal of social good, but this does not change the fact that their motivation is a desperate attempt to avoid the vegetative forms of meaninglessness. Crusaders are caught up in the vitality, drama, and group cohesiveness inherent in causes and movements.

From the point of view of the present theory we can understand the social attractiveness of a cause because it duplicates, if only on a transient basis, many of the essential social features of the primate troop. It provides individual recognition, a common cause, and some human concern for other members of the group.

Much of the attractiveness of juvenile gangs arises from the same basic social factors. The gangs supply a sense of social acceptance, combined with the shared experience of danger and adventure which satisfies some of the innate needs of the young primate male.

The failure to find meaning is one of the most important and pervasive problems in our personal lives. The present theory suggests that the problem arises from the failure to satisfy or even to stimulate the fundamental human

* The descriptions of the syndromes given here are summarized from Maddi's analysis. For a related analysis see also Frankl (2, 3).

"social values" that are intended to motivate the human decision system. When these social or psychological needs are neither stimulated nor satisfied the decision system simply *idles* without meaning or purpose. The victim is aware that he is dissatisfied, bored, and depressed but he does not know why. In a primitive society, where the social motives are more routinely stimulated, the individual gradually learns what activities he enjoys. In a modern society, because of a lack of relevant social experience, he may fail to understand or respond to the guidance of the human motivation system.

In Chapter 9, when we discussed the basic theory of motivation, we found that activities could be motivated either by positive or by negative values. When we fail to engage in social activities or to satisfy our social value system the negative values begin to dominate. There is a general drift of *all* relevant values in the negative direction. The emotion of joy versus sorrow, which is the prime motivator of social activity, drifts toward sorrow. The emotion of pride versus shame drifts toward shame. The emotion of interest versus boredom, which is intended to motivate intellectual activity, drifts toward boredom. The individual becomes depressed, loses his sense of pride and personal worth, and becomes chronically bored. In effect he loses his sense of meaning and significance in life. The extent of the hunger for real human relations is dramatically demonstrated in the present success of the "sensitivity" training concept. It is tragic that the instinctively social human animal has been so psychologically deprived that he must be taught to touch, to talk, and to communicate his emotions.

One of the most important functions of a practical personal ethics is to provide the individual with an understanding of this problem and how to deal with it. Our analysis of the structure of human values suggests that human social and intellectual needs are very complex. Although a number of separate "motives" are involved, most of the motives take on their full significance only within a cohesive social environment where each person is treated as an individual, where the quality of each person as an individual is significant, and where reputations and sentiments can be developed. Unfortunately, for a large fraction of our urban population this essential social context is missing. People need guidance and assistance in developing the essential social context.

The individual at present is at a serious disadvantage in his effort to find or create such a social environment. The activity of big corporations and big government serves unknowingly to disrupt the development of a stable social environment. Obviously we need to reevaluate this larger social system carefully to see how it can be better adapted to the support of real human needs. Many possible personal solutions are being tried, but there may be no really satisfactory solution within the present structure of society.

Some of the most innovative concepts have come from the so-called counterculture movements. The urban commune with intimate group living and economic interdependence represents one interesting approach to the problem. The development of tightly knit communities of small businesses within an urban area based on crafts, services, and "natural" foods is another creative approach to the problem. Still others are experimenting with rural communes built around cooperative effort in farming and gardening. Each of these "radical" concepts in group living reflects the need to solve the main existential problem that afflicts our society. Each of these experiments is experiencing a variety of problems, both social and economic. So it is likely that no one has found any really satisfactory solutions, but people deserve the freedom to experiment and look for solutions that will meet their personal needs.

Many others are experimenting with more conventional solutions. Some are looking for stable neighborhoods in order to build meaningful lives around community activities. Others are looking for jobs in small businesses or are trying to develop their own small business in which they can create a stable human society. Some are focusing their efforts in the professional sphere to try to develop a personal reputation and lasting relationships within a professional community. But in most cases the real motives and purposes are not really understood, and the efforts are much less successful than they might be with a better appreciation of the essential problem.

Although the theory cannot yet provide any real guidance about which of the alternatives is likely to be best, it does make it clear that none of the solutions is likely to be successful unless it involves more than superficial human relations. Success depends on the existence of a cohesive social group, close personal relations, commonly shared goals and objectives, a real caring about others within the group, and sufficient stability to allow significant sentiments to develop.

THE SEXUAL WILDERNESS

Recent technological and social developments have produced profound changes in human sexual relationships. Of course, the impact of contraceptives has been widely recognized. But this is only one of many factors underlying the revolutionary changes now in progress. Some of the most important trends seem to be a consequence of subtle psychological changes in the urban social environment. Vance Packard (7) aptly described the resulting confusion as a "sexual wilderness."

Our review of human evolutionary history indicates that nature has provided certain innate human desires intended to bind male and female in a durable working partnership. In a primitive society the marriage operates quite naturally as a total partnership. The partnership encompasses economics, comradeship, sex, children, and social status. The durability of the primitive marriage cannot be attributed to any of these factors individually; rather it is a consequence of the ensemble of factors working together. Modern urban society has fragmented these traditional forces so that they no longer operate in unison to support the institution of marriage. The most serious damage has been done not to the physical attraction or emotional infatuation that bring a couple together, but rather to the enduring emotional bonds that serve to maintain a lasting relationship. As a consequence, the innate psychological forces which traditionally held the marriages together no longer seem adequate to the task. For some, the obvious solution may be to recreate a social environment in which the traditional institution of marriage can survive. For others, the solution may be to redesign the institution of marriage to bring it into harmony with the realities of modern society. The issues involved are complex, and the proper answers are far from obvious.

In primitive hunting and agricultural societies the man and woman tended to work together toward common objectives. The woman could see and appreciate the efforts of her husband; and the man could see and appreciate the labors of his wife. This awareness of the efforts of the other contributed to mutual understanding, as well as to a feeling of economic interdependence. But the total impact of this close daily contact was probably far more significant than one might expect on purely rational grounds. Man and woman are linked by innate emotional bonds that serve to hold them together. It seems likely that daily proximity and mutual awareness serve to reinforce these emotional bonds, so that the sense of affection and love is increased. It seems probable that in the primitive environment the couple are not only more economically dependent on each other, but that they feel a stronger commitment of emotional ties and thus are more strongly motivated to want to stay together.

In a modern urban family, the man's efforts to support the woman occur beyond her sight and often involve work that she does not understand. Her feeling for the man is likely to be both rationally and emotionally attenuated. She may even experience a sense of anger and frustration that the man has deserted her all day. The man who is away all day cannot share her experiences raising the children, so he lacks a real appreciation of her efforts. When he returns at the end of a day, he may be greeted with the pent-up anger of a woman who feels socially isolated as well as deserted.

In a primitive society the woman has frequent contact with relatives and neighbors, as well as her children, as she goes about her daily tasks. Thus, the woman's social desires can be met and satisfied within the neighborhood and the home. Unfortunately in many modern homes the woman may feel isolated, almost as if she were in solitary confinement. A woman who feels trapped in this way may choose to go to work for purely social reasons, regardless of economic need or career ambitions.

In many species of primates the males appear to have an innate inhibition against taking a female who belongs to another male. It is possible that a similar inhibition may have operated in earlier human societies. But the structure of modern urban society is such that even if this innate inhibition exists, it has little chance to operate. When the woman is working in an office she is surrounded by men who rarely (if ever) see her with her husband. In a purely rational sense they know she is married, but they do not see her with her man and the innate value system may not respond emotionally as if she were "taken."

The modern office tends to bring mature "high status" (and therefore sexually attractive) men into close contact with young attractive women. This circumstance can be disruptive to the existing marriage commitments of both the man and the woman. The young woman is likely to be attracted to the mature man. She can see and respect his professional efforts and will treat him with a respect and admiration that he does not receive at home. The man in turn finds an attractive woman who feeds his ego and regenerates his sense of self-respect. In short, the emotional bonds that are designed to hold man and wife together are inadvertently invoked in the modern office to pull them apart.

In a primitive society, when the men are away from the village they tend to be in all-male groups. Thus, there is little opportunity for sexual temptation. Life within the village is organized so that social encounters between men and women tend to occur primarily when the man and wife are together. Emotionally as well as rationally they were seen by others as belonging to each other.

In primitive societies the economic interdependence of men and women also served as a practical bond. The physical strength and daring of the man in hunting and in battle were needed by the woman. The patience, manual dexterity, and mothering skills of the woman were essential to the home. Thus, even in an economic sense neither partner was complete without the other. In the specialized economic structure of modern society each can be independent. Economic benefits previously available only within the institution of marriage can be obtained through the medium of monetary exchange. Each individual still contributes his specialized talents—but to society at large. The diversified economic needs of each individual are fulfilled not by his mate but by society at large through the medium of exchange. In this way the traditional economic basis of marriage has been weakened.

It is sometimes suggested that the increased frequency of marriage failures reflects a deteriorating moral climate that may result from the decreased influence of established religions. But the important factors appear to be much more fundamental. There have been fundamental changes in the social structure that have weakened the emotional, social, and economic foundations of marriage. Of course because of the decreased influence of the church, people may feel more free to terminate a bad marriage, but the high frequency of marriage failures must be blamed on other causes. The basic forces which push a marriage apart are largely social, economic, and emotional.

Our objective here is not to try to suggest solutions to the problems, but rather to show how the origin of some of the problems seems to be related to innate values that are part of our genetic inheritance. If we are to be successful in analyzing the problem and in finding appropriate solutions, it will be necessary to define both our objectives and the structure of the problem in terms of the innate values that are our human inheritance. As yet there has been almost no effort to look at the problem from this fundamental point of view.

SEX ROLES AND EQUAL RIGHTS

The issue of equal rights for women is one of the most controversial current social issues. There is little disagreement about the objective of equality, but there is a great deal of controversy about what constitutes equality. Unfortunately the issue is extremely complex. Although the natural social principle defines a theoretical approach to the problem, it does not yet provide any answers. It may nevertheless be useful to define some of the issues and thus illustrate the essential complexity of the problem.

The "social benefit function" treats all adult citizens equally, so the value-theory approach begins with absolutely equal treatment of men and women. The theory says that policy should be chosen to maximize the social benefit function. But there are important differences between the innate values of the average man and the average woman. For this reason identical treatment is inappropriate and could be achieved only at the disadvantage of both sexes. According to the social benefit function men and women should be treated differently but in a way that is most beneficial and equitable for both. If there are areas where the interests of men and women are in conflict, the social principle would seek the most mutually beneficial compromise.

Unfortunately it is much easier to agree on this as a working principle than it is to agree on the consequences of the principle. We do not even know enough to be sure which alternatives would be preferred by either sex alone, so the problem of selecting a suitable compromise only complicates the problem. But the theory does raise very serious questions about the validity of many arguments now being used in the debate. Many of the arguments are based on the premise that equal rights necessarily implies the elimination of sex distinctions and identical treatment in all areas of activity. Because men and women are fundamentally different, an effort to treat them in identical ways can introduce new and artificial areas of unfairness and inequality. The proper goal has to be fairness, justice, and equity rather than identical treatment.

THE CURSE OF SPECIALIZATION

The modern social and technical environment confronts us in hundreds of different ways with demands for intensive specialization that narrow our life-style and limit our ability to experience the joy and beauty of life. The human animal is equipped by evolution with a value system which motivates a *diversity* of activity. The appetitive characteristics of the innate value system motivate a balance of intellectual, physical, and emotional activity. In our primitive past such a balance was desirable for survival. It also protected the individual from the damage of excessive concentration on any single activity. Efficiency in the modern society is achieved as a consequence of extreme specialization. Although the specialization contributes to economic efficiency it entails a great human cost.

Some individuals are confined to office jobs with almost no physical activity. Others are involved in excessive physical labor with almost no intellectual challenge. Both extremes are damaging to body and mind. A job requiring creative intellectual effort can be very rewarding, but even such a job can become a burden if it is not relieved by social or physical interludes of less demanding intellectual effort.

In our present specialized economy almost everyone works too intensely at a job that is too narrowly specialized. The psychic rewards of the job might be considerable if it were experienced in moderation, but when the job is experienced in excess it becomes a burden. While specialization undoubtedly contributes to efficiency in the traditional economic and materialistic sense, it also involves substantial human costs. When benefits are measured in *human* rather than monetary units, excessive specialization can be very *inefficient*.

MALDISTRIBUTION OF EDUCATION

The same error of specialization that is committed in the job environment is carried over into the field of education. Education can be a very rewarding experience, but the concept that it should be taken in one continuous dose beginning at age six and ending after completion of a higher education seems most inappropriate. In terms of human values we have a serious maldistribution of educational experience. The children and youth get far too much, the young and mature adults get far too little. The result is a disadvantage to both groups.

It seems probable that both the individual and society would be better served if the educational process were better distributed. This suggests that many individuals may find a better balance if they interrupt their formal education at the end of the high school experience. Like many other young primate males, most young men at this age feel an acute need to develop independence and learn what type of person they really are. Some youthful years spent in comparative freedom from responsibilities could be very rewarding to most men and to many women. It would provide a much needed opportunity for adventure. It would allow them to develop a realistic appreciation of their own potentialities and of the opportunities and responsibilities of the human experience.

The example illustrates another characteristic of the natural ethical perspective. The traditional parent might oppose such a shift in the educational schedule because it will postpone the time when the young man or woman can begin the climb toward "success and achievement" in business, and thus may entail some loss in final "level achievement" at age forty or fifty. The value-theory perspective does not focus solely on such ultimate objectives but rather on the satisfactions and joys of the journey. It assigns importance to the satisfactions of youth as well as the achievements of age.

SPECIALIZATION IN EDUCATION

Much of today's education is too narrowly specialized. It is directed more to economic than to human objectives. There is, of course, a need to learn a trade and to learn professional skills. But when education is limited to such material, students can acquire a very distorted view of life. There is a tendency to think and act in terms of narrow economic objectives and to try to squeeze the fullness of life into the rigid confines of a scientific or mathematical discipline. It is not surprising that the students complain about the lack of relevance of the educational curriculum.

One of the serious problems with our modern technical education is that it tends to emphasize a mental attitude of objectivity and detachment. Obviously objectivity has an important place in science and engineering, but when detachment is carried over into personal life, into relations with other people, and into the political arena, it is destructive of human values. A satisfying personal life is built not on objectivity and detachment, but on concern, commitment, and faith. To appreciate the warmth of love or the joy of success, we must care enough to be willing to work and to sacrifice. To suppress our concern for human values is to suppress the essence of life.

Much more attention needs to be given to the life sciences and the humanities, but even more important, attention needs to be given to the art of living. Students need to develop an understanding of the importance and relevance of human values. They need to learn the importance of following the dictates of the heart, as well as the logic of the intellect. They need to be exposed to literature dealing with human wisdom, past and present, rather than just the "know-how" of the present.

THE SPECTATOR SOCIETY

To realize the fullest human experience we need to stimulate and satisfy the full ensemble of human values. The previous sections dealt primarily with issues based on the spectrum of Social Values. But we should not ignore the role and importance of the Intellectual Values. The availability of communication media such as television, radio, hi-fi, movies, magazines, newspapers, and books has made high quality (and low quality) cultural activities available to the public as never before. Although much of the material is banal and trivial, it is almost all of professional quality.

But this torrent of professional literature, drama, art, music, sports, and dancing is not an unmixed blessing. Although it allows the individual to observe professional quality cultural events, it also tends to inhibit amateur participation in such activities. The professional quality of the activities becomes accepted as a universal norm, and we discriminate critically with regard to small differences in the ability and quality of the professionals. Against such standards an amateur performance is hopelessly inadequate. Rather than reveal such inadequacy, the average individual retreats from active participation. In doing so, he misses the joy of personal participation that is necessary to fully stimulate the innate intellectual values. Participation in cultural activities becomes limited to professionals and spectators. Again, we are faced with the curse of specialization. The professional is so intensely involved in his activity that much of the joy of participation is lost. The nonprofessional is denied the satisfaction of even modest participation.

We need to change our attitude toward cultural achievement. Although it is appropriate to judge professional performances by professional standards, we must avoid applying professional standards to amateur participation. The prevailing atmosphere of scorn for amateur quality performances is one of the most destructive attitudes in our culture. It exaggerates what is already a difficult cultural problem. We should learn to enjoy and applaud amateur performance in friendly gatherings for its own distinctive emotional quality, and for the achievement it represents. A response to a teen-ager's performance which says in effect "he will never be a professional folk singer" is irrelevant and destructive. A more appropriate response might be ". . . and do you know he is *also* a first-rate basketball player."

The individual who is actively involved in developing his own skills, testing his limits of creativity, and exploring his potential for excellence is far more likely to find life enjoyable and meaningful. Our tendency to degenerate into a spectator society is one more reason for the sense of apathy and loss of meaning that seems so common in our society.

THE NEED FOR COMMITMENT

The spectator philosophy, combined with the scientific philosophy of objectivity and detachment, is destructive of the vitality of society as well as the meaning of our personal lives. It is fashionable to avoid expressing moral judgments, to be objective, cynical, tough, and unbiased. But we cannot suppress moral judgment without also suppressing our humanity. When human values are damaged, we should give vent to our moral outrage.

Human society was designed by evolution to operate in response to moral suasion. When we claim that it does not matter what others think, we are denying essential human values. In evolution's design, the wrong-doer was to be shamed into doing right. When in our sophistication we suppress such responses or refuse to recognize the validity of social standards of behavior, we erode the foundation of human civilization. Just such a sophisticated amorality proved to be the fatal flaw of the Nixon presidency. If our society is to regain its health, we need to place more, not less emphasis on the essential features of human morality. We need leaders who are realistic; but we also need leaders who believe that there is a real distinction between right and wrong.

THE CONFLICT BETWEEN ECONOMIC AND HUMAN VALUES

The "natural" social principle immediately focuses attention on one of the central problems of social policy in the United States. The real goals of social policy are defined in terms of human values, but the economy is driven and guided by purely *economic* forces (or economic values). If the economy is to become an efficient servant of human values, then some way must be found to align the economic forces so that they properly serve human values. Of course, the present theory cannot specify how this should be done, but it does focus attention on a number of important issues that must be addressed if economic policy is to serve human rather than purely economic objectives.

The tendency to give priority to economic rather than human objectives can be explained in part as a consequence of the way we have traditionally defined social objectives. The traditional goals of social policy have centered on individual freedom, equality of opportunity, freedom from foreign oppression, and freedom from poverty. Although these are obviously high-priority objectives, they do not set very high goals for social policy. Like the personal goals that flow from the psychological doctrine of "basic needs," these traditional social objectives are focused negatively on mininum acceptable standards rather than positively on maximum achievement.

For many years this negative conception of *social* objectives has existed side by side with economic objectives that were positively stated in terms of economic values. If a proposed public policy could contribute to economic growth without actually transgressing the minimum standards of social policy, it was likely to be adopted regardless of any adverse effects on human values. The cumulative effects of such decisions have become progressively more oppressive to human values, so that within the last few years there has been a growing outcry of concern. The present theory provides a badly needed positive definition of the goals of social policy which should help in focusing this concern.

We are so used to measuring policy alternatives against economic criteria that we are often blind to the social cost of our "economic efficiency." Obviously we cannot afford to be economically inefficient, but we need to be sure that the economic machine is efficiently serving human values and human needs. In general the free-market economy does a remarkably efficient job of fulfilling the physical needs of the people. But often this is accomplished at a high cost in terms of other human values. The present economic system within the United States includes many features that seem unnecessarily damaging to human values.

JOB SECURITY

The sudden turns of federal policy produce a great deal of unnecessary hardship for individuals. The problem is particularly acute for workers in specialized competitive industry. Current trends in this respect are not encouraging. The fraction of economic activity connected with government contracts is increasing. The size of the corporations is increasing, and the size of the contracts at stake is increasing. The large shifts of government financial support from one company to the other, as the companies compete for large contracts, produces an unnecessary loss of personal security as well as unnecessary mobility in the population.

Both job insecurity and forced mobility are damaging to human values. Job insecurity contributes to a sense of hopelessness and lack of control over one's destiny. It proves to the individual that he is being treated only as a cog in an impersonal economic machine. This job insecurity (which is completely unrelated to the job performance of the employee) contributes unnecessarily to the general destruction of personal dignity which is so characteristic of the present industrial system. The resultant unnecessary transience of the workers and their families adds unnecessarily to the faceless character of the society. Families are uprooted and the opportunity to develop a sense of belonging in a stable neighborhood is destroyed.

These cruel disruptions of families and careers are tolerated in the name of "competition" and "efficiency" in a "free-enterprise" economy. There is a dogma that efficiency can be maintained only through unrestrained economic competition. But when the competition occurs between corporate giants and for huge government contracts, thousands of families are reduced to helpless pawns in a futile economic game. The purpose of competition is supposed to be efficiency, but we ignore the economic and human costs of such disruptive competition.

The rules of this raw competitive game are deliberately enforced by Congress to ensure that the taxpayer gets "maximum value" for his tax dollar. In fact the actual incentives produced by these rules are often contrary to the interests of the taxpayer. Contract selection procedures require each contract award to be treated as a separate and unrelated decision, essentially independent of the history of the company and without regard to the impact on the work force. Consequently, contracts are awarded on the basis of elaborate proposals which provide optimistic estimates of both the quality of the product and the probable production costs. Because the past performance of a corporation is treated as if it is of minor importance for the awarding of new contracts, there is little motivation to waste management talent to improve performance on existing contracts. The effort on existing contracts tends to concentrate on explaining why cost overruns are the responsibility of the government and not of the contractor.

THE COMPETITIVE TREADMILL

Our present economic system tends to concentrate money, power, and responsibility in the hands of a small number of talented and hard-working people. Typically, these people are compulsive workers; they carry far more than their share of the burdens of society. They also enjoy such an excessive share of excitement, responsibility, and prestige that these excess social "benefits" can become a burden rather than a benefit. Others in the society, who may be almost equally talented and ambitious, are denied access to these benefits. Both groups lose.

There is usually no way within the system for the hard-working executive to pause and rest. If he gets off the treadmill he may never be able to get back on. Thus the system goes on, with some people unhappy because they have too much responsibility, others unhappy that they have too little. With a little attention to the importance of human values it seems that it should be possible to provide a more equitable distribution of leisure and responsibility.

THE LOSS OF QUALITY

The overemphasis on purely economic criteria has led to a deterioration not only in the quality of life but also in the quality of products. In an earlier period when products were simpler the customer usually could judge quality either by examining the product or by considering the reputation of the craftsman. At present the consumer has no way of judging the quality or durability of most of what he buys. But he can clearly identify differences in price. Because of the inability of the consumer to recognize quality or durability in a product, quality products cannot survive in the marketplace. The consumer spends his money for products that perform poorly and require frequent replacement or repair. In terms of simple economic measures it is all for the best. Corporate sales and profits are increased. The consumer spends more and has to work harder. The GNP increases.

But in terms of human values, the trend is most unfortunate. Toys disintegrate within the first few days and children learn that nothing has lasting value. The general neglect of quality in all products adds to the instability and transience of our lives and adds one more way in which the individual is unable to plan or control his destiny. The rapid turnover of low-quality products accelerates consumption of irreplaceable natural resources and hastens the deterioration of the environment.

HUMAN VALUES AND THE "SCIENCE" OF ECONOMICS

As is typical in our present educational system, modern economics is a science of "means" without an "end." It tells us how to achieve economic growth but it cannot tell us whether economic growth is desirable. It treats consumption as if it were a goal in and of itself, but it cannot tell us what consumption is desirable and what consumption is undesirable. It treats irreplaceable natural resources as if they were of zero value, and leads to economic decisions that maximize the rate of destruction of these natural resources. It ignores the value of clean air, clean water, and an attractive physical environment. It ignores the social consequences of economic decisions. It treats human labor as if it is a commodity to be sold and bartered, and it ignores the value of the human dignity that comes from honest creative labor (8).

There is a need for a new science of economics, one that can relate economic means to human objectives. The real goal of a science of economics should be to align the economic structure so that it is as efficient as possible in the support of human objectives. There is no reason why such a science cannot be developed, but it will require us to recognize human values as the primary criterion for economic policy.

The purpose of a theory of values such as that developed here is to help define appropriate objectives for society. In effect the theory focuses attention on certain social problems that need to be solved; it does not provide any ready-made solutions.

The mistakes that have been made in the management of economic policy have been both errors of omission and errors of commission. On one hand we have failed to develop any systematic approach for introducing human value objectives into the operation of the economy. On the other hand too many decisions about economic policy have been made in terms of "efficiency" as measured by purely economic criteria, rather than in terms of efficiency as measured by "human values."

We have an economic system that is both powered and steered by economic motives. Like a dog chasing its tail, it goes faster and faster in a whirl of energy and effort, serving no useful purpose. If the system is to be effective in serving human values, we must harness it and steer toward human objectives.

There can be no doubt about the urgency of a reevaluation of economic goals and objectives. Our wasteful methods of production are rapidly destroying irreplaceable natural resources: oil, coal, iron ore, and all types of minerals. Unless we begin to plan ahead we may soon be faced with a dramatic reduction in our standard of living. A society in which economic wealth is based almost entirely on the rapid consumption of capital resources can have no permanence. If we are to build an enduring social system we must build it on a principle of equilibrium in which any consumption of capital is matched by a regeneration of capital resources.

THE NEED FOR EDUCATION IN VALUES

If we are to evolve toward a future where life will be more worth living we must begin to give priority to human values. This will almost certainly require a very broad reorientation of the goals and objectives for our society. In a democratic society such a reorientation cannot be imposed by the government. It must be an outgrowth of a change in public opinion and such a change of opinion can occur only through the education of an informed public.

Our civilization is almost unique in its concentration of education on production know-how without a parallel education in values and objectives. In almost all historical societies education with regard to the "values" of the society has been one of the most important aspects of the educational process. To avoid interference with freedom of religion we have deliberately taken instruction in human values out of the schools and have left that responsibility to the families and the churches. But with the decline in the influence of the churches the practical effect has been that we have almost completely defaulted in the responsibility to provide an education in values.

The concentration of public education on the purely mechanical and scientific aspects of life leads quite naturally to a cynical materialistic world view. The resulting vacuum of knowledge about human values leaves the youth as easy prey for the commercial philosophy. Our most effective education with regard to human values is provided by commercial television, which very deliberately and effectively reinforces a philosophy of conspicuous consumption. Thus demand is created for the consumption of a larger volume of irrelevant products. By default, our education in human values has fallen into the hands of commercial interests whose sole purpose is the creation of unnecessary demand. Thus, the economic system teaches a distorted view of human values which sacrifices human welfare and natural resources to the false objectives of commercial profit and accelerating "economic growth." If we are to improve our ability as a democratic society to make good decisions concerning human value issues, there is a need for deliberate education in human values to counteract the distorted image provided by commercial television.

THE NEED FOR PLANNING

In the past, we have been able to proceed fairly satisfactorily without any long-range planning because resources seemed unlimited. There was a continent to conquer, and individual initiative could be given freedom to address the problems.

Now we face an entirely different situation. Land and resources are limited. As economic activity approaches the limits of available land and resources, we must give careful consideration to how the resources are to be used (1). Without deliberate social guidance the free enterprise economy will continue to consume and destroy natural resources without concern for the future consequences.

In the past, government policy has operated on the assumption that the proper objective is "economic growth." People have operated as if their objective was to "conquer nature." It is obvious that these are no longer appropriate objectives. We have already conquered nature, we are in the process of destroying it. The advantages of economic growth need to be weighed against the social costs of growth.

Many of the problems we face in achieving meaning and purpose in our personal lives arise because of the present structure of our economic and social system. The problems of ecology and resource conservation that we foresee in the future can be addressed only through collective action of the entire economic and social system. If we are to achieve constructive changes in the system we must begin by thinking through our goals and objectives for change. There is a critical need for creative long-term planning on a national basis.

THE NEED FOR LEADERSHIP

One of the purposes of long-range research and planning is to identify realistic national objectives that make sense in terms of human values. When there is agreement on common goals there is a natural human willingness to pull together toward the achievement of the objectives. Thus the objectives have a healing and unifying effect on the society.

But the public will not respond to social objectives unless they are presented in a way that highlights the emotional appeal inherent in the *human values* of the objective. The development and presentation of goals so that the human values are dramatized is one of the major responsibilities of political leadership. The ability to present objectives in this way was the real genius of the Kennedy administration.

The specific programs of John F. Kennedy were little better or worse than those of his successors or predecessors, but he had a genius for presenting ideas so that they captured the imagination, particularly of the youth of the country. His appeal was not to selfish personal interests or to personal security but to basic human values: to justice, to adventure, to the chance to participate in great causes, to the opportunity to contribute something worthwhile to society. It was his ability to appeal to the innate altruistic emotions of youth that made him such an effective leader. It was because of this appeal that his death was so deeply mourned. There exists within our society a great reservoir of energy and good will. It awaits only the definition of valid goals and objectives and the development of creative leadership.

There is no shortage of great goals to be pursued. In the haste of technical progress, we have ignored some of the fundamental principles of long-term survival within the finite environment of planet earth. These oversights have brought us simultaneously into three great crises: (1) In our ignorance of human nature we have created a technological and social environment which is oppressive to the human spirit. (2) Because of our disregard for the environment, we are facing a crisis of pollution which is damaging to human health and threatens the extinction of many species. (3) Finally, in our wasteful approach to production we are rapidly destroying our irreplaceable stock of fossil energy and mineral resources.

It is widely believed that modern technology has solved the problem of production and that we are entering a new age of leisure and affluence. This view is based on a fundamental misunderstanding of the achievements of modern technology. Technology has not yet solved the basic problem of production. It has only developed efficient techniques for exploiting capital resources that have accumulated on the planet over millions of years. The apparent economic security of modern society is based on rapid consumption of accumulated capital. It does not reflect any comparable improvement in real economic productivity. An economic system that is built on the consumption of capital can have no permanence. It faces an inevitable collapse when the accumulated capital is exhausted. (8) If we are to build a society of permanence we must find ways to operate the economy in harmony and equilibrium with the environment. This defines an entirely new technological challenge which is essential to the survival of our society.

We have been living in an age of rapid technological change. Alvin Toffler's remarkably successful book *Future Shock* (9) provides an excellent description of the destructive human impact of this excessive pace of change. But Toffler's disturbing conclusion seems to be that we should learn to live with and adapt to rapid change. When we reexamine this problem from the perspective of human values we seem to be led to quite a different conclusion. The transience and impermanence of our society is destructive to human values. Rather than attempting to adapt human nature to an inhuman pace of change, we might do better to harness technology so that it works in support of the enduring human values.

Indeed if our social policy for the future is developed with wisdom, then it must inevitably lead toward more stability and toward a reduction in the pace of change. This comforting conclusion follows automatically as a consequence of two inescapable principles which should ultimately define the path of social policy. First, the proper objectives for social policy are defined by the enduring structure of our innate human values. Second, the feasible policy alternatives are bounded by the finite resources of planet earth. If human policy is to be compatible with these two fundamental considerations, it must inevitably lead to a reduction in the pace of change and to a new modern life-style that is more satisfying and more humane. It should lead to a *proper* level of economic activity and a *proper* population size. It surely will not lead to an accelerating rate of change and more rapid economic and population growth.

These fundamental considerations point to some challenging new directions for our society that await only the catalyst of an informed and creative leadership.

SUMMARY

The main objective of this chapter has been to show that the value-theory perspective can be used to highlight many of the existing evils of our society. It is not a way of justifying the status quo. It offers a new conceptual approach for thinking about social problems which promises to be particularly effective both in identifying problems and in suggesting changes in social policy.

The discussion in this chapter deliberately avoided making any specific recommendations for changes in social policy. In a book of this type it is better to leave such initiatives to the reader and to future research on social policy. However, it should be obvious that the discussion in the chapter could easily be extended to lead to some rather specific and constructive policy recommendations, both for social policy and for the development of our personal lifestyles.

REFERENCES

1. Ewald, William R., Jr. *Environment and Change.* Bloomington: Indiana University Press, 1968.
2. Frankl, Victor Emil. *The Doctor and the Soul.* New York: Knopf, 1955.
3. Frankl, Victor Emil. *Man's Search for Meaning.* Boston: Beacon Press, 1962.
4. Leys, Wayne A. R. *Ethics for Policy Decisions: The Art of Asking Deliberative Questions.* New York: Prentice Hall, 1952.
5. Maddi, Salvatore R. "The Search for Meaning." Nebraska Symposium on Motivation, 1970.
6. Muller, Herbert J. *Uses of the Future.* Bloomington: Indiana University Press, 1974.
7. Packard, Vance Oakley. *The Sexual Wilderness.* New York: McKay, 1968.
8. Schumacher, E. F. *Small Is Beautiful: Economics As If People Mattered.* New York: Harper and Row, 1973.
9. Toffler, Alvin. *Future Shock.* New York: Random House, 1970.

Chapter 18

Beyond the Beginning

Yes, but
the answer spoilt
the question!

from the comic strip
Pogo by WALT KELLY

THE MAIN OBJECTIVE of this book has been to lay the foundations for a new understanding of human values. To develop a coherent theory we have looked at the subject from many different points of view. Each new perspective has contributed additional concepts needed to complete the total picture.

As we have progressed through our discussion of human values, we have encountered a sequence of shifts of perspective. Part I begins in the field of decision science, with an analysis of the design principles for automatic decision systems. This theoretical analysis is followed by a discussion of the way these same principles have been incorporated within the design of the human brain. Part I ends with two important shifts of perspective as we examine the brain, first physiologically and then introspectively, to be sure that the model is consistent with reality. Part II analyzes the actual human values as a problem in evolutionary "system design." The analysis shifts back and forth among three separate points of view: first, the broad perspective of primate behavior, drawing on the resources of primatology; second, the more limited perspective of human behavior, drawing on anthropology, psychology, and sociology; and finally the very limited perspective of personal introspection. In Part III we shift back to the field of decision science and begin to apply the results. The perspective shifts from personal ethics and personal philosophy to social ethics and finally to public policy. In many ways the essence of the theory lies in this continuously shifting perspective as we address the subject of values from many different but complementary points of view.

Before closing the discussion we must call attention to a final shift in perspective that will be needed to apply the theory in our personal lives. Throughout, our point of view has been scientific, rational, and objective. This approach was essential, for if we are to have a solid theoretical foundation for the development of personal ethics and public policy, it is necessary for the framework to be logical and consistent. A coherent theoretical framework can be developed only through rational and objective analysis. Moreover, the theory is most likely to be useful for public policy if it can be applied in a similar rational and scientific manner.

But if we were to try to apply the concepts to our personal lives solely in terms of objective rationality we would miss one of the major messages of the theory. Some of the most important human values are emotional rather than intellectual. If we become too narrowly intellectual and analytical in our personal lives there is a risk of shackling emotions, and numbing the zest for life. To gain the full benefit of the theory as a social concept and a personal philosophy it will be necessary to explore the symbolic and emotional side of the theory. This final shift of perspective cannot be accomplished by a scientific approach. It requires emotional and artistic techniques; and it is a task for creative artists, writers, musicians, and moral leaders.

In developing the theory from the objective and scientific perspective, we have addressed many of the long-standing "mysteries of life" and, from a purely objective scientific perspective, have provided certain solutions and answers. But we must not allow the scientific answers to spoil the philosophical questions. From a subjective and emotional perspective the fundamental mysteries of human existence remain as a profound personal enigma—with the result that we enjoy the ambiguity and mystery of literary and poetic speculation.

Within such artistic expression there is emotional truth. The scientific truth moves only the intellect, the artistic truth moves the heart. If we were to lose our reverence for cosmic mysteries of human existence we would damage some of our most important human values. Although our approach has been rational and objective, the quest for human values leads ultimately back to our innate and intuitive human spirit as the supreme guide to human policy.

With the rapid pace of change, humanity is once again in need of creative spiritual and emotional guidance. Because of the scientific emphasis in our society, effective new leadership will have to be compatible with a rational scientific foundation, but the leadership must go beyond the limits of coldly rational science. It must ultimately be based on an inspired appeal to the human spirit.

The modern educational emphasis on the scientific method, combined with the deemphasis of spiritual, religious, and philosophical concepts, has produced a distortion in our culture in which "rational" materialistic values are exaggerated and the spiritual and emotional values are largely ignored or suppressed by the "rational" mind.

One of the objectives of this book is to help redress this balance; to restore to the human spirit its rightful role as the ultimate guide to human goals and aspirations. In an earlier age, this message could have been carried in song and poetry direct to the human heart. Today, our lives have become so intellectualized that the message must be addressed to the head in order to reach the heart.

Index